*Methods of
Experimental Physics*

VOLUME 12

ASTROPHYSICS

PART B: Radio Telescopes

METHODS OF EXPERIMENTAL PHYSICS:

L. Marton, *Editor-in-Chief*

Claire Marton, *Assistant Editor*

1. Classical Methods
   Edited by Immanuel Estermann
2. Electronic Methods, Second Edition (in two parts)
   Edited by E. Bleuler and R. O. Haxby
3. Molecular Physics, Second Edition (in two parts)
   Edited by Dudley Williams
4. Atomic and Electron Physics—Part A: Atomic Sources and Detectors
   Part B: Free Atoms
   Edited by Vernon W. Hughes and Howard L. Schultz
5. Nuclear Physics (in two parts)
   Edited by Luke C. L. Yuan and Chien-Shiung Wu
6. Solid State Physics (in two parts)
   Edited by K. Lark-Horovitz and Vivian A. Johnson
7. Atomic and Electron Physics—Atomic Interactions (in two parts)
   Edited by Benjamin Bederson and Wade L. Fite
8. Problems and Solutions for Students
   Edited by L. Marton and W. F. Hornyak
9. Plasma Physics (in two parts)
   Edited by Hans R. Griem and Ralph H. Lovberg
10. Physical Principles of Far-Infrared Radiation
    L. C. Robinson
11. Solid State Physics
    Edited by R. V. Coleman
12. Astrophysics—Part A: Optical and Infrared
    Edited by N. Carleton
    Part B: Radio Telescopes, Part C: Radio Observations
    Edited by M. L. Meeks

Volume 12

Astrophysics

PART B: Radio Telescopes

Edited by

M. L. MEEKS

*Massachusetts Institute of Technology
Cambridge, Massachusetts
and
Northeast Radio Observatory Corporation
Haystack Observatory
Westford, Massachusetts*

1976

ACADEMIC PRESS · **New York** **San Francisco** **London**
A Subsidiary of Harcourt Brace Jovanovich, Publishers

COPYRIGHT © 1976, BY ACADEMIC PRESS, INC.
ALL RIGHTS RESERVED.
NO PART OF THIS PUBLICATION MAY BE REPRODUCED OR
TRANSMITTED IN ANY FORM OR BY ANY MEANS, ELECTRONIC
OR MECHANICAL, INCLUDING PHOTOCOPY, RECORDING, OR ANY
INFORMATION STORAGE AND RETRIEVAL SYSTEM, WITHOUT
PERMISSION IN WRITING FROM THE PUBLISHER.

ACADEMIC PRESS, INC.
111 Fifth Avenue, New York, New York 10003

United Kingdom Edition published by
ACADEMIC PRESS, INC. (LONDON) LTD.
24/28 Oval Road, London NW1

Library of Congress Cataloging in Publication Data

Main entry under title:

Astrophysics.

    (Methods of experimental physics ; v. 12)
    Includes bibliographical references.
    CONTENTS: pt. B. Radio telescopes.
    1.   Astrophysics—Methodology.   I.   Meeks, Marion Littleton,   (date)   II.   Series.
QB461.A773   523.01   75-34188
ISBN 0–12–475952–1 (v. 12, pt. B)

PRINTED IN THE UNITED STATES OF AMERICA

CONTENTS

CONTRIBUTORS xi

FOREWORD xiii

PREFACE xv

CONTENTS OF VOLUME 12, PARTS A AND C xvii

CONTRIBUTORS TO VOLUME 12, PARTS A AND C xxi

1. Radio Telescopes

   1.1. Essentials of Radiometric Measurements 1
      by M. L. MEEKS

       1.1.1. Introduction 1
       1.1.2. Antenna Considerations 2
       1.1.3. Radiometer Considerations 5

   1.2. Types of Astronomical Antennas 7
      by W. J. WELCH

       1.2.1. Introduction 7
       1.2.2. Pencil-Beam Antennas 10
       1.2.3. Aperture Synthesis 24

   1.3. Analysis of Paraboloidal-Reflector Systems 29
      by W. V. T. RUSCH

       1.3.1. Introduction 29
       1.3.2. Analysis of the Paraboloid with a Prime-Focus Feed 30
       1.3.3. Analysis of Multireflector Systems 48

1.4. Feed Systems for Paraboloidal Reflectors. 64
     by JOHN RUZE

     1.4.1. Introduction 64
     1.4.2. Basic Feed Types 64
     1.4.3. Horn Feeds 65
     1.4.4. Dipole Feeds 71
     1.4.5. Loop Feed 73
     1.4.6. Log-Periodic Dipole Array 73
     1.4.7. Helix Feed 74
     1.4.8. Conical Spiral 76
     1.4.9. Flat Spirals 77
     1.4.10. Tertiary Reflector Systems 77
     1.4.11. Summary of Feed Types 81

1.5. Antenna Calibration 82
     by R. WIELEBINSKI

     1.5.1. Introduction 82
     1.5.2. Natural Radio Sources 83
     1.5.3. Determination of the Flux Scale 84
     1.5.4. The Calibrating Sources 86
     1.5.5. Mechanical Measurements 94
     1.5.6. Pointing Theory 96

1.6. Practical Problems of Antenna Arrays 98
     by J. C. JAMES

     1.6.1. Introduction 98
     1.6.2. A 16-Dipole Array 98
     1.6.3. Array Patterns 101
     1.6.4. Grating Lobes 102
     1.6.5. Considerations When Building an Array 104
     1.6.6. Practical Suggestions 110
     1.6.7. Miscellaneous Notes 117

2. Atmospheric Effects

   2.1. The Ionosphere 119
        by TOR HAGFORS

        2.1.1. Introduction 119
        2.1.2. Propagation in the Regular Ionosphere 120

2.1.3.	Ionospheric Absorption of Electromagnetic Waves	125
2.1.4.	Faraday Rotation	126
2.1.5.	Phase and Group Delays in the Ionosphere	127
2.1.6.	Refraction Effects	128
2.1.7.	Propagation in the Irregular Ionosphere	130
2.1.8.	The Effect of Small-Scale Irregularities	131
2.1.9.	The Effect of Large-Scale Irregularities	133
2.1.10.	Ionospheric Measurements	134

2.2. Structure of the Neutral Atmosphere 136
by R. K. CRANE

2.2.1.	Introduction	136
2.2.2.	Atmospheric Temperature	136
2.2.3.	Atmospheric Pressure	138
2.2.4.	Atmospheric Water Vapor	139
2.2.5.	Propagation Effects	140

2.3. Absorption and Emission by Atmospheric Gases 142
by J. W. WATERS

2.3.1.	Introduction	142
2.3.2.	Radiative Transfer at Microwave Frequencies	142
2.3.3.	Microwave Spectral Line Absorption	145
2.3.4.	The Microwave Spectrum of the Terrestrial Atmosphere	172

2.4. Extinction by Condensed Water 177
by R. K. CRANE

2.4.1.	Introduction	177
2.4.2.	Solution to the Scattering Problem	177
2.4.3.	Effects of Drop-Size Distributions	178
2.4.4.	Models for Attenuation Computations	180
2.4.5.	Single-Scattering Albedo	182
2.4.6.	Multiple Scattering	183
2.4.7.	Measured Attenuation	184

2.5. Refraction Effects in the Neutral Atmosphere 186
     by R. K. Crane

     2.5.1. Introduction 186
     2.5.2. Radio Refractivity 186
     2.5.3. Bending 190
     2.5.4. Path Length 196

3. Radiometers

3.1. Radiometer Fundamentals 201
     by R. M. Price

     3.1.1. Introduction 201
     3.1.2. Types of Signals in Radio Astronomy 201
     3.1.3. Measurement of Radio Astronomy Signals 202
     3.1.4. The Basic Receiver System 210
     3.1.5. Practical Receiver Configurations 212
     3.1.6. Special-Purpose Receivers 219
     3.1.7. Present Trends in Receiver Systems 223
     3.1.8. Considerations in Radiometer System Design . . . 223

3.2. Parametric Amplifiers 225
     by Jochen Edrich

     3.2.1. Fundamentals of Nonlinear Reactances 225
     3.2.2. Fundamentals of Parametric Amplifiers 228
     3.2.3. Design Considerations and Practical Parametric Amplifiers 233

3.3. Maser Amplifiers 246
     by K. Sigfrid Yngvesson

     3.3.1. Basic Properties of Maser Amplifiers 246
     3.3.2. Systems and Operational Considerations 257
     3.3.3. Data for Specific Maser Amplifiers and Systems . . 260
     3.3.4. Millimeter Wave Masers 262

3.4.	Multichannel-Filter Spectrometers		266
	by HAYS PENFIELD		
	3.4.1.	General Description	266
	3.4.2.	Filter Characteristics	269
	3.4.3.	Detectors	272
	3.4.4.	Integrators	272
	3.4.5.	Special Operating Features	275
	3.4.6.	Output Devices	276
	3.4.7.	Calibration	277
3.5.	Autocorrelation Spectrometers		280
	by B. F. C. COOPER		
	3.5.1.	Introduction	280
	3.5.2.	Sampling and Quantizing Considerations	281
	3.5.3.	Computation of the Power Spectrum	284
	3.5.4.	Standard Deviation of Spectral Estimate	285
	3.5.5.	Spectrum Denormalization	285
	3.5.6.	Prefilters and Video Converters	287
	3.5.7.	Digital Correlator Logic	288
	3.5.8.	A Scheme for Optional One-Bit or Two-Bit Correlation	294
	3.5.9.	Extension to Cross-Correlation Spectrometry	295
	3.5.10.	Some Examples of Correlation Spectrometers	297

INDEX FOR VOLUME 12, PART B 299

INDEX FOR VOLUME 12, PART C 304

CONTRIBUTORS

Numbers in parentheses indicate the pages on which the authors' contributions begin.

B. F. C. COOPER, *Division of Radiophysics, Commonwealth Scientific and Industrial Research Organization, Epping, NSW, Australia* (280)

R. K. CRANE, *Lincoln Laboratory, Massachusetts Institute of Technology, Lexington, Massachusetts* (136, 177, 186)

JOCHEN EDRICH, *Department of Electrical Engineering and Denver Research Institute, University of Denver, Denver, Colorado* (225)

TOR HAGFORS,* *Lincoln Laboratory, Massachusetts Institute of Technology, Lexington, Massachusetts* (119)

J. C. JAMES, *Teledyne Brown Engineering, Huntsville, Alabama* (98)

M. L. MEEKS, *Haystack Observatory, Northeast Radio Observatory Corporation, Westford, Massachusetts* (1)

HAYS PENFIELD, *Center for Astrophysics, Harvard College Observatory, Cambridge, Massachusetts* (266)

R. M. PRICE, *Physics Department and Research Laboratory of Electronics, Massachusetts Institute of Technology, Cambridge, Massachusetts* (201)

W. V. T. RUSCH, *Department of Electrical Engineering, University of Southern California, Los Angeles, California* (29)

JOHN RUZE, *Lincoln Laboratory, Massachusetts Institute of Technology, Lexington, Massachusetts* (64)

J. W. WATERS,† *Research Laboratory of Electronics, Massachusetts Institute of Technology, Cambridge, Massachusetts* (142)

W. J. WELCH, *Radio Astronomy Laboratory, University of California, Berkeley, California* (7)

* Present address: Norges Tekniske Hogskole, Trondheim, Norway.
† Present address: Jet Propulsion Laboratory, California Institute of Technology, Pasadena, California.

R. WIELEBINSKI, *Max-Planck Institut für Radioastronomie, Bonn, German Federal Republic* (82)

K. SIGFRID YNGVESSON, *Electrical and Computer Engineering Department, University of Massachusetts, Amherst, Massachusetts* (246)

FOREWORD

I have already had the pleasure to welcome to our treatise Part A of the Methods of Astrophysics, edited by Professor N. Carleton; Parts B and C, edited by Dr. M. L. Meeks, are introduced here. They contain a presentation of the methods used in radio astrophysics, a branch of physics which in a short time developed to a degree that in some respects it surpasses the present methods used in optical astrophysics.

In earlier volumes I used the opportunity offered by the Foreword to announce expected additions to this series. Professor D. Williams, who edited both the first and second editions of our Molecular Physics volume, kindly consented to edit a volume on Spectroscopy. Professor G. Weissler is organizing a volume on Vacuum Technology methods, and discussions are underway to cover several other areas of physics.

My warmest thanks to Dr. Meeks and his authors for a remarkable and great effort in putting together these volumes. I hope the scientific community will agree with my judgment on its outstanding quality.

L. MARTON

PREFACE

Many different fields have contributed to the methods of radio astronomy, for example, antenna theory, the engineering of radiometers and digital systems, and computer programming for data analysis and display. We intend that these volumes provide a guide to the various methods currently used in radio astronomy whether the reader be an astronomer, a graduate student, a physicist working in a related field, or an engineer developing equipment for a radio telescope.

Although most of the thirty-one authors represented are radio astronomers, the others are specialists who have contributed directly to the advancement of radio astronomy through work in their own field. Interdisciplinary collaboration, particularly between electrical engineers and astronomers, has been fruitful in the development of new techniques and devices in this field. We hope that these volumes will encourage further collaboration by bringing together descriptions of the essential methods from many different points of view. The dissimilar backgrounds of the contributors has, however, presented some editorial problems. It has not proven feasible to standardize the notation throughout the volumes; for example, frequency is designated by v in some chapters and f in others.

Volumes 12B and 12C were conceived as a single-volume companion to Professor N. Carleton's Volume 12A dealing with the optical and infrared regions of the spectrum. When all chapters for the radio volume were completed and edited, the decision was made to publish the material in two volumes rather than one.

Let me take this opportunity to thank all the contributors for their efforts, cooperation, and patience in the preparation of this volume.

M. L. MEEKS

CONTENTS OF VOLUME 12, PARTS A AND C

PART A

1. Photomultipliers: Their Cause and Cure
   by ANDREW T. YOUNG

   1.1. Introduction
   1.2. An Idealized Photomultiplier
   1.3. Basic Physics of Photomultipliers
   1.4. Real Photomultipliers
   1.5. Photomultipliers and System Components

2. Other Components in Photometric Systems
   by ANDREW T. YOUNG

   2.1. Optical Systems
   2.2. Calibration Problems and Standard Sources
   2.3. Principles of Photometer Design

3. Observational Technique and Data Reduction
   by ANDREW T. YOUNG

   3.1. Atmospheric Extinction
   3.2. Transformation to a Standard System

4. Reshaping and Stabilization of Astronomical Images
   by DONALD M. HUNTEN

   4.1. Reshaping of Images
   4.2. Stabilization of Images

5. Detective Performance of Photographic Plates
   by D. W. LATHAM

   5.1. Introduction
   5.2. Photographic Photometry
   5.3. Signal-to-Noise and Detective Quantum Efficiency
   5.4. Detective Performance of Kodak Spectroscopic Plates, Types IIa-0, 103a-0, and IIIa-J

6. Two-Dimensional Electronic Recording
    - 6.1. Phosphor Output Image Tubes   by E. J. Wampler
    - 6.2. Electrographic Tubes   by Gerald E. Kron
    - 6.3. Television Systems for Astronomical Applications   by John L. Lowrance and Paul Zucchino

7. X-Ray and Gamma-Ray Detection by Means of Atmospheric Interaction: Fluorescence and Čerenkov Radiation
   by G. G. Fazio
    - 7.1. Introduction
    - 7.2. Detection of Cosmic X Rays by Atmospheric Fluorescence
    - 7.3. Detection of Cosmic Gamma Rays by Atmospheric Čerenkov Radiation

8. Polarization Techniques
   by K. Serkowski
    - 8.1. Introduction
    - 8.2. Analyzers for Linearly Polarized Light
    - 8.3. Retarders
    - 8.4. Depolarizers
    - 8.5. Optimum Design of an Astronomical Polarimeter
    - 8.6. Instrumental Corrections
    - 8.7. Astronomical Polarimetry in the Future: Television and Image Tube Techniques

9. The Instrumentation and Techniques of Infrared Photometry
   by F. J. Low and G. H. Rieke
    - 9.1. Introduction
    - 9.2. Detectors
    - 9.3. Associated Apparatus
    - 9.4. Telescope Design
    - 9.5. Modulation and Space Filtering Techniques
    - 9.6. Atmospheric Limitations
    - 9.7. The Infrared Photometric System
    - 9.8. Observing Procedure

10. Diffraction Grating Instruments
    by Daniel J. Schroeder
    - 10.1. General Spectrometer Considerations
    - 10.2. Diffraction Gratings

- 10.3. Grating Spectrometers
- 10.4. Echelle Spectrometers
- 10.5. Concluding Comments

11. Fourier Spectrometers
    by HERBERT W. SCHNOPPER and RODGER I. THOMPSON

    - 11.0. Introduction
    - 11.1. Historical Background
    - 11.2. Theory of Fourier Transform Spectroscopy
    - 11.3. Fourier Spectroscopy in Practice
    - Appendix A

12. Fabry–Perot Instruments for Astronomy
    by F. L. ROESLER

    - 12.0. Introduction
    - 12.1. The Ideal Fabry–Perot Interferometer
    - 12.2. Application of the Fabry–Perot Interferometer as a Spectrometer
    - 12.3. Multiple Fabry–Perot Spectrometers
    - 12.4. Observation of Astronomical Sources with Fabry–Perot Spectrometers
    - 12.5. Examples of Basic Fabry–Perot Spectrometer Design for Astronomical Observations
    - 12.6. Adjustment and Calibration of Fabry–Perot Spectrometers
    - 12.7. Comparison with Other Instruments

PART C

4. Single-Antenna Observations

    - 4.1. Observations of Small-Diameter Sources   by JOHN R. DICKEL
    - 4.2. Fundamentals of Spectral-Line Measurements   by D. R. W. WILLIAMS
    - 4.3. Measurements with Radio-Frequency Spectrometers   by J. A. BALL
    - 4.4. Measurements of Galactic 21-cm Hydrogen   by CARL HEILES and G. T. WRIXON
    - 4.5. Pulsar Observing Techniques   by G. RICHARD HUGUENIN
    - 4.6. Lunar Occultation Measurements   by C. HAZARD
    - 4.7. Scintillation Measurements   by L. T. LITTLE

5. Interferometers and Arrays

    5.1. Theory of Two-Element Interferometers by A. E. E. ROGERS
    5.2. Connected-Element Interferometry by GUY POOLEY
    5.3. Very Long Baseline Interferometer Systems by J. M. MORAN
    5.4. Frequency and Time Standards by ROBERT F. C. VESSOT
    5.5. Very Long Baseline Interferometric Observations and Data Reduction by J. M. MORAN
    5.6. The Estimation of Astrometric and Geodetic Parameters by IRWIN I. SHAPIRO

6. Computer Programs for Radio Astronomy

    6.1. Radial-Velocity Corrections for Earth Motion by M. A. GORDON
    6.2. The Fast Fourier Transform by NORMAN BRENNER
    6.3. Data Presentation Techniques
        6.3.1. Contour Mapping by NORMAN BRENNER and STANLEY H. ZISK
        6.3.2. Ruled-Surface Mapping by NORMAN BRENNER
        6.3.3. Gray-Scale Mapping by NORMAN BRENNER

Appendixes A–K

CONTRIBUTORS TO VOLUME 12, PARTS A AND C

PART A

G. G. FAZIO, *Smithsonian Astrophysical Observatory, Cambridge, Massachusetts*

DONALD M. HUNTEN, *Kitt Peak National Observatory, Tucson, Arizona*

GERALD E. KRON, *US Naval Observatory, Flagstaff Station, Flagstaff, Arizona*

D. W. LATHAM, *Smithsonian Astrophysical Observatory, Cambridge, Massachusetts*

F. J. LOW, *Lunar and Planetary Laboratory, University of Arizona, Tucson, Arizona*

JOHN L. LOWRANCE, *Princeton University Observatory, Princeton University, Princeton, New Jersey*

G. H. RIEKE, *Lunar and Planetary Observatory, University of Arizona, Tuscon, Arizona*

F. L. ROESLER, *Department of Physics, University of Wisconsin, Madison, Wisconsin*

HERBERT W. SCHNOPPER, *Department of Physics, Massachusetts Institute of Technology, Cambridge, Massachusetts*

DANIEL J. SCHROEDER, *Thompson Observatory, Beloit College, Beloit, Wisconsin*

K. SERKOWSKI, *Lunar and Planetary Laboratory, University of Arizona, Tucson, Arizona*

RODGER I. THOMPSON, *Steward Observatory, University of Arizona, Tucson, Arizona*

E. J. WAMPLER, *Lick Observatory, Board of Studies in Astronomy and Astrophysics, University of California, Santa Cruz, California*

ANDREW T. YOUNG, *Department of Physics, Texas A & M University, College Station, Texas*

PAUL ZUCCHINO, *Princeton University Observatory, Princeton University, Princeton, New Jersey*

PART C

J. A. BALL, *Center for Astrophysics, Harvard College Observatory and Smithsonian Astrophysical Observatory, Cambridge, Massachusetts*

NORMAN BRENNER, *Department of Earth and Planetary Sciences, Massachusetts Institute of Technology, Cambridge, Massachusetts*

JOHN R. DICKEL, *University of Illinois Observatory, Urbana, Illinois*

M. A. GORDON, *National Radio Astronomy Observatory, Green Bank, West Virginia*

C. HAZARD, *Institute of Astronomy, University of Cambridge, Cambridge, England*

CARL HEILES, *Astronomy Department, University of California, Berkeley, California*

G. RICHARD HUGUENIN, *Department of Physics and Astronomy, University of Massachusetts, Amherst, Massachusetts*

L. T. LITTLE, *Electronics Laboratory, University of Kent, Canterbury, Kent, England*

J. M. MORAN, *Center for Astrophysics, Harvard College Observatory and Smithsonian Astrophysical Observatory, Cambridge, Massachusetts*

GUY POOLEY, *Mullard Radio Astronomy Observatory, University of Cambridge, Cambridge, England*

A. E. E. ROGERS, *Haystack Observatory, Northeast Radio Observatory Corporation, Westford, Massachusetts*

IRWIN I. SHAPIRO, *Department of Earth and Planetary Sciences and Department of Physics, Massachusetts Institute of Technology, Cambridge, Massachusetts*

ROBERT F. C. VESSOT, *Center for Astrophysics, Harvard College Observatory and Smithsonian Astrophysical Observatory, Cambridge, Massachusetts*

D. R. W. WILLIAMS, *Radio Astronomy Laboratory, University of California, Berkeley, California*

G. T. WRIXON,[*] *Bell Telephone Laboratories, Crawford Hill Laboratory, Holmdel, New Jersey*

STANLEY H. ZISK, *Haystack Observatory, Northeast Radio Observatory Corporation, Westford, Massachusetts*

---

[*] Present address: Department of Electric Engineering, University College, Cork, Ireland.

1. RADIO TELESCOPES

1.1. Essentials of Radiometric Measurements*

1.1.1. Introduction

Radio astronomy covers a frequency range extending from a few megahertz up to a frequency of about 300 GHz or equivalently a wavelength range from roughly a hundred meters down to 1 mm. The low-frequency limit of this band is set by the transmission properties of the ionosphere and the high-frequency limit by the onset of extensive absorption by water vapor in the atmosphere. Over this entire region amplifiers are available which preserve the phase of the received signal (coherent amplifiers) as contrasted with the infrared region of the spectrum where power detectors are presently used. Hence it is possible to consider the radio domain as a whole in discussing equipment, observing techniques, and methods of analysis.

A radio telescope in its simplest form consists of three elements: (1) an antenna that selectively collects radiation from a small region of sky, (2) a radiometric receiver, referred to as a "radiometer," that amplifies a restricted frequency band from the output of the antenna, and (3) an indicator that registers the radiometer output so that it may be recorded by the observer. Radio emission from astronomical sources is noiselike in character, and the signal must be measured in the presence of several kinds of extraneous noise such as thermal emission from the surroundings and noise generated within the radiometer itself.

In making an observation the observer typically steers the antenna so that it points alternately toward a radio source and toward a comparison region where there is no radio source but where the background noise is otherwise identical. The difference between indicator readings on and off source represents the power received from the astronomical source.

Radiometers are designed to be linear in power so the received signal can be calibrated in terms of a reference noise source. The reference or calibration signal is superimposed on the signal from the antenna terminals and is usually injected while the antenna is pointing toward the comparison region.

The radiation measured with radio telescopes can be related to thermal

---

* Chapter 1.1 is by M. L. Meeks.

radiation at specific temperatures. Consider blackbody radiation from a surface with absolute temperature T. The *specific intensity* of thermal radiation per unit bandwidth is a function of frequency v, and will be designated I_v. The Rayleigh–Jeans approximation gives

$$I_v = 2kTv^2/c^2 \tag{1.1.1}$$

provided that $hv/k \ll 1$, where k is Boltzmann's constant and c is the velocity of light. For $T = 10°\text{K}$ and $v \ll 200$ GHz the condition for validity of this approximation is satisfied, and so Eq. (1.1.1) applies to nearly all radio observations. It is important to note that I_v is thus proportional to the absolute temperature, and we can characterize a particular specific intensity I_v by a corresponding equivalent brightness temperature T_B by means of Eq. (1.1.1). This correspondence can be used to define an equivalent brightness temperature for nonthermal radiation.

Now let us consider the problem of calibrating a radio telescope, and the units to be used for measuring the received power. If we surround an antenna with an enclosure at a uniform temperature T_A, or equivalently replace the antenna with a matched load at temperature T_A, then it can be shown[1] that the input power amplified by the radiometer will be given by

$$P_v \, \Delta v = kT_A \, \Delta v, \tag{1.1.2}$$

where P_v is the input power per unit bandwidth and Δv the radiometer bandwidth. Equation (1.1.2) in this case provides a relationship between P_v and the temperature T_A, the so-called *antenna temperature*. With two or more matched loads at well-established temperatures one can determine the change in equivalent antenna temperature that the calibration signal represents. Thus a scale can be established for measuring noise power in terms of antenna temperature in degrees Kelvin. This practice conveniently makes the measurements independent of radiometer bandwidth.

1.1.2. Antenna Considerations

Antennas in radio astronomy are characterized by their receiving pattern, their polarization properties, and their frequency response. Conceptually it is simpler to consider first antennas as radiators rather than receivers of radio waves. Let us begin by defining the *directivity function* $D(\theta, \phi)$, where θ and ϕ are the angles in a spherical coordinate system attached to the antenna. The directivity is defined in terms of the radiated power $P(\theta, \phi)$ as

$$D(\theta, \phi) = \frac{P(\theta, \phi)}{P_T/4\pi}, \tag{1.1.3}$$

---

[1] J. D. Kraus, "Radio Astronomy," pp. 97–101. McGraw-Hill, New York, 1966.

1.1. ESSENTIALS OF RADIOMETRIC MEASUREMENTS

where P_T is the power radiated by the antenna. For a lossless, isotropic antenna the directivity becomes $D = 1$, and in general the integral over all solid angles Ω is thus

$$\iint_{4\pi} D(\theta, \phi) \, d\Omega = 4\pi. \tag{1.1.4}$$

The power gain function $G(\theta, \phi)$ is defined similarly as

$$G(\theta, \phi) = \frac{P(\theta, \phi)}{P_S/4\pi} \tag{1.1.5}$$

except that P_S is the power supplied to the antenna terminals. The ratio $G(\theta, \phi)/D(\theta, \phi)$ is the so-called *radiation efficiency* $\eta_R = P_T/P_S$. Ohmic losses in the antenna account for η_R being less than one.

Antennas usually have gain patterns with a main beam in which most of the radiation is concentrated and a distribution of much weaker sidelobes in various other directions. The radiation pattern may be characterized by the angular dimensions of the main beam, usually measured to the half maximum power contour, and by the relative intensity of the strongest sidelobe. A complete description is provided by the *antenna pattern* $f(\theta, \phi)$, a function normalized to the maximum value of the directivity or gain, D_0 or G_0, respectively, so that

$$f(\theta, \phi) = \frac{D(\theta, \phi)}{D_0} = \frac{G(\theta, \phi)}{G_0}. \tag{1.1.6}$$

Figure 1 shows two representations of a typical antenna pattern for a pencil-beam antenna. This pattern was measured by scanning the antenna over an intense radio source whose angular size was small in comparison with the size of the main beam.

Considering an antenna as a collector of radio waves, we think in terms of an aperture with an *effective area* $A(\theta, \phi)$, the capture area for plane waves traveling in directions specified by θ and ϕ. The intercepted power is then given by

$$P_\nu \, \Delta\nu = \frac{\Delta\nu}{2} \iint_{4\pi} I_\nu A(\theta, \phi) \, d\Omega. \tag{1.1.7}$$

The factor $\tfrac{1}{2}$ in the above equation reflects the fact that an antenna can respond to only one kind of polarization, and I_ν represents an unpolarized intensity containing both polarizations, for example, right and left circular. From thermodynamic considerations it can be shown[1] that the angular distributions $G(\theta, \phi)$ and $A(\theta, \phi)$ are identical except for a constant factor

$$G(\theta, \phi) = (4\pi/\lambda^2) A(\theta, \phi). \tag{1.1.8}$$

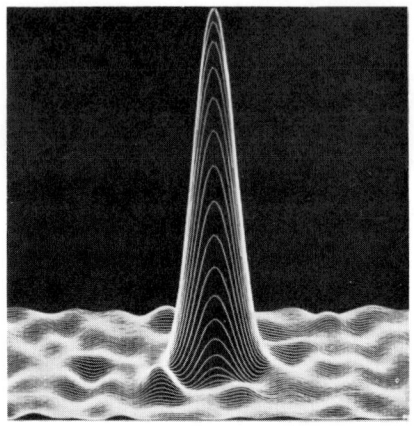
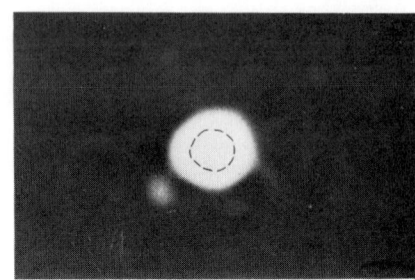

Fig. 1. Two representations of the antenna pattern of the Haystack antenna, a 120-ft diameter paraboloidal reflector measured at a frequency of 15.5 GHz. The radio source Virgo A was used to obtain this pattern, and the two images represent the same data displayed by means of computer graphics. A small coma-type sidelobe may be seen above the background noise.

We can rewrite Eq. (1.1.7) in terms of measured antenna temperature T_A and the distribution of brightness temperature T_B over the sky as specified by Eqs. (1.1.1) and (1.1.2). This procedure gives the following important equation as a result:

$$T_A = \frac{1}{\lambda^2} \iint_{4\pi} T_B(\theta, \phi) A(\theta, \phi) \, d\Omega. \qquad (1.1.9)$$

We will consider one special case involving the above equation. Other cases are discussed in Chapter 4.1. If the angular extent of the source region is very much smaller than the beamwidth of the antenna (a "point source") and if the beam is centered on the source, then the maximum effective area of the antenna may be taken as a constant A_0 so that

$$T_A = (A_0/2k) S_\nu, \qquad (1.1.10)$$

where S_ν, the so-called *flux density*, is given by

$$S_\nu = \iint_{\text{source}} I_\nu \, d\Omega. \qquad (1.1.11)$$

The unit of flux density recently adopted is the *jansky*, abbreviated Jy. This unit represents 10^{-26} Wm^{-2} Hz^{-1} and replaces the previous designation of *flux unit* for this quantity.

1.1. ESSENTIALS OF RADIOMETRIC MEASUREMENTS

A figure of merit for an antenna system called the *aperture efficiency* η is defined in terms of effective area A_0 along the axis of the main beam and the geometric area of the antenna A_{geo} as:

$$\eta = A_0/A_{geo}. \qquad (1.1.12)$$

The effective area and aperture efficiency of an antenna may be measured by observing a small-diameter radio source of known flux density and making use of Eqs. (1.1.10) and (1.1.12). In practice, however, it is necessary to take into account absorption by the atmosphere when making observations, particularly at short radio wavelengths. The various effects of the atmosphere are discussed in Part 2. The chapters that follow in Part 1 consider various aspects of antenna systems used in radio astronomy.

1.1.3. Radiometer Considerations

Since the emission from astronomical sources is noiselike in character, the problem of observing these signals involves detecting and measuring very small changes in noise level against a background of extraneous noise, including noise generated by the radiometer and noise picked up by the antenna from the ground and from the atmosphere. Because noise is generated in the radiometer, a fluctuating output would be obtained even if a matched load at absolute zero were connected to the input terminals. It is customary to specify the radiometer noise in terms of an equivalent noise power referred to the input terminals of the radiometer. In other words a noise power per unit bandwidth P_N injected into an ideal noiseless amplifier would produce the same output noise level that is observed from the actual radiometer. This equivalent input noise power can be expressed as

$$P_N \Delta v = kT_N \Delta v. \qquad (1.1.13)$$

This equation serves to define the *noise temperature* T_N of the radiometer.

A radio telescope pointed toward a source that produces an antenna temperature T_{source} would receive a total input noise temperature given by

$$T_{source} + T_{system} = T_{source} + (T_N + T_{ant} + T_{atmos})$$

Here the *system temperature* T_{system} includes the effects of noise picked up by the antenna through sidelobes directed toward the ground T_{ant} and the noise contribution of the atmosphere T_{atmos}. The radiometer output will be a fluctuating signal, and from general statistical considerations[2] the root mean square (rms) fluctuation ΔT will be given by

$$\Delta T = (\text{const}) \frac{T_{source} + T_{system}}{(\tau \Delta v)^{1/2}}, \qquad (1.1.14)$$

[2] T. Hagfors and M. L. Meeks, *Astron. J.* **69**, 447 (1964).

where τ is the effective integration time and $\Delta \nu$ is the radiometer bandwidth. The constant factor will have a value between one and roughly three depending on details of the radiometric system (see Section 3.1.2). Equation (1.1.14) shows that with a given system the rms noise may be reduced by increasing the integration time, but the decrease goes as the square root of τ. Similarly if the bandwidth can be increased, the sensitivity will be increased as the square root of $\Delta \nu$. In the case of spectral-line observations, however, the bandwidth is limited by the linewidth of the astronomical signal. Line observations require a low system noise to obtain sensitivity, but for continuum observations wide bandwidth may compensate for a higher noise temperature. Part 3 of this book discusses various radiometer systems, and Chapter 3.1 provides a detailed introduction to the subject.

Single-antenna observations are considered in Part 4, and this portion of the book deals with techniques of measurement primarily, rather than with equipment. Part 5 is concerned with interferometers and arrays; discussions of the equipment, observing techniques, and data analysis are included. Finally Part 6 deals with computer programs that are generally useful in radio astronomy.

1.2. Types of Astronomical Antennas*

1.2.1. Introduction

The variety in the equipment which has been built for radio astronomical observations is probably greater than in any other branch of astronomy. This variety is due at least in part to the large range of wavelengths over which observations are made: from a 100-m wavelength down to 1 mm, a range of about 10^5. Whereas quasi-optical antennas such as the parabolic reflector are appropriate for millimeter wavelengths, multielement arrays of elementary dipole antennas are suitable for the decameter wavelengths. The choice of instrumentation depends also, of course, on the particular observational goals. For example, whenever the goal has been image formation, particularly with high angular resolution, multiple antenna systems have generally been used.

Every antenna has the following general characteristics: (a) *input impedance*: the impedance appearing at its input terminals when it is coupled to a transmitter; (b) *polarization*: the sense of polarization that it radiates or receives in every direction. This may be linear, circular, or elliptical; (c) *radiation pattern*: the field strength that the antenna radiates in every direction. The simplest antenna normally radiates most of its energy in one direction, with the angular width of this main lobe or pencil beam determined by the size and design of the antenna. Weaker secondary maxima in other directions are called *sidelobes*; (d) *gain* or *directivity*; the radiated power in the direction of the main beam relative to what would be radiated by an isotropic antenna in that direction; (e) *effective collecting area* A_e; the product of A_e with the flux of radiation from a given direction is the power delivered to a matched receiver at the antenna terminals. The effective collecting area in a given direction is proportional to the gain in that direction; (f) the *bandwidth* or frequency dependence of all the above parameters. A final characteristic which is often a mechanical rather than electrical property concerns the pointing capability of the antenna. It is the accuracy and reproducibility with which the main lobe can be directed toward a particular direction in the sky.

One important property of any antenna operating in a linear reciprocal medium is that its radiation pattern when it is used as a transmitter is the same as its receiving pattern This is a consequence of the fact that electro-

---

* Chapter 1.2 is by W. J. Welch.

magnetic fields obey a simple reciprocity principle. Even though a radio astronomical antenna will only be used for reception, it is often easier to visualize and calculate its pattern by considering it in the transmitting mode. This practice is so common that it permeates the terminology used to describe antennas. For example, a dipole placed at the focus of a paraboloid to receive the signal from a distant source is generally called the antenna "feed," as though it were coupled to a transmitter rather than a receiver.

The receiving pattern of the elementary paraboloidal antenna is best understood by visualizing it in the transmitting mode. Figure 1 illustrates

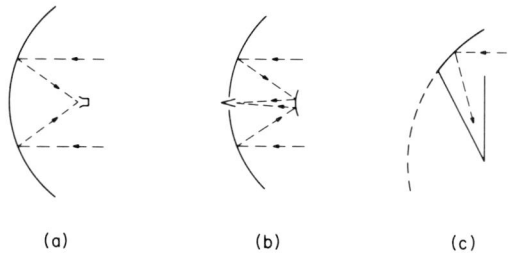

Fig. 1. Paraboloidal antennas: (a) with the feed at the prime focus; incoming rays are shown; (b) the Cassegrain antenna; the secondary reflector is a hyperboloid; (c) the horn reflector.

three types of paraboloidal antennas. Because of the geometrical properties of the parabola, the field radiated by an elementary dipole at the focus and reflected from the parabola has a plane phase front at the aperture plane of the reflector. The antenna therefore radiates as though it were an aperture in an opaque screen illuminated by a plane wave. The radiation pattern is then just the same as the diffraction pattern of a circular aperture having the same diameter as the reflector. A cross section of this familiar pattern appears in Fig. 2. The angular half-width of the central lobe or "pencil beam" is inversely proportional to the diameter of the aperture divided by the operating wavelength. The secondary lobes or sidelobes are also evident in the figure. The angular resolution of the paraboloid is limited by the width of the central lobe. This so-called "Rayleigh limit" may be theoretically less than 1/10 arc sec for an optical telescope, but it is typically 1 arc min or more for even the largest paraboloidal antennas. This example illustrates the general fact that the angular resolving power of any antenna is inversely proportional to its maximum "aperture" dimension measured in wavelengths.

One characteristic difference between radio telescopes and optical telescopes is the way in which the signal detector is coupled to the telescope. For

1.2. TYPES OF ASTRONOMICAL ANTENNAS

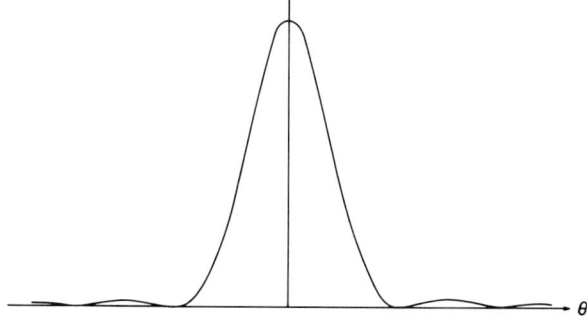

FIG. 2. Cross section of the pattern of a circular-aperture antenna. The angular width of the main lobe is proportional to the diameter of the aperture divided by the operating wavelength.

the former there is no simple equivalent of the retina of the eye or the photographic plate which may be placed in the focal plane to form a detailed image of an astronomical object. The output of an antenna directed at an astronomical radio source is simply a voltage at a pair of terminals which is coupled to a linear amplifier (often through a transmission line) and finally detected and recorded. The photographic plate may be regarded as a detector of a large number of spatial modes as compared with the single spatial mode represented by the single terminal-pair output of the antenna. The diffraction pattern or receiving pattern of the antenna is the projection of this single mode on the celestial sphere. A map of the radio source is obtained in sequence by scanning the pencil beam across the source and recording the changes in detected output. Note that the single mode contains only one sense of polarization, whether it be linear, circular, or elliptical, and the map will be in only one sense of polarization. In principle, an instantaneous image of the radio source may be formed by detecting the outputs of an array of receiving dipoles located in the focal plane of the antenna. Because each voltage must be coupled to a separate amplifier and receiving system, it is impractical to obtain more than a few spatial modes in this way. In fact, most single-reflector, radio-telescope antennas operate with a single-output terminal pair.

Another characteristic difference between radio telescopes and optical telescopes is that the resolution of the former is determined by the diffraction (Rayleigh) limit whereas for the latter it is generally determined by atmospheric "seeing" or imperfections in the figure of the reflector. There are several reasons for the difference. Whereas the Rayleigh limit for a large optical telescope may be 1/10 arc sec or smaller, atmospheric turbulence generally restricts resolution to about 1 arc sec or greater. Because of this, the figure of the mirror is usually smoothed to provide a resolution of no better than about 0.3 sec. Because of the multimode character of the photographic plate, these effects produce an extended image of a star (a point source) on the plate, and none of the starlight is lost. The situation for the radio telescope is different. Atmospheric "seeing" effects are negligible at radiowavelengths. The telescope resolution is then determined by either diffraction or by roughness of the antenna construction. Because of the single-mode character of the antenna feed, the effect of antenna roughness is to scatter the radiation arriving from a point source into modes which are not received by the "feed" at the focus. This radiation is lost. Hence it is the general practice to build the radio telescope so that surface roughness or construction imperfections do not significantly affect the operation of the instrument at the shortest intended operating wavelength. The criterion for this is that the rms surface deviation from the ideal figure be not greater than about 1/20 of the shortest operating wavelength.

Of the various characteristics of antennas, the two which are probably the most important in radio astronomical applications are the angular resolution of the antenna and its collecting area. Quite a wide range of these two parameters has been achieved in practice, with collecting areas as large as 10^6 m^2 and angular resolution better than 10^{-3} arc sec.

Existing systems generally fall into two categories: pencil- or single-beam antennas and interferometers. A further division which applies to the former category is into filled and unfilled aperture systems. For the filled aperture antenna, the angular resolution and the collecting area are related. For the unfilled aperture, these two parameters may be specified independently, a possibility which offers a number of advantages. In the following sections we describe briefly a number of existing radio-telescope antennas of various types.

1.2.2. Pencil-Beam Antennas

The pattern of the pencil-beam antenna has one main lobe or maximum with a single-output terminal pair, or perhaps a few main lobes each with its own separate output. The output at a single terminal pair corresponds to one main lobe for only one sense of polarization: linear, circular, or elliptical. The same pattern but in an orthogonal polarization may be brought out to another terminal pair. Sweeping the pencil beam or beams over a portion of the sky produces a map of the region. Pointing the beam to different directions is accomplished in some cases by mechanically changing the antenna orientation and in other cases by changing the relative electrical phase of the antenna elements. Pencil-beam antennas are of two basic types: the filled aperture and the unfilled aperture.

1.2.2.1. Filled Apertures. Antennas in this category have no substantial gaps; energy is collected over the entire aperture area. The most common and simplest radio telescopes are in this group.

1.2.2.1.1. THE SIMPLE HORN. Figure 3 shows the cross section of either a

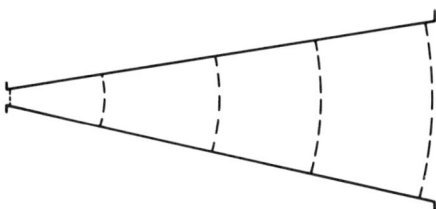

FIG. 3. Cross section of a conical or pyramidal horn antenna. The dotted lines show wavefronts within the horn when it is transmitting.

pyramidal or conical horn antenna. The "antenna terminals" are the throat of the horn where it smoothly joins to either rectangular or circular waveguide. The curved dotted lines show the phase fronts of the field in the horn when it is transmitting. Evidently, the radiation pattern of the horn is approximately the same as the diffraction pattern of an aperture in an opaque screen (the same size as the horn mouth) illuminated by a nearby point source. Because of the curvature of the phase front, the main lobe is slightly broader than that of the paraboloid of the same size discussed above (see Fig. 2).

The curvature of the phase front or phase "error" at the mouth of the horn restricts the maximum mouth dimension to less than about 15 wavelengths in practice. Larger apertures would require impractically long horns. Maximum practical horn gain is therefore about 1000, and the corresponding minimum beamwidth is 5–10°. The effective area $A_e(0, 0)$ for a horn designed for maximum gain is typically about half the mouth area. The principal use of the single-horn antenna is to serve as a gain standard for the calibration of larger radiotelescopes. The simple geometry permits calculation of the horn gain or collecting area with better than 5% accuracy.[1] A careful measurement will yield the gain with 1% accuracy.

Horns are also used for a few direct astronomical observations where high angular resolution is not important but either accurate knowledge of the antenna gain or low level of backlobes (characteristic of horns) is. For example horns have proven very valuable in measurements of the isotropic

[1] S. A. Schelkunoff and H. T. Friis, "Antennas: Theory and Practice," Chapter 16. Wiley, New York, 1952.

microwave background radiation, where the low level of back lobes is important to eliminate pick-up of radiation from the ground.[2]

1.2.2.1.2. PARABOLOID ANTENNAS. The most common antenna used for radio astronomy is the parabolic reflector with the feed horn or dipole located at the parabolic focus. Figure 1a shows a cross section of this type of antenna. Various mountings for such antennas are discussed in a later paragraph. Typically, the mounting allows the reflector to be directed toward any part of the sky.

One principal advantage of this antenna is the ease with which the receiver may be coupled to it. The input terminals are at the feed horn or dipole. Operation over a wide range of wavelengths is simple; to change from one band of wavelengths to another requires only the change of the feed. In practice, the receiver for a given band is located behind the feed at the prime focus, and it is changed along with the feed when operation is changed from one band to another. The shortest operating wave length is determined by the smoothness of the reflector surface and is usually about 20 times the rms surface roughness. The longest operating wavelength depends on the mean spacing between the feed support legs. When this spacing is less than about one-half wavelength, the legs shield the feed from the reflector. At each wavelength the receiving pattern is essentially that shown in Fig. 2 with the lobe width approximately inversely proportional to the reflector diameter divided by the operating wavelength. The precise beamwidth, relative size of sidelobes, etc. depend on the details of the feed pattern (see Section 1.3.2).

In contrast to optical reflectors, the radio telescope paraboloid has a short focal ratio (the ratio of focal distance to telescope diameter). The highly curved reflector and closely mounted feed form a mechanically rigid structure. Usually only one or two feeds will be located at or near the focus, so the poor image-forming quality of the short-focal-length system is unimportant.

For large fully steerable reflectors, current manufacturing practice will produce a reflector whose ratio of rms surface inaccuracy to reflector diameter is about 10^{-5} or greater. The inaccuracy is largely due to gravitational deflections of the surface when the structure is moved and to the effects of wind. The factor 10^{-5} means that at the shortest operating wavelength the beam width is about 1 min of arc, and this is the maximum resolution attainable by this type of antenna system. The *homologous design principle* of von Hoerner[3] promises to extend this limit. An antenna constructed according to this principle is allowed to deform under gravity as the antenna orientation is changed, but it must always change from one paraboloid to another with a different focus. Thus it is only necessary to be able to refocus the antenna to be able to use it successfully in all orientations.

[2] P. G. Roll and D. T. Wilkinson, *Phys. Rev. Lett.* **16**, 405 (1966).
[3] S. von Hoerner, *J. Struct. Div. Proc. Amer. Soc. Civil Eng.* **93**, 461 (1967).

1.2. TYPES OF ASTRONOMICAL ANTENNAS

The largest fully steerable paraboloid currently in operation is the 100-m telescope at Bonn.[4] This antenna is intended to operate at wavelengths as short as a few centimeters.

The simple prime-focus, fully steerable, paraboloidal antenna has some disadvantages. Two which have been mentioned above are the poor image-forming quality of the short-focus system and the limit on resolution due to mechanical imperfections. There are several additional difficulties: (a) the aperture blocking of the feed and feed-support legs produces some loss in gain and, more important, relatively high sidelobes; (b) the broad pattern of the feed extends beyond the edge of the reflector and picks up some thermal radiation from the ground; (c) the short-focus system has somewhat high cross-polarization sensitivity away from the telescope axis; (d) only the largest systems have adequate space for locating receiving equipment at the prime focus. A few modifications have been introduced which overcome some of these difficulties.

The horn-reflector antenna developed at the Bell Telephone Laboratories overcomes a few of the weaknesses of the simpler prime-focus antennas.[5] This antenna, shown in cross section in Fig. 1c is essentially an off-axis-fed section of a parabola with hornlike walls guiding the signal between the feed and the reflector. Because there is no aperture blocking, the gain is high, and the side- and backlobes are very low. The collecting area is typically 70% of the aperture area as compared with the 50–60% that is common with the simple on-axis prime-focus systems. The horn throat couples to the input waveguide in the same way it does for the simple horns described above. Changing operation to shorter wavelengths is accomplished by adding a short section to the horn which carries it down to the dimensions of the appropriate smaller waveguide. The presence of the walls shields the "feed" from direct thermal radiation from the ground. The horn walls provide some mechanical support to the parabolic section. The weight of the horn section, however, restricts the overall practical size of such structures. An aperture as large as that of the Bonn telescope would be impractical for the horn reflector. Horn reflectors have been used successfully for low-gain absolute radio-source calibrations, for studies of the isotropic background radiation, and for spectroscopic studies of the galaxy.[6]

A modification to the simple prime-focus antenna borrowed from standard optical practice is the two-reflector Cassegrain system. Figure 1b shows a cross section of a Cassegrain antenna. One focus of a hyperboloid is located at the focus of the paraboloid. The focal point of the composite two-reflector system is at the second hyperboloid focus, which is usually near the vertex

---

[4] R. Wielebinski, *Nature (London)* **228**, 507 (1970).
[5] A. B. Crawford, D. C. Hogg, and L. E. Hunt, *Bell Syst. Tech. J.* **40**, 1095 (1961).
[6] A. A. Penzias and R. W. Wilson, *Astrophys. J.* **142**, 419 (1965).

of the primary paraboloid. The effective focal ratio at the Cassegrain focus is much larger than that of the primary reflector alone, and, as with an optical telescope, the image quality is much improved. This means that feeds may be located off-axis in the focal plane with less loss of antenna gain than with prime-focus antennas.

There are three other important practical advantages of the Cassegrain radio telescope. One is that the long effective focal ratio provides a pattern which has lower cross-polarization sensitivity off-axis. Another advantage is that the spill-over from the feed at the Cassegrain focus is directed more toward the sky rather than at the ground, so that less interfering thermal radiation from the ground is picked up. A third advantage is largely mechanical. Because the Cassegrain focus is near the vertex of the primary reflector, bulky receiving equipment can be conveniently located behind the primary reflector. Heavy equipment at the prime focus of a single reflector antenna creates excessive tension in the feed-support struts resulting in deflection of the reflector. Because of the aperture blockage of the secondary reflector and its support struts, the side-lobe characteristics of the Cassegrain antenna are similar to those of the prime-focus fed antennas.

Cassegrain antennas are in wide use, particularly at the shorter wavelengths where the narrower beamwidth required of the feed at the Cassegrain focus is relatively easy to achieve.[7] Because of the low feed spill-over, the lowest-noise systems use either Cassegrain antennas or the horn reflector described above.[8]

1.2.2.1.3. RADIOTELESCOPE MOUNTINGS. Mountings for radiotelescopes are generally of three types: (a) polar or equatorial, (b) azimuth elevation, and (c) transit. Figure 4 shows examples of types (a) and (b).

The polar or equatorial mount, shown in Fig. 4a, is so named because one of the two orthogonal axes about which it can rotate is parallel to the axis of the earth. Motions about the two axes are in the local-hour-angle and declination coordinates of the celestial sphere. Thus tracking a source as the earth rotates requires rotation of the mount only about its polar axis (at a constant rate). This simplicity is the chief virtue of the equatorial mount and the reason why it is universally used for optical telescopes. Most of the smaller radiotelescopes, with diameters of 25 m or less, use equatorial mounts. The 42.7-m (140-ft) telescope at the National Radio Astronomy Observatory (NRAO) is the largest reflector to use this type of mounting.[9] Generally, structures as large as the 42.7-m (140-ft) reflector or larger are too heavy to be easily supported by an equatorial mount.

[7] J. R. Cogdell et al., *IEEE Trans. Antennas Propagat.* **AP-18**, 515 (1970).

[8] S. A. Rocci, in *Deep Space Missile Tracking Antenna: An Aviat. Space Symp. Held Conjunction ASME Winter Ann. Meeting, New York, Nov. 27–Dec. 1, 1966.* Amer. Soc. Mech. Eng., New York, 1966.

[9] O. Struve, R. M. Emberson, and J. W. Findlay, *Publ. Astron. Soc. Pac.* **72**, 439 (1960).

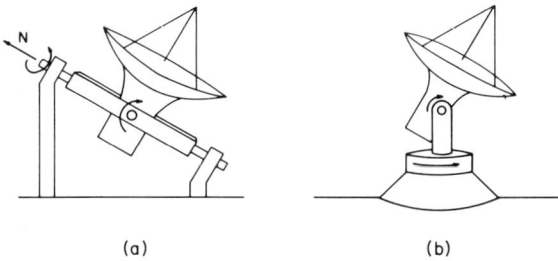

FIG. 4. Telescope mountings: (a) equatorial with one axis parallel to the pole of the earth; (b) azimuth elevation with one axis pointing to the local zenith.

When a reflector is operated near its shortest wavelength limit, its gain is often affected by the gravitational deflections of the reflector as its orientation is changed from one part of the sky to another. Because gravity acts downward, the deflections are a function of both the declination and hour-angle orientations of the antenna, and it is often difficult to calibrate these effects and correct for them for the polar mount.

The other most common mounting is the azimuth elevation or "az-el" mounting shown in Fig. 4b. As the name implies, one of its axes is parallel to the local vertical, and motions about it change the azimuth of the antenna orientation. Rotations around the other axis change elevation. All the largest reflecting telescopes use this mounting[10] because it is capable of carrying much more weight. One practical difficulty with this mount occurs because its axes do not coincide with celestial coordinates, the positional coordinates of the radio sources. It becomes necessary to use computer control systems to make the required coordinate conversions. With the presently available fast and inexpensive computers, this requirement no longer presents a serious problem. One advantage of this mount is that the effects of gravitational deflection are a function of only one coordinate axis, the elevation, and they are therefore easier to calibrate. Pointing corrections to compensate for refraction are also only a function of the elevation and are more easily made with this mount than with the equatorial mount.

One characteristic of the az-el mounted reflector is that the projection of its aspect on the celestial coordinates depends on the orientation of the antenna. This means, for example, that the polarization angle of the antenna rotates with respect to the source as the source is being followed. If the circumstances of the observation require that this angle remain constant, the feed must be rotated to compensate. On the other hand, this characteristic is sometimes an advantage. One may use the rotation of a fixed linearly polarized feed with

[10] E. G. Bowen and H. C. Minnett, *Proc. IRE (Aust.)* **24**, 106 (1963).

respect to the source as the source is tracked across the sky to measure the magnitude and orientation of the linear polarization of the source.

The transit mounting is simpler than either the equatorial or az-el mounts; it allows the reflector to move about only one axis, i.e., elevation. The elevation axis is fixed along an East–West direction. Hence, sources are observed only as they transit the local meridian. Because all visible sources do transit the meridian as the earth rotates, full sky maps may be made with this arrangement. Sources are, however, observed briefly only once each day, and it is difficult to accumulate the large amounts of observation time needed for good signal–noise on the weakest sources. The great advantage of this mounting is its simplicity. The 91.4-m (300-ft) NRAO transit telescope at Green Bank, West Virginia, is the largest instrument using this type of mounting.[11]

1.2.2.1.4. RADOMES. The 36.6-m (120-ft) reflector of the Haystack Radio Observatory at Westford, Massachusetts is housed within a radome, an approximately spherical, radio-transparent, protective dome.[12] The radome, as shown in Fig. 5, protects the antenna from the deleterious effects of sunshine and severe weather. More important, it protects the reflector from

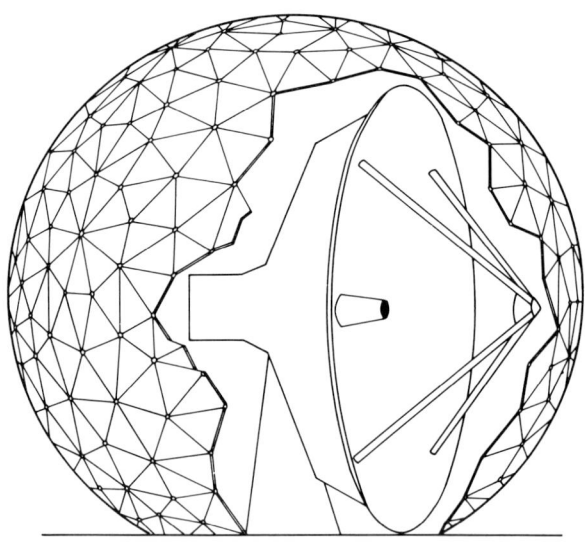

FIG. 5. A cutaway view of a radome showing the antenna within.

[11] J. W. Findlay, *Sky Telesc.* **25**, 68 (1963).
[12] M. L. Meeks, J. A. Ball, and A. B. Hull, *IEEE Trans. Antennas Propagat.* **AP-16**, 746, (1968).

deflection due to the wind, normally a very serious problem with large reflectors. As a result, the antenna is operable at shorter wavelengths and in higher winds than comparable antennas which are unprotected. The principal advantage of a radome is the improvement in pointing accuracy in the presence of wind.

The radome consists of many thin dielectric panels joined at their edges to the metal space frame. The aperture blocking of the metal frame results in a small loss in gain. In addition, the radome places upper and lower operating wavelength limits on the system. The long-wavelength limit occurs when the dimensions of the frame openings are on the order of one-half wavelength. Longer wavelengths are completely reflected by the metal space frame. The short wavelength limit is set by the resonant reflectivity and losses of panels. As the panel thickness approaches one-quarter wavelength in the dielectric, the transmission through the plastic drops off sharply.

For very large reflectors, the radome permits a considerable cost saving by eliminating the wind-survival requirement for the reflector structure. It is quite likely that if fully steerable antennas greater than 100-m diameters are built in the future, they will be housed in radomes.

1.2.2.1.5. KRAUS-TYPE TELESCOPE. Kraus has developed a transit instrument which provides large collecting area economically.[13] The antenna system, shown in Fig. 6, consists of a fixed standing section of a paraboloid

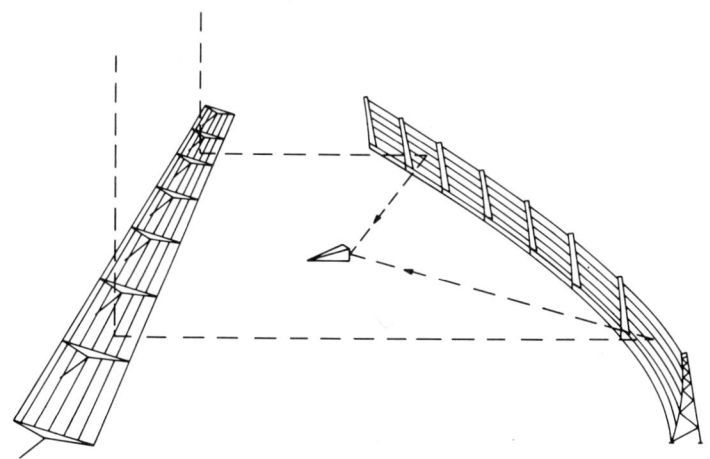

FIG. 6. The Kraus antenna. The angle of the flat plate may be changed to change the direction of the beam.

[13] J. D. Kraus, *Sky Telescope* **26**, 12 (1963).

with its horizontal beam reflected by a tiltable flat plate toward different directions in the meridian plane. The relatively inexpensive section of paraboloid uses the rigidity of the ground to maintain its shape. Being flat, the figure of the movable plate is easier to maintain as it is tipped to steer the beam. The ground between the parabolic section and the feed is a flat conducting sheet providing an image of both the reflector and the feed. This allows the feed to have one-fourth the height required in the absence of the ground plane and also eliminates much of the thermal noise that would otherwise be picked up by the feed from the ground.

The primary advantage of this antenna is that it is much cheaper to build than a fully steerable parabolic reflector of comparable collecting area. Another advantage is its large focal ratio which permits image formation and allows tracking of sources for a short period around the meridian. It also permits locating several feeds for different frequencies near the focus.

The principal disadvantage is that it is basically a transit instrument. Also because the design gains great advantage from the rigidity of the ground, the standing reflector is long and narrow producing a highly elliptical beam on the sky.

The largest instrument of this type is at Nançay, France.[14] Unlike the Kraus antenna at Ohio State University, this telescope does not employ a ground plane. Also, its standing reflector has a circular shape in the horizontal plane permitting tracking of sources for about 1 hr.

1.2.2.1.6. SPHERICAL REFLECTORS. Spherical reflectors have so far enjoyed only limited use in radio astronomy, although they have a few important advantages over parabolic reflector systems. Unlike the paraboloid, the spherical reflector has no fixed optical axis, and a fixed spherical reflector may be used to gather signals from different directions. The reflector may be fed by a line source, or, as is sometimes done in optical systems, it may be equipped with a correcting secondary reflector which focuses the rays to a point. Two such possibilities appear in Fig. 7.

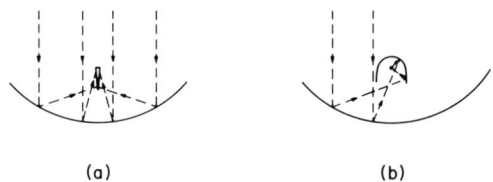

(a) (b)

FIG. 7. The spherical reflector: (a) with a line-source feed; and (b) with a correcting secondary reflector which focuses the rays to a point.

[14] E. J. Blum, A. Boischot, and J. Lequeux, *Proc. IRE (Aust.)* **24**, 208 (1963).

The largest spherical reflector system presently in use is the 304.8-m (1000-ft) reflector at Arecibo, Puerto Rico.[15] This reflector is a section of a sphere of radius 265 m (870 ft) and lies fixed in a natural depression in the ground with its aperture plane horizontal. Its principal feed is a line source (Fig. 7a) fastened to a carriage suspended by cables above the reflector. The carriage allows the feed to be steered about the center of the sphere permitting the main beam to be pointed toward any direction within 20° of the zenith. The feed consists of a tapered section of slotted waveguide with different sections of the guide receiving signals reflected from different annular sections of the sphere. Portions of a plane wave reflected from different parts of the mirror arrive in phase at the top of the guide. While the wave guide feed is simple in principal, it has the disadvantage of being very narrow band.

In addition, point-source feeds are in use on the Arecibo telescope. Operation at long wavelengths and illumination of only a part of the reflector minimize the deleterious effects of spherical abberation for the point-source feeds.

One important advantage of this particular reflector is that it is supported everywhere by the ground. This rigid support should permit operation at short wavelengths, and the surface of the Arecibo reflector is currently being figured for operation down to a 10-cm wavelength.

1.2.2.1.7. PARABOLIC CYLINDER REFLECTOR. Reflectors in the shape of a parabolic cylinder as shown in Fig. 8 are employed in radio astronomy both as

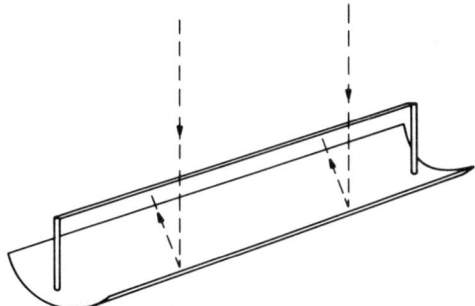

FIG. 8. The cylindrical reflector with its line feed.

individual telescopes and also as components of multiantenna systems. The focus of this reflector is a line parallel to the reflector surface as compared with the point focus of a paraboloid of revolution. A line feed is therefore required. Steering the beam in the cross-sectional plane of the reflector requires tipping the reflector as is the case for the paraboloid of revolution.

[15] W. E. Gordon and L. M. LaLonde, *IRE Trans. Antennas Propagat.* **AP-9** 17 (1961).

Steering the beam in the plane containing the focal line may, however, be accomplished by introducing a progressive phase shift along the line feed. This possibility of steering the beam electrically offers important advantages, particularly for large reflectors.

The largest single-reflector telescope of this type is the 600-MHz telescope of the University of Illinois.[16] The reflector in this system, 122 m (400 ft) wide and 183 m (600 ft) along its axis, is formed from a natural gully which is smoothed and covered by a reflecting mesh. The orientation of the reflector is fixed with its axis lying in the meridian plane, and it is used as a transit instrument. The feed is a line array of 276 elements 130 m (425 ft) long and located 46.6 m (153 ft) above the vertex of the reflector. With the elements of the array in phase, the beam is directed toward the zenith. Introduction of different progressive phase shifts along the array steers the beam to different directions within $\pm 30°$ of the zenith. The feed array has two novel features. A nonuniform distribution of elements permits both a reduction in the necessary number of elements and also low side lobes. In addition, the individual elements are circularly polarized so that appropriate rotation of each element produces the progressive phase shift needed to steer the telescope beam. Being able to steer the beam over a large angle without having to rotate the reflector is an economic advantage. The disadvantage of this system is the necessary complexity of the feed. The line feed must contain many elements, and redirecting the beam requires adjustment of the phase of each. The major difficulty is the loss of sensitivity which results from the long cable runs to the individual elements. Elimination of this loss requires separate preamplifiers at each feed (or group of feeds).

Another large telescope of this type is the steerable cylindrical reflector operated by the Tata Institute of Fundamental Research near Ootacamund, India.[17] The reflector is 529 m long and 30 m wide. The latitude of the site is 11° and the reflector is located on a hill of the same slope so that the main axis of the reflector is parallel to the earth's pole. Sources are therefore easily tracked by means of rotating the antenna about this axis. The surface consists of closely spaced wires stretched over 24 parallel parabolic frames. The antenna operates at 326.5 MHz. The feed consists of a linear array of dipoles parallel to the main axis of the reflector. The dipole outputs are combined to form 12 simultaneous contiguous beams separated by about 3 arc min in declination. Each beam has a width of about 3.6 min in declination and 2° in right ascension. By appropriate phasing of the dipole outputs, the beams may be steered in declination from $-36°$ to $+36°$.

1.2.2.1.8. FILLED ARRAYS. At low frequencies the filled aperture antenna is a planar array of separate but contiguous dipoles, their separate outputs

[16] G. W. Swenson, Jr. and Y. T. Lo, *IRE Trans. Antennas Propagat.* **AP-9**, 9 (1961).
[17] G. Swarup *et al.*, *Nature (London)* **230**, 185 (1971).

brought together by cables. Examples of this type of antenna are the solid 26-MHz array of the Clark Lake Radio Observatory[18] and the 50-MHz array of the Jicamarca Radar Observatory.[19] Separate dipoles may be used because at long wavelengths the dipoles are long and have sufficient cross section that a filled array of large size may be built with a reasonable number of elements. In addition, the signals may be brought large distances by cable from each element because cable losses are low. Preamplification at each element is unnecessary because the level of cosmic noise at each dipole far exceeds the noise introduced by cable loss or by the receiver.

The Jicamarca array consists of 9216 crossed dipoles over a flat reflecting screen. It is 288 m on a side and has an area of approximately 8.9 ha (22 acre). With all the elements connected in phase the beam is directed toward the zenith. Inserting the appropriate lengths of cable into lines feeding different parts of the array introduces the necessary progressive phase shift across the array to steer the beam away from the zenith. The practical difficulty with a large array of this kind is the large number of cable changes or switching operations required to redirect the beam.

1.2.2.2. Unfilled Apertures. The full aperture antenna of diameter D has a beam width which is proportional to λ/D and an effective collecting area which is proportional to D^2. Hence the resolution and collecting area or sensitivity are related and cannot be independently specified. There are many circumstances, however, in which high angular resolution is essential but in which the concomitant large collecting area and sensitivity of a filled aperture is either uneconomical or unnecessary. A number of antenna systems have been devised which have the pencil-beam characteristics of a very large filled aperture but which have only modest collecting areas. These systems generally operate by special processing of the signals received by different parts of the antenna.

1.2.2.2.1. THE CROSS ANTENNA. The cross system consists of two long narrow apertures or linear arrays arranged in the shape of a cross as shown in Fig. 9a. The reception pattern of either arm (Fig. 9b) consists of an elliptical central lobe with weak elliptical secondary lobes. If L is the length and W the width of an arm, then the angular widths of the narrow and broad dimensions of the main lobe are proportional to λ/L and λ/W, respectively. The pencil beam results from multiplying the outputs of the two arms of the cross. The product pattern is large only in those directions in which the patterns of both arms are large. The cross-hatched area in Fig. 9c is the resultant main beam. The secondary lobes along the principal axes can be kept small by suitable tapering of the element sensitivities toward the ends of the arms. The angular resolution of the cross is the same as that of a filled aperture of dimensions L

[18] W. Ericson, *IEEE Trans. Antennas Propagat.* **AP-13**, 422 (1965).
[19] W. K. Klemperer, G. R. Ochs, and K. L. Bowles, *Astron. J.* **69**, 22 (1964).

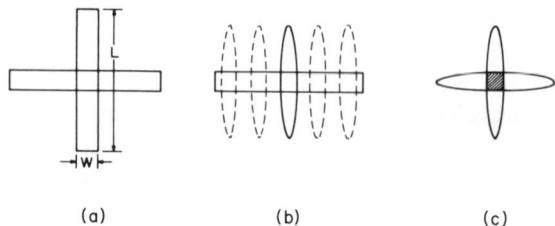

FIG. 9. (a) The multiplying cross antenna; (b) one arm of the cross and its pattern. The solid line shows the principal lobe, the dotted lines the secondary lobes; (c) The principal lobes of the two arms. The cross-hatched area is the main lobe which results from the multiplication of the outputs of the two arms.

by L. The collecting area can be independently specified by appropriate selection of the arm widths W.

A number of cross systems have been built and used successfully. The largest is the one-mile cross of the University of Sydney.[20] One arm of this instrument lies along an East–West (E–W) line and the other is North–South (N–S). Each arm consists of a single, long reflector in the shape of a parabolic cylinder approximately 12.2 m (40 ft) wide. The N–S reflector is fixed, and its main lobe is tipped in elevation by introduction of a progressive phase shift along its line feed. Because this telescope is a transit instrument, no electrical steering of the E–W feed is required. Mechanical rotation of the E–W reflector about its axis puts the broad maximum of its N–S lobe in the same direction as the maximum of the N–S arm. The instrument operates at 408 MHz and 111.5 MHz with beam widths of 3 and 10 arc min, respectively. The range of the beam is $\pm 55°$ from the vertical. The outputs of the N–S feed array are actually combined simultaneously with different progressive phase shifts to produce several outputs at different elevations. At 408 MHz there are 11 beams separated by 1.5 arc min and at 111.5 MHz, 3 beams separated by 7 arc min. Thus at 408 MHz the instrument scans a strip of about 16 arc min width with its 11 contiguous beams.

This telescope has very high angular resolution and a collecting area of about 37,000 m² (400,000 ft²). Because it is a transit telescope, it is best suited for survey observations.

1.2.2.2.2. THE CROSSED-GRATING ANTENNA. The transit-type cross system is not well suited for detailed studies of individual sources which require tracking the sources across the sky. The crossed-grating telescope was introduced primarily for solar studies. Its design permits observation of the sun at a wide range of hour angles. Instead of continuous or filled arms this cross has discrete spaced elements in its arms. Unlike the full arm which has

[20] B. Y. Mills, A. G. Little, K. V. Sheridan, and O. B. Slee, *Proc. IRE* **46**, 67 (1958).

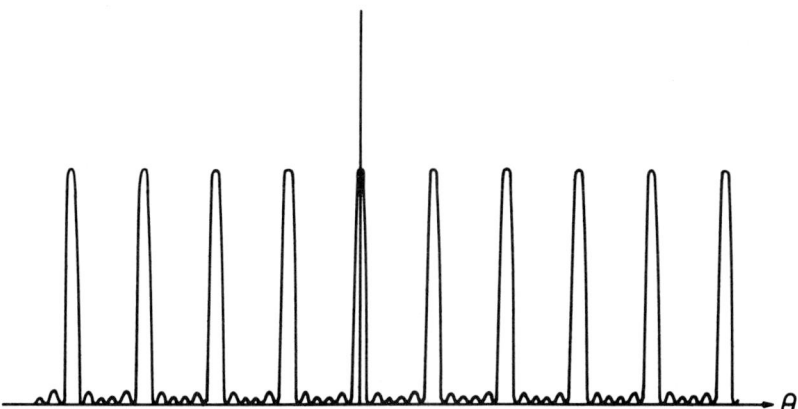

FIG. 10. The pattern of a grating array.

only one principal maximum in its pattern, the arm with spaced elements has an array of principal maxima or grating lobes as shown in Fig. 10 in one dimension. If the spacing between the elements is d, then the grating lobes are spaced by λ/d rad. The sun may be studied over a large range of hour angles as it successively transits each grating lobe. The lobes are separated by about 1° or more so that only one lobe is on the sun at a time. Because the sun is so bright, there is no danger of confusion from other radio sources in other lobes. The individual elements are usually small dishes with some resolution of their own, and hence they must track the sun keeping it on the maxima of their beams. Multiplication of the outputs of the two arms produces a pattern with a grid of lobes. In order to permit observation of a particular solar latitude at a particular time, the N–S interference pattern must be movable by means of an adjustable progressive phase shift along the N–S arm.

Examples of this type of system are the 21-cm wavelength cross-grating solar telescope at Fleurs near Sydney, Australia,[21] and the Stanford University 10-cm wavelength Spectroheliograph.[22] The Fleurs telescope has 32 5.8-m (19-ft) parabolic dishes in each arm. The arms are 366 m (1200 ft) in length, producing a grid of grating lobes, the lobes approximately 2 arc min in diameter and spaced by 1°. The beams of the individual antennas are about 2.5° and therefore include about 9 of the interferometer lobes. Because of this multilobe characteristic, these antennas are limited to solar observations.

1.2.2.2.3. THE COMPOUND-GRATING ANTENNA. There have been a number of experiments with multiplying antenna systems aimed at eliminating the multilobe characteristic of the grating array. In one such experiment the output of the Fleurs E–W array was multiplied by the output of a 18.3-m

---

[21] W. N. Cristiansen and R. F. Mullaly, *Proc. IRE (Aust.)* **24**, 165 (1963).
[22] R. N. Bracewell and G. Swarup, *IRE Trans. Antennas Propagat.* **AP-9** 22 (1961).

(60-ft) dish located at the end of the array.[23] Because the array element spacing is 12.2 m (40 ft), the pattern of the large reflector is narrower than the grating lobe spacing, and the product pattern contains only one grating lobe. The result is an elliptical fan beam whose N–S extent is given by the beamwidth of the 18.3-m (60-ft) dish at 21 cm, about 45 arc min, and whose E–W extent is 2 arc min as determined by the grating. The large reflector tracks the source under investigation and successive transits of the grating lobes are observed. The disadvantage of this particular system is the highly elliptical beam.

1.2.2.2.4. THE CIRCULAR ARRAY. The Culgoora radio heliograph is a large circular array with associated complicated data processing equipment designed principally for studies of the sun at 3.75-m wavelength.[24] It consists of 96 steerable parabolic antennas of 13-m diameter evenly distributed around a circle of 3-km diameter.

The annular aperture fed in-phase produces a beam pattern with one-main central lobe having about the same width as the lobe of the corresponding filled circular aperture but with much higher sidelobes. If the antennas are connected with a progressive phase shift of $2n\pi$ ($n = 1, 2, 3, \ldots$) per circuit, other axially symmetric patterns result which have a null on axis. An appropriate sum of such patterns for different values of n will, when subtracted from the in-phase pattern, produce a composite pattern with low side lobes. It is, in fact, possible to reproduce the pattern of a full circular aperture in this way.

The Culgoora system employs such a synthesis by rapidly making observations in sequence with the different phase connections and then combining them to yield a composite observation for a particular beam direction. In addition to this beam synthesis, the outputs from the antennas are fed to different receivers through different phase paths to produce a simultaneous line or array of beams on the sky. There are 48 of these beams, 3.5 arc min in diameter, spaced by 2.1 arc min along a N–S line. Finally, this row of beams is rapidly scanned from East to West providing a full picture of the sun in about 1 sec.

Because the array has discrete elements, it has secondary grating lobes. In the Culgoora instrument the spacings are such that grating lobes are about 2° apart, and the sun is observed with an isolated lobe. The principal limitatation of this instrument is that because of the grating lobes only strong isolated sources may be studied with it. One other limitation is that its size cannot easily be increased as can that of a cross, for example.

1.2.3. Aperture Synthesis

Whereas all the preceding antenna systems are basically designed to provide a single beam for finding radio source positions and for mapping their bright-

[23] N. R. Labrum, E. Harting, T. Krishnan, and W. J. Payten, *Proc. IRE (Aust.)* **24**, 148 (1963).
[24] J. P. Wild, *Proc. IRE (Aust.)* **28**, 279 (1967).

ness distributions, the interferometer collects this kind of data in a rather different way. The outputs of pairs of small antennas are recorded sequentially with different spacings and orientations of the antennas. When all of the spacings and orientations which would fill an aperture of area A on the ground have been used, the data is combined to give a map of the sky with the same resolution that a filled aperture of area A would have. It is the sequential synthesis of the aperture that distinguishes the interferometer synthesis telescope from the pencil beam telescope.

1.2.3.1. The Two-Element Interferometer. The two-element system is the most basic and most widely used interferometer. In general, both the spacing and orientation of the two antennas may be varied. For example, the interferometer of the Owens Valley Observatory consists of two 27.5-m (90-ft) steerable paraboloids which can be located at various stations along a 488-m (1600-ft) E–W baseline and at stations along a 488-m (1600-ft) N–S baseline.[25] A wide range of spacings and orientations may be selected for observations.

The outputs of the two antennas are multiplied and the resulting system pattern looks like a sine curve as shown in Fig. 11a. In Fig. 11b the finite

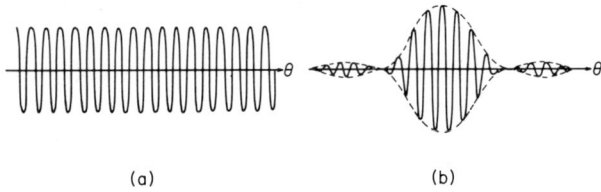

(a)   (b)

FIG. 11. Pattern of the two-element interferometer: (a) the sinusoidal pattern of the two elements; (b) the composite pattern including the patterns of the individual antennas.

beamwidth of the individual antennas is included. During observations the antennas track the region of interest. The strategies for measuring source positions and brightness distributions are different.

1.2.3.1.1. POSITION MEASUREMENT. If an isolated radio source (isolated by the resolution of the individual antennas) passes through the interferometer pattern of Fig. 11 as the earth rotates, the receiver output will be a low-frequency sinusoid. One period corresponds to passage of one lobe of the pattern. If the antenna separation is s, then the angular width of the lobes at the instrumental meridian is λ/s. If a nearby calibration source of known position is observed before or after the source of unknown position, the relative phase of the outputs gives the relative positions of the sources with an ambiguity of $n\lambda/s$ ($n = 1, 2, 3, \ldots$). The ambiguity may be removed by means

[25] Richard B. Read, *IRE Trans. Antennas Propagat.* **AP-9**, 31 (1961).

of another observation with shorter spacing. In this way positions relative to standard sources may be obtained with angular accuracies on the order of 1/10 of a lobe width or less. For example, the NRAO interferometer at Green Bank, West Virginia, operating at 11 cm with a maximum baseline of 2700 m has obtained positions of a number of small sources with accuracies better than 1 arc sec.[26] Other observatories have reported similar accuracies.

1.2.3.1.2. SOURCE MAPPING. If the source has a finite extent with respect to the sinusoidal pattern of Fig. 11, then the resulting interferometer output will be proportional to the Fourier spatial-frequency component of the source brightness distribution with the period of the lobe pattern. In effect the interferometer with spacing s measures the Fourier spatial frequency component of the source with period λ/s. Observations of the source on successive days with all spacings from zero up to the maximum available spacing L yields the Fourier transform of the source brightness distribution with spatial frequencies up to L/λ. The inverse transform then yields a map with the same resolution that would be obtained with a fan beam antenna of length L. In this way the antenna of length L is synthesized by the interferometer. If all the spacings and orientations that are contained within an aperture of side L are employed, then the pattern of a square pencil-beam antenna of side L is synthesized.

If the source to be studied is known to have finite angular extent Θ, then the required spacings for the synthesis are multiples of λ/Θ.[27] The antenna synthesized in this way is clearly a grating, and the synthesized pattern will have grating lobes separated in angle by Θ.

The preceding discussion is appropriate for sources observed near the local meridian. If a source is followed for many hours while it is above the horizon, the rotation of the earth varies the aspect of a fixed interferometer baseline as seen from the direction of the source, varying both its orientation and its effective length. This use of earth rotation permits synthesis of full apertures with relatively limited baseline settings. In particular, an E–W baseline interferometer will provide full maps of sources which are not too close to the celestial equator. The Cambridge 1600-m (1-mile) interferometer, which operates entirely on an E–W baseline, is an example of an earth-rotation synthesis telescope.[28]

1.2.3.1.3. VERY LONG-BASELINE INTERFEROMETRY. The highest-resolution radio observations have been made by very long-baseline interferometry (VLBI) (see Chapter 5.5). In this technique signals received at two widely separated antennas are tape recorded independently and later multiplied in a computer to produce the sinusoidal interference pattern. The technique relies on stable atomic oscillators at the two antennas which

[26] C. M. Wade, *Astrophys. J.* **162**, 381 (1970).
[27] R. N. Bracewell, *Proc. IRE* **46**, 97 (1958).
[28] A. Hewish, *Proc. IRE (Aust.)*, **24**, 225 (1963).

permit the high-frequency signals received from the source to be translated by heterodyne methods to a frequency band that is low enough to be recorded with the coherence of the signals maintained. Both the continuum radiation from quasars[29] and the line emission from interstellar molecular masers have been observed by this method.[30] With intercontinental baselines, observations at centimeter wavelengths have discovered source sizes smaller than 1/1000 arc sec.

1.2.3.2. Multielement Interferometers. The multielement interferometer consists of a number of two-element interferometers operating simultaneously. That is, the outputs of all the antenna elements are combined in pairs and recorded at the same time. The result is a system that can complete the aperture synthesis for a given region of the sky in a time shorter than that required by a simple two-element telescope. The Cambridge 1600-m (1-mile) interferometer, which has three antennas, is an example.[31] Its three antennas lie on an E–W line. Two have a fixed spacing of 762 m (2500 ft); the third is mounted on rails and can be moved from adjacent to one antenna to a distance of 762 m (2500 ft). The output of the movable antenna is combined with that of each fixed antenna. A point of the sky is observed with the movable antenna at each of about 50 locations along its track providing data at all spacings up to 1524 m (5000 ft). At each location the observation lasts for 12 hr so that a two-dimensional aperture is synthesized by the earth-rotation method discussed in Section 1.2.3.1.

With the addition of more elements, a more efficient system results if a good choice of all the possible simultaneous two-antenna spacings is made. Ideally, there should be no duplications in antenna spacing when all of the antenna outputs are combined in pairs, and, at the same time, every spacing should be present (in units of the smallest spacing) up to the largest spacing. Such an array would be the most efficient because it has no redundant spacings. Studies of linear arrays show that zero redundancy is possible only if the number of antenna elements is 4 or less.[32] In substantially larger arrays, the number of redundant spacings in a "minimum-redundancy" array is typically 20–30% of the total number. One example of a low-redundancy eight-element array has the following spacings: · 1 · 3 · 6 · 6 · 2 · 3 · 2 ·. The numbers are units of the smallest spacing. In this array, 22% of the pair spacings are duplicates.

The five-element interferometer of the Stanford University Radio Astronomy Institute is an example of a fixed, multielement interferometer with

[29] M. H. Cohen, W. Cannon, G. A. Purcell, D. B. Schaffer, J. J. Broderick, K. I. Kellermann, and D. L. Jauncey, *Astrophys. J.* **170**, 207 (1971).
[30] K. J. Johnston, *et al.*, *Astrophys. J.* **166**, L21 (1971).
[31] A. Hewish, *Proc. IRE (Aust.)* **24**, 225 (1963).
[32] A. T. Moffet, *IEEE Trans. Antennas Propagat.* **AP-16**, 172 (1968).

minimum redundancy spacings.[33] Five 18.3-m (60-ft) paraboloids are arranged on an E–W line with the following spacings: · 1 · 1 · 4 · 3 ·. All spacings, in units of 22.9 m (75 feet), are present up to the maximum of 206 m (675 feet) with only the unit spacing duplicated. The antennas are fixed. Tracking of a point in the sky permits earth-rotation synthesis of a two-dimensional aperture. From one 10-hr observation of a region of the sky a complete map of the region is synthesized. Because the outputs of nine independent spacings are being recorded at once, this instrument is nine times faster than a movable baseline two-element system covering the same total baseline, although it has only $2\frac{1}{2}$ times as many elements. Operating at 2.8-cm wavelength, the Stanford interferometer has a resolution of about 20 arcsec and a field of view, determined by the beamwidths of the individual dishes of about 7 arc min.

The Westerbork synthesis telescope of Leiden University is another complex instrument.[34] Ten fixed and two movable 25-m dishes are located along an E–W line. The output of each movable antenna is multiplied by that of each fixed one, and these 20 products are recorded simultaneously. The largest baseline is 1500 m. The operating wavelength is 20 cm, so that the beam synthesized by different locations of the movable antennas as well as by earth rotation has a resolution of about 20 arc sec. The field of view set by the resolution of the individual antennas is 30 arc min. However, because the smallest spacing used is 75 m the synthesized maps have grating lobes separated by about 10 arc min.

---

[33] R. N. Bracewell, R. S. Colvin, K. M. Price, and A. R. Thompson, *Sky Telescope* **42**, 4 (1971).

[34] P. C. van der Kruit, *Astron. Astrophys.* **15**, 110 (1971).

1.3. Analysis of Paraboloidal-Reflector Systems*

1.3.1. Introduction

The rf performance of paraboloidal reflectors may be analyzed using various techniques of diffraction theory. Such techniques generally involve the integration of the dyadic Green's function over the currents induced on the reflector (A1–A6).† These induced currents are usually estimated using geometrical optics (B1–B5). In principle, however, they may also be calculated using moment-method techniques (C1–C9) or by applying classical results to appropriate boundary-value problems (D1–D9). The currents having been determined, evaluation of the fields becomes straightforward although the computations may be lengthy and laborious (E1–E9). Computers are generally needed for all but a few special cases that may be integrated in closed form. In the event that the integrals have stationary points, the integrals can be evaluated in terms of one or more simple, closed-form expressions (F1–F6) which frequently lend themselves to greatly simplified geometrical ray-tracing interpretation (G1–G13).

By its very nature, radio astronomy employs the antenna as a receiving device, i.e., as a transducer of the incident electromagnetic wave into charges, currents, and voltages in the receiving electronics. Consequently, the receiving parameters of the antenna (e.g., effective aperture as a function of angle of incidence) are needed for system analysis and design, and for interpretation of the data. Unfortunately, direct mathematical evaluation of these receiving parameters is extremely difficult except in a few special circumstances. On the other hand, reciprocity may be used to relate the receiving parameters with the corresponding transmitting parameters (H1–H8), and the transmit mode is generally more amenable to mathematical evaluation. Consequently, the analysis described in this section will be primarily analysis of the transmitting properties of the antenna.

---

* Chapter 1.3 is by W. V. T. Rusch.

† Letter and number combinations refer to the list of references at the end of the text of this chapter.

1.3.2. Analysis of the Paraboloid with a Prime-Focus Feed

The analysis is based on the assumption that the reflector lies in the far-field region of the feed.† The fields of the feed may then be represented by[1,2]:

$$\bar{E}_f(\rho, \psi, \xi) \equiv [E_\psi(\psi, \xi)\bar{a}_\psi + E_\xi(\psi, \xi)\bar{a}_\xi]e^{-jk\rho}/\rho, \qquad (1.3.1)$$

where

$$E_\psi(\psi, \xi) = \sum_m [A_m(\psi) \sin m\xi + B_m(\psi) \cos m\xi] \qquad (1.3.2a)$$

and

$$E_\xi(\psi, \xi) = \sum_m [C_m(\psi) \cos m\xi + D_m(\psi) \sin m\xi] \qquad (1.3.2b)$$

The feed-reflector geometry is shown in Fig. 1, where (ρ, ψ, ξ) constitute the

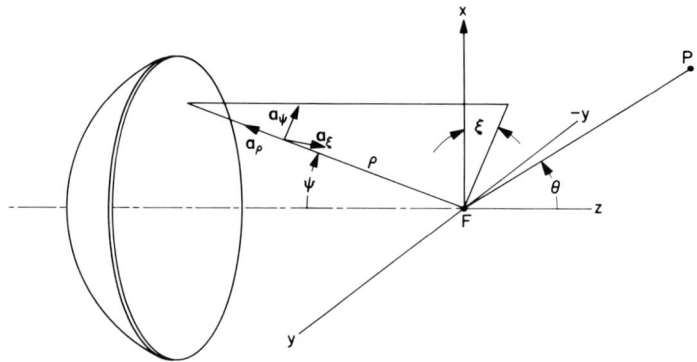

FIG. 1. Geometry of a paraboloid with a prime-focus feed.

polar coordinates of a right-handed coordinate system directed toward the vertex of the reflector along the negative z-axis.

1.3.2.1. Aperture Efficiency.
The conventional paraboloidal reflector is illuminated by a feed located at or near the prime focus of the reflector. The aperture efficiency η of a prime-focus reflector illuminated by the fields of

---

[1] W. V. T. Rusch and P. D. Potter, "Analysis of Reflector Antennas," pp. 74–81. Academic Press, New York, 1970.

[2] A. C. Ludwig, JPL Tech. Rep. 32-1430, Jet Propulsion Lab., Pasadena, California, February 15, 1970.

---

† Should this assumption not be valid, the mathematical complexity will be considerably increased although the principles of the analysis will remain unchanged.

1.3. ANALYSIS OF PARABOLOIDAL-REFLECTOR SYSTEMS

Eqs. (1.3.1) and (1.3.2) can be computed using physical optics.[1,3,4,B5] The result is

$$\eta = \frac{\pi \cot(\Psi/2)}{2Z_0 P_T} \left\{ \left| \int_0^\Psi [A_1(\psi) + C_1(\psi)] \tan\left(\frac{\psi}{2}\right) d\psi \right|^2 \right.$$
$$\left. + \left| \int_0^\Psi [B_1(\psi) - D_1(\psi)] \tan\left(\frac{\psi}{2}\right) d\psi \right|^2 \right\}, \quad (1.3.3)$$

where Z_0 is the impedance of free space and P_T is the total power radiated by the feed. (Joule heating losses have been neglected, and the effects of aperture blockage are considered in Section 1.3.2.3.) Because of the axial symmetry of the reflector, only the $m = 1$ components of the feed radiation contribute to the aperture efficiency.

The theoretical pattern of an open-ended, reflection-free, circular waveguide excited in the dominant TE_{11} mode has been chosen to illustrate the application of Eq. (1.3.3). This pattern has been selected because it represents a relatively useful class of feeds, and its pattern and total power are known in closed form.[5] Only the $m = 1$ mode is radiated; $A_1(\psi)$ is the E-plane pattern and $C_1(\psi)$ is the H-plane pattern, the only nonzero components of the feed radiation. These theoretical patterns are plotted in Figs. 2a and b, respectively, as functions of the waveguide radius (in free-space wavelengths).

The aperture efficiency is dependent on several interrelated effects, of which the two dominant effects are (1) illumination efficiency, the degree to which the aperture is illuminated as uniformly as possible, and (2) spillover, the fractional feed power not intercepted by the reflector. As a general rule, an improvement in one of these factors will cause a degradation in the other, e.g., a highly tapered feed pattern will virtually eliminate spillover while simultaneously reducing illumination efficiency. A truncated paraboloidal reflector will therefore yield maximum aperture efficiency for a feed pattern which effectively compromises the illumination efficiency and spillover.

The pattern of an open-ended circular waveguide located with its phase center at the prime focus of the paraboloid was inserted in Eq. (1.3.3), and the aperture efficiency was determined as a function of F/D and a/λ, where F is the paraboloid focal length, D the paraboloid diameter, a the waveguide radius, and λ the free-space wavelength. The results are plotted in Fig. 3. It is evident that each F/D combination yields a maximum efficiency for a

---

[3] A. C. Ludwig (ed.), JPL Tech. Rep. 32-979, p. 69, Jet Propulsion Lab., Pasadena, California, April 1967.

[4] A. C. Ludwig, SPS No. 37-26, Vol. IV, p. 200, Jet Propulsion Lab., Pasadena, California, April 1964.

[5] S. Silver, "Microwave Antenna Theory and Design." pp. 336–341. McGraw-Hill, New York, 1949.

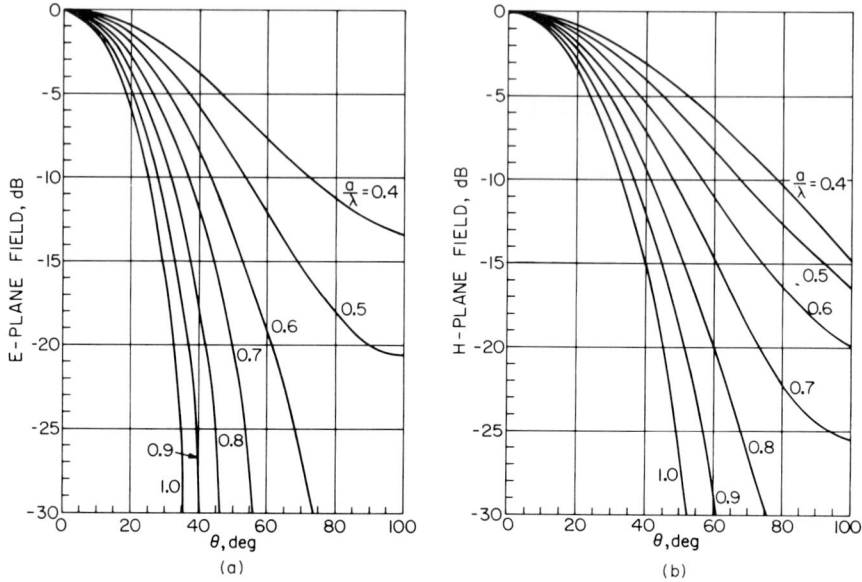

FIG. 2. (a) E-plane field of open-ended, reflection-free, circular waveguide in TE_{11} mode. (b) H-plane field of open-ended, reflection-free, circular waveguide in TE_{11} mode.

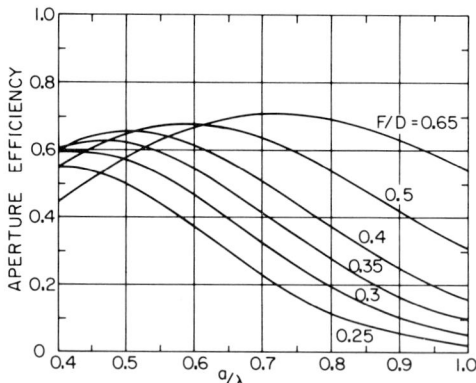

FIG. 3. Aperture efficiency of paraboloid illuminated by TE_{11} feed.

1.3. ANALYSIS OF PARABOLOIDAL-REFLECTOR SYSTEMS

particular waveguide radius/wavelength ratio, although the maxima are relatively broad.

The paraboloid edge angle, i.e., the angle between the vertex and the rim as viewed from the focus, is related to the F/D ratio by

$$\Psi = 2 \tan^{-1}(D/4F). \tag{1.3.4}$$

Furthermore, the distance from the focus to the reflector rim is greater than the distance from the focus to the vertex. This additional space loss that an inverse-distance electromagnetic wave undergoes when it travels to the edge of the reflector is given by

$$\text{space loss} = 40 \log_{10}\left(\sec \frac{\Psi}{2}\right) \tag{1.3.5}$$

Equations (1.3.4) and (1.3.5) are presented graphically in Fig. 4.

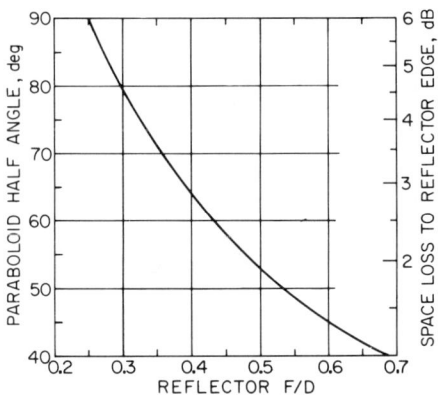

FIG. 4. Paraboloid half-angle and space loss versus reflector F/D, using Eqs. (1.3.4) and (1.3.5).

The radius/wavelength ratio of the waveguide illuminator yielding maximum aperture efficiency is tabulated in Table I as a function of F/D and the corresponding edge angle. The E- and H-plane pattern tapers† corresponding to the reflector edge angle are tabulated. In general, the E-plane pattern of the open-ended waveguide is tapered more strongly at the edge of the reflector than the H-plane pattern. In the final two columns of the table the space loss has been added to the pattern taper, thus yielding the actual taper on the illumination across the circular aperture.

---

† Pattern taper = $20 \log_{10} |E(0)/E(\Psi)|$.

TABLE I. Pattern Characteristics of Open-Ended, Reflection-Free, Circular Waveguide in TE_{11} Mode for Maximum Efficiency

Reflector F/D	Reflector edge angle (deg)	Space loss (dB)	a/λ for maximum efficiency	E-plane pattern edge taper (dB)	H-plane pattern edge taper (dB)	E-plane edge taper with space loss (dB)	H-plane edge taper with space loss (dB)
0.65	42.1	1.2	0.72	14.4	8.3	15.6	9.5
0.60	45.2	1.4	0.67	13.9	8.3	15.3	9.7
0.55	48.9	1.6	0.62	13.7	8.4	15.4	10.0
0.50	53.1	1.9	0.58	13.6	8.3	15.6	10.2
0.45	58.1	2.3	0.54	13.6	8.9	15.9	11.2
0.40	64.0	2.9	0.50	13.4	9.3	16.3	12.1
0.35	71.1	3.6	0.47	13.6	10.0	17.2	13.6
0.30	79.6	4.6	0.42	12.5	10.8	17.1	15.3
0.25	90.0	6.0	0.40	12.6	12.6	18.6	18.6

1.3.2.2. Radiation Pattern.
The far-zone radiated field of the illuminated paraboloid is[1]

$$\bar{E}(P) = \frac{-j\omega\mu_0}{4\pi R} e^{-jkR} \int [\bar{J}_S - (\bar{J}_S \cdot \bar{a}_R)\bar{a}_R] e^{jk\bar{\rho}\cdot\bar{a}_R} \, dS, \quad (1.3.6)$$

where $\bar{J}_S$ is the induced surface-current density on the reflector. The physical optics approximation of $\bar{J}_S$ for a linearly polarized illuminator is then

$$\bar{J}_S = 2\left(\frac{\varepsilon_0}{\mu_0}\right)^{1/2} \frac{e^{-jk\rho}}{\rho} \left\{ [A_1(\psi) - C_1(\psi)] \cos\frac{\psi}{2} \sin\xi \cos\xi \, \bar{a}_x \right.$$

$$\left. - [A_1(\psi) \sin^2\xi + C_1(\psi) \cos^2\xi] \cos\frac{\psi}{2} \bar{a}_y + A_1(\psi) \sin\xi \sin\frac{\psi}{2} \bar{a}_z \right\}.$$

$$(1.3.7)$$

Inserting Eq. (1.3.7) into Eq. (1.3.6) and carrying out the azimuthal integration yields the two vector components of the far-zone radiated field:

$$E_\theta(P) = jkF \sin\phi \, \frac{e^{-jkR}}{R} \int_0^\Psi \exp[-jk\rho(\cos\theta\cos\psi + 1)]$$

$$\times \left\{ A_1(\psi) \cos\theta[J_0(\beta) - J_2(\beta)] + C_1(\psi) \cos\theta[J_0(\beta) + J_2(\beta)] \right.$$

$$\left. - 2j \sin\theta \tan\frac{\psi}{2} J_1(\beta) A_1(\psi) \right\} \tan\frac{\psi}{2} \, d\psi, \quad (1.3.8a)$$

$$E_\phi(P) = jkF \cos\phi \, \frac{e^{-jkR}}{R} \int_0^\Psi \exp[-jk\rho(\cos\theta\cos\psi + 1)]$$

$$\times \{A_1(\psi)[J_0(\beta) + J_2(\beta)] + C_1(\psi)[J_0(\beta) - J_2(\beta)]\} \tan\frac{\psi}{2} \, d\psi, \quad (1.3.8b)$$

where J_0, J_1, and J_2 are the Bessel functions of order 0, 1, and 2, respectively, and $\beta = k\rho \sin\theta \sin\psi$. Integration of Eqs. (1.3.8) then yields $E_\theta(P)$ and $E_\phi(P)$ which can be combined in any azimuthal plane to determine both the normally polarized and the cross-polarized components of the radiation field. In the E-plane† $E_\theta(P)$ constitutes the entire field; in the H-plane†

---

† In a linearly polarized system the E-plane is the principal plane containing the E-vector of the radiated field and the H-plane is the principal plane containing the H-vector.

$E_\phi(P)$ constitutes the entire field. Such integrations have been carried out by several authors.[6-13] For angles near to the reflector axis (boresight) the longitudinal components of current may be neglected and the surface integral can be converted into an equivalent aperture integral.[14,15] This aperture method introduces a path-length phase error which is less than 1/16 wavelength provided[16]:

$$\sin \frac{\theta}{2} \leq \left[\frac{1}{2}(F/D)(\lambda/D)\right]^{1/2}. \qquad (1.3.9)$$

At wide angles the pattern is determined primarily from contributions from the edges of the reflector.[F3,F4,F6,G9]

In order to illustrate the application of Eqs. (1.3.8), the illumination function of the TE_{11} open-ended circular waveguide shown in Figs. 2a and b has been selected. The patterns were obtained for paraboloidal reflectors illuminated by the circular waveguide which yielded optimum efficiency for a particular F/D [cf. Fig. 3 and Table I]. Results for $F/D = 0.4$ and $a/\lambda = 0.5$ are plotted in Fig. 5 together with the reference pattern of a uniformly illuminated circular aperture. The abscissa is the normalized quantity

$$u = \pi(D/\lambda) \sin \theta. \qquad (1.3.10)$$

For values of θ such that $\sin \theta \ll \cos \theta$, the radiation pattern is independent of the reflector diameter when it is plotted versus u. Consequently, the patterns are "universal" curves, valid for all but the smallest D/λ ratios. The E-plane pattern is broader than the H-plane pattern, reflecting the more tapered E-plane illumination. Both patterns are wider than the pattern for uniform illumination. Only the first two sidelobes are included in the range of u shown, since the detailed sidelobe structure at wide angles is not generally

[6] J. P. Schouten and B. J. Beukelman, *Appl. Sci. Res.* (B) **4**, 137 (1954).

[7] D. Carter, *Convent. Rec. IRE 1954 Nat. Convent. New York* **1**, 60 (1954).

[8] R. L. Pease, Tech. Rept. No. 184, MIT Lincoln Lab., Lexington, Massachusetts, August 5, 1958.

[9] L. B. Tartakovskii, *Radiotekh. Elektron.* **4**, 920 (1959).

[10] S. S. Sandler, *IRE Trans. Antennas Propagat.* **AP-8**, 368 (1960).

[11] W. V. T. Rusch, Tech. Rep. No. 32-434, Jet Propulsion Lab., Pasadena, California, May 1963.

[12] M. S. Afifi, *in* Electromagnetic Wave Theory, Part 2, *Proc. Symp. Delft, The Netherlands, September 1965* (J. Brown, ed.)., p. 669. Pergamon, Oxford, 1967.

[13] W. V. T. Rusch, *in* Tech. Rept. No. 32-979 (A. C. Ludwig, ed.), p. 41, Jet Propulsion Lab., Pasadena, California (April 1967).

[14] S. Silver, "Microwave Antenna Theory and Design," pp. 169-199. McGraw-Hill, New York, 1949.

[15] J. Ruze, *IEEE Trans. Antennas Propagat.* **AP-13**, 660 (1965).

[16] W. V. T. Rusch and P. D. Potter, "Analysis of Reflector Antennas," pp. 91-92, Academic Press, New York, 1970.

1.3. ANALYSIS OF PARABOLOIDAL-REFLECTOR SYSTEMS

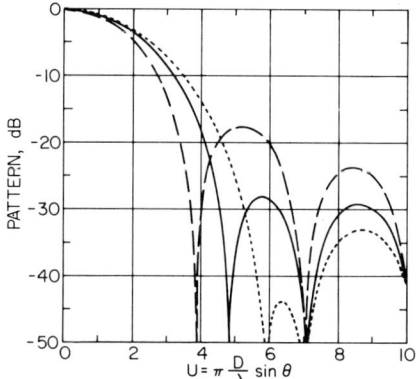

FIG. 5. Pattern of paraboloid illuminated by TE_{11} feed with $F/D = 0.4$ and $a/\lambda = 0.5$. The solid line represents the H-plane, the dotted line the E-plane, and the dashed line uniform illumination.

of interest to the radio astronomer. The H-plane sidelobes are considerably lower than the sidelobes for uniform illumination, and the E-plane sidelobes are still lower.

A comparison of the H-plane patterns for four optimized F/D values is plotted in Fig. 6. As the F/D decreases from 0.65 to 0.25, the edge taper increases [cf. Table I]. Consequently, the main beam broadens and the first sidelobe drops until it gradually merges into the second sidelobe.

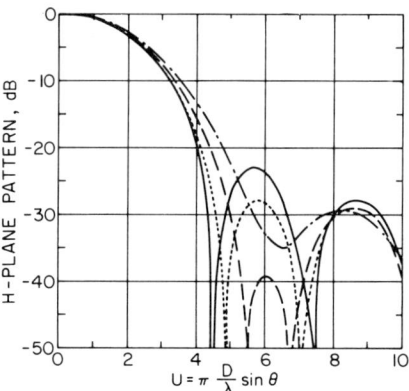

FIG. 6. H-plane pattern of paraboloid illuminated by TE_{11} feed: —, $a/\lambda = 0.72$, $F/D = 0.65$; ···, $a/\lambda = 0.50$, $F/D = 0.40$; ---, $a/\lambda = 0.42$, $F/D = 0.30$; -·-, $a/\lambda = 0.40$, $F/D = 0.25$.

The half-power beamwidth for a particular illumination function is inversely proportional to the D/λ ratio. This relationship can be expressed as

$$\text{HPBW} = K_B \lambda/D \tag{1.3.11}$$

Values of K_B (in degrees) are tabulated in Table II for the optimized TE_{11}

TABLE II. Beamwidth Characteristics of Paraboloid Illuminated by TE_{11} Feed

Reflector F/D	TE_{11} Feed a/λ	E-plane HPBW × (D/λ) (deg)	H-plane HPBW × (D/λ) (deg)
0.65	0.72	71.0	64.5
0.50	0.58	71.8	65.0
0.40	0.50	72.5	67.0
0.35	0.46	73.0	68.0
0.30	0.42	73.5	69.2
0.25	0.40	75.3	72.5

mode illuminated paraboloids. The values range from 64.5 to 72.5 (H-plane) and 71 to 75.3 (E-plane). The generally broader E-plane beamwidths are due to the increased E-plane illumination taper.

1.3.2.3. Aperture Blocking. Figure 7 shows an aperture projection of the

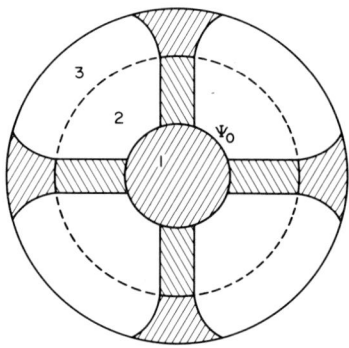

FIG. 7. Aperture projection of blocked portions of a paraboloidal reflector.

blocked portions of a paraboloidal reflector when it is transmitting an idealized geometrical beam. Region 1 represents the geometrical shadow of a centrally located member of the feed structure cast by the beam as it leaves the reflector. Similarly, region 2 is the shadow of the feed support struts cast by the outgoing beam. Region 3 represents the geometrical shadow of the support struts cast on the reflector by the diverging beam as it emerges from the focal region of the antenna.

In principle, simple geometrical concepts are insufficient to describe such diffractive antenna performance characteristics as aperture blocking at microwave frequencies. The blocking obstacles in Region 1 are, however, usually large relative to a wavelength, and their effects can be accurately described using the geometrical blocking approximation provided that the angles of observation are not far from boresight.[17-22] This approximation effectively assumes that the projection of the blocking obstacle onto the reflecting surface cancels contributions to the radiated field from currents on these blocked portions of the surface. Thus the radiation pattern associated with the blocked aperture is the superposition of the pattern of the unblocked aperture and the pattern of the blocked portion of the aperture excited 180° out of phase. The broad pattern of the blocked portions reduces the main beam and the even sidelobes while it raises the first, third, etc., sidelobes. In practice, the central blocking is included in the field calculation by inserting an appropriate angle Ψ_0 in the lower limits of Eqs. (1.3.3) and (1.3.8). H-plane patterns are plotted in Fig. 8 for a paraboloid with $F/D = 0.4$ illuminated by an open-ended TE_{11}-mode circular waveguide of 0.5-wavelength radius. A centrally located circular obstacle of diameter d blocks the aperture. The patterns are plotted for d/D ratios of 0 (unblocked), 0.10, 0.15, 0.20. It is evident that the first sidelobe rises nearly 10 dB over the range of blocking.

The loss in axial gain due to central blocking may amount to a decibel or more. In general, this loss can be expressed as

$$\text{blockage loss (dB)} = -20 \log_{10}[1 - a(d/D)^2] \approx 8.7a(d/D)^2. \quad (1.3.12)$$

For the parameters of Fig. 8 the values of a varied from 2.2 to 2.3. The parameter a has been estimated[22] to be

$$a \cong G_0/32(F/D)^2, \quad (1.3.13)$$

where G_0 is the gain of the feed.

[17] S. Silver, "Microwave Antenna Theory and Design," pp. 169–199. McGraw-Hill, New York, 1949.
[18] P. D. Potter, JPL Rep. No. TR32-149 (September 1961).
[19] R. C. Hansen, "Microwave Scanning Antennas," Vol. 1. Academic Press, New York, 1964.
[20] S. Altshuler, L. M. Frantz, and M. I. Sancer, TRW Rep. Prepared Under Contract F33 (675)-67-CO121, Redondo Beach, California (September 1967).
[21] R. E. Collin, and F. J. Zucker, "Antenna Theory," p. 48. McGraw-Hill, New York, 1969.
[22] W. V. T. Rusch and P. D. Potter, "Analysis of Reflector Antennas," pp. 82–86, 95–98, 119. Academic Press, New York, 1970.

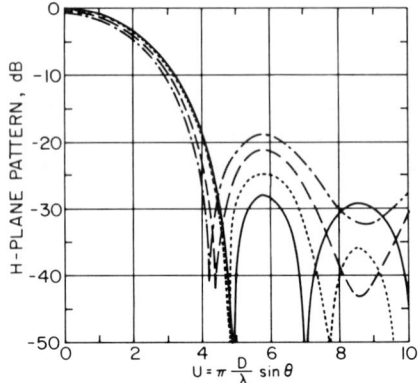

FIG. 8. *H*-plane pattern of blocked paraboloid illuminated by TE_{11} feed with $F/D = 0.4$, $a/\lambda = 0.5$, and $d/D = 0.00$ (—), 0.10 (···), 0.15 (– – –), and 0.20 (– · –).

The blocking effects of the feed-support struts are considerably more difficult to evaluate than central blocking because (1) their transverse dimensions are usually comparable to or smaller than a wavelength, and (2) they are illuminated both by the plane-type wave emerging from the antenna aperture and also by the spherical-type wave emerging from the feed. No completely satisfactory theory to describe the blocking effects of these struts has been developed. In principle, a moment-method solution is feasible, but because of the numerical complexity and large storage requirements, only a two-dimensional moment-method solution of the feed-support problem has been reported.[C9] Several approximate techniques to evaluate the strut blocking effects have been undertaken.[23-28] If the illuminating fields of the feed† are assumed to be determined independently, the radiating currents on the antenna structure may be decomposed into the following partial current distributions:

(1) the currents on the main reflector due to the feed which, together with the source currents of the feed, determine the fields of the unblocked antenna aperture;

[23] E. Everhart and J. W. Kantorski, *Astron. J.* **64**, 455 (1959).
[24] C. L. Gray, *Microwave J.* **7**, No. 3, 88 (1964).
[25] J. H. Wested, Tech. Rep. P2118, Microwave Lab., Danish Acad. of Tech. Sci., Copenhagen, Denmark (March 1, 1966).
[26] F. I. Sheftman, Lincoln Lab. Tech. Rep. 416 (September 23, 1966).
[27] J. Ruze, *Microwave J.* **11**, No. 12, 76 (1968).
[28] W. V. T. Rusch, *1971 G-AP Int. Symp. Digest*, Los Angeles 211 (1971).

† Either prime focus or Cassegrain.

1.3. ANALYSIS OF PARABOLOIDAL-REFLECTOR SYSTEMS

(2) the currents on the central feed structure due to the presence of the main reflector and the feed-support struts: these currents determine the central blocking fields;

(3) the currents induced on the struts due to the presence of the main reflector;

(4) the currents induced on the struts due to the direct fields of the feed; and

(5) the perturbation currents induced on the main reflector due to the presence of the struts: these currents determine the blocking effects in region 3, as well as perturbing the current distribution in regions 1 and 2.

These five current distributions must be determined simultaneously inasmuch as only the total distribution satisfies the boundary conditions. Consequently, a moment-method solution represents the only feasible procedure for a completely rigorous solution. It appears, however that the strut currents [distributions (3) and (4)] can be approximated in a straightforward cause and effect manner.[C9] For example, the currents induced on the struts due to the presence of the main reflector can be approximated by the currents that would flow on that portion of an infinite cylindrical structure of the same cross section in free-space immersed in an infinite plane wave, where the wave amplitude and polarization are locally the same as for the wave emerging from the antenna at a corresponding point on the strut. Having approximated the distribution (3) component of the total strut current in this manner, the corresponding component of the total radiated field can be determined using the free-space dyadic Green's function. Because distribution (3) appears to constitute a major component of the blocking fields, the material below describes the cross polarization and loss in boresight gain caused by this component of the total strut current.

The blocking effects of the plane-wave component of the total strut current are conveniently described in terms of the E-wave and H-wave *induced field ratios*, IFR_E and IFR_H.[29,C9] These induced field ratios are characteristic properties of an infinite cylindrical structure immersed in an incident plane wave in the same way that the various scattering cross sections are characteristic properties. The IFR is defined as the ratio of the forward-scattered field to the field radiated by an aperture of the same width. It is a complex number which depends on the cross section, wavelength, polarization, and direction of the incident plane wave. The optical blocking approximation sets the IFR of the blocking object equal to -1.

The values of IFR_E and IFR_H for a right circular cylinder immersed in a normally incident plane wave are plotted in Fig. 9. The cylinder radius/wavelength (a/λ) ratio is indicated as a parameter along the E-wave and H-

[29] A. F. Kay, *IEEE Trans. Antennas Propagat.* **AP-13**, 188 (1965).

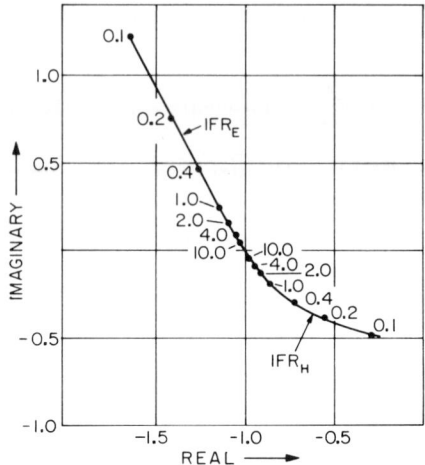

FIG. 9. Complex IFR_E and IFR_H plotted as a function of a/λ for a right-circular cylinder.

wave branches of the curve. As this ratio increases, both IFR_E and IFR_H approach the optical limit of $-1.0 + j0.0$.

The loss in boresight gain due to central blockage and the plane-wave component of strut current for a linearly polarized antenna is

$$\text{loss (dB)} = -20 \log_{10} \left| 1 - \frac{\int_0^{2\pi} \int_0^{\beta} F(t, \phi) t \, dt \, d\phi}{\int_0^{2\pi} \int_0^{1} F(t, \phi) t \, dt \, d\phi} \right.$$

$$+ \sum_{i=1}^{\substack{\text{number} \\ \text{of} \\ \text{struts}}} \left\{ \left(W_i \frac{D}{2} \right) \left(IFR'_{Ei} \cos^2 \gamma_i + IFR'_{Hi} \sin^2 \gamma_i \right) \right.$$

$$\left. \left. \times \frac{\int_\beta^{\beta_i} F(t, \phi_i) \, dt}{\left(\frac{D}{2}\right)^2 \int_0^{2\pi} \int_0^{1} F(t, \phi) t \, dt \, d\phi} \right\} \right|, \quad (1.3.14)$$

where t is the normalized radial aperture coordinate, ϕ the azimuthal aperture coordinate, $F(t, \phi)$ the aperture distribution function, β the fractional diameter blocking by central blockage, W_i the width of ith strut, D the aperture diameter, β_i the fractional radius blocking by ith strut, γ_i the angle between electric vector and ith strut, α_i the angle between ith strut and aperture plane, IFR'_{Ei} the IFR_E for ith strut cross section with linear dimensions

1.3. ANALYSIS OF PARABOLOIDAL-REFLECTOR SYSTEMS

reduced by cos α_i, and IFR$'_{Hi}$ the IFR$_H$ for ith strut cross section with linear dimensions reduced by cos α_i.

The second term represents the optical approximation to axially symmetric central blockage. The third term represents a summation over the plane-wave component of strut currents on each strut in terms of the parameters defined in Fig. 10. The strut term in Eq. (1.3.14) can also be derived by replacing the

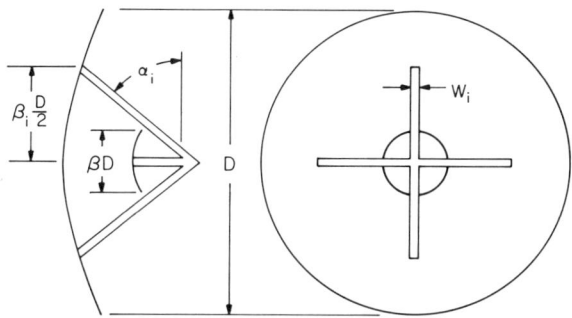

FIG. 10. Strut geometry.

struts with equivalent electric and magnetic line sources, the intensities of which are proportional to the IFRs and the aperture field. These equivalent line sources are particularly useful in the determination of sidelobe properties.[28]

For a uniformly illuminated aperture Eq. (1.3.14) reduces to

loss (dB)
$$= -20 \log_{10} \left| 1 - \beta^2 + \sum_i \left\{ \frac{A_i}{(\pi D/2)^2} (\text{IFR}'_{Ei} \cos^2 \gamma_i + \text{IFR}'_{Hi} \sin^2 \gamma_i) \right\} \right|,$$
(1.3.15)

where A_i is the projected area of the ith strut on the aperture plane. To illustrate the application of Eq. (1.3.15), numerical values were obtained for a rim-mounted, principal-plane quadripod lying in the aperture plane: if $\beta = 0.1$, and the four struts are circular cross section of diameter $2a = 1.0$ wavelength, the strut loss is 0.93 dB for a 25-wavelength diameter aperture and 0.50 dB for a 50-wavelength diameter aperture. If square struts of width $w = 1.0$ wavelength are used and the flat faces are parallel to the aperture plane, the strut losses are 1.17 and 0.61 dB, respectively, for the 25- and 50-wavelength apertures. These results reflect the relatively larger IFR magnitudes of the square struts. From a mechanical point of view, however, the struts with square cross section have far superior mechanical properties (e.g., compressional strength, moment of inertia, radius of gyration, etc.) than circular struts of the same width. Consequently, equivalent mechanical properties can be achieved with square struts of smaller physical dimensions.

In addition to degrading boresight gain, a feed-support strut will generate cross polarization if it is not aligned parallel to or perpendicular to the electric field in the aperture. The boresight cross polarization level due to the plane-wave component of strut currents is

$$\mathrm{CP\ (dB)} = 20 \log_{10} \left| \sum_i \left\{ \left(W_i \frac{D}{2}\right)(\mathrm{IFR}'_{Hi} - \mathrm{IFR}'_{Ei}) \sin \gamma_i \cos \gamma_i \right.\right.$$
$$\left.\left. \times \frac{\int_\beta^{\beta_i} F(t, \phi_i)\, dt}{\left(\dfrac{D}{2}\right)^2 \int_0^{2\pi} \int_0^1 F(t, \phi) t\, dt\, d\phi} \right\} \right|. \qquad (1.3.16)$$

Clearly, if $\sin \gamma_i \cos \gamma_i$ is zero (e.g., principal-plane struts), cross polarization is not generated. Other strut configurations, e.g., an equiangular tripod, will also not generate boresight cross polarization because of cancellation properties of the total geometry, although each strut individually may generate a cross-polarized component.

1.3.2.4. Feed Defocusing. Displacement of the feed from the focus of the paraboloid will result in

(1) a change in the peak gain (virtually always a decrease);
(2) a broadening of the main pencil beam;
(3) a rearrangement of the sidelobe structure, which may or may not be accompanied by an increase in the average sidelobe level.

These effects are considered detrimental to the operation of antennas designed for maximum gain with minimum beamwidth. Under special circumstances, however, it may be highly desirable to operate in such "defocused" conditions.

For example, it is well known that a lateral component of defocusing produces a shift in the direction of the main pencil beam. If the amount of lateral defocusing is not excessive, the antenna will continue to operate properly in the pencil-beam mode with only slight degradation in performance.[10,15,19,30-33] A second example requires the feed to be axially

[30] S. Silver and C. S. Pao, Radiat. Lab. Rep. No. 479, MIT, Cambridge, Massachusetts (1944).

[31] F. B. Hilderbrand, Radiat. Lab. Rep. No. 1078, MIT, Cambridge, Massachusetts (February 20, 1946).

[32] K. S. Kelleher and H. P. Coleman, NRL Rep. 4088, Naval Res. Lab., Washington, D.C. (December 31, 1952).

[33] Y. T. Lo, *IRE Trans. Antennas Propagat.* **AP-8**, 347 (1960).

1.3. ANALYSIS OF PARABOLOIDAL-REFLECTOR SYSTEMS

defocused in order to achieve broader target coverage than would ordinarily be possible for a given aperture size.[E6, 34-36]

To illustrate the effects of axial defocusing, the illumination function of a half-wavelength radius TE_{11} open-ended circular waveguide was used to illuminate a 25-wavelength diameter reflector with $F/D = 0.4$. The resulting H-plane patterns are shown in Fig. 11 for axial defocusing values of 0.0

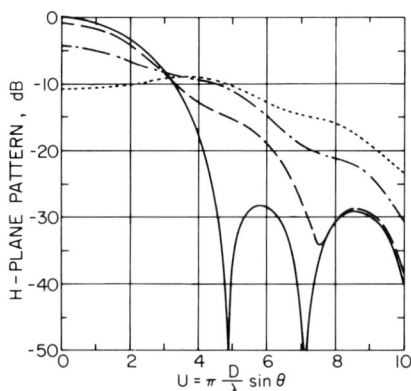

FIG. 11. H-plane pattern of paraboloid illuminated by axially defocused TE_{11} feed with values of 0.0 (—), 0.5λ (– – –), 1.0λ (– · –), and 1.5λ (· · ·) for $D/\lambda = 25$, $F/D = 0.4$, and $a/\lambda = 0.5$.

(focused), 0.5λ, 1.0λ, and 1.5λ beyond the focus (away from the reflector). Similar patterns are obtained for comparable defocusing toward the reflector, although accurate studies reveal that axial defocusing is not symmetrical about the focus.[36] Figure 11 reveals that axial defocusing causes the beamwidth to increase, the sidelobes to rise, and the nulls to fill. For the extreme value of 1.5λ defocusing, the boresight field is slightly below the field on either side. This bifurcated effect is more pronounced for less tapered feeds. If the relatively untapered pattern of an infinitesimal electrical dipole is used to illuminate the paraboloid, the bifurcation becomes a complete axial null for defocusing values of

$$Z = m\frac{\lambda}{2}\left[1 + \left(\frac{4F}{D}\right)^2\right], \qquad m = \pm 1, \pm 2, \ldots . \qquad (1.3.17)$$

[34] H. A. Wheeler, *IRE Trans. Antennas Propagat.* **AP-10**, 573 (1962).
[35] H. W. Redlien, Jr., *IEEE Trans. Antennas Propagat.* **AP-16**, 415 (1968).
[36] P. G. Ingerson and W. V. T. Rusch, *IEEE Trans. Antennas Propagat.* **AP-21**, 104 (1973).

Reciprocity yields the identical property for the focal-region field of a paraboloid receiving a normally incident plane wave.[37]

Lateral displacement of the feed of a paraboloidal antenna causes the pencil beam to scan on the opposite side of the reflector axis. The ratio of beam scan angle to feed scan angle (feed squint) is defined as the *beam deviation factor*, generally of the order of 0.7 to 0.9. The *Petzval surface*, a term from classical optics, is the surface of best focus in an optical system in the absence of astigmatism. For a single mirror the radius of curvature of the Petzval surface is one-half the radius of curvature of the mirror. In the primary microwave reference to the subject, the Petzval surface of a paraboloidal mirror is derived by Ruze[15] to be another paraboloid of half the focal length, tangent to the focal plane at the focus, and described by the equation.

$$\rho_{\text{Petz}}^2 = 2FZ_{\text{Petz}} \tag{1.3.18}$$

where ρ_{Petz} is the distance from the reflector axis and Z_{Petz} is the axial distance beyond the focus.

To illustrate the scanning properties of a microwave scanning reflector, the half-wavelength radius TE_{11} open-ended circular waveguide feed was scanned in the focal region of a 25-wavelength reflector with $F/D = 0.4$. The feed was translated in such a manner that its beam peak remained parallel to the reflector axis. By trial-and-error, the contour to achieve maximum scan gain for a series of scan angles to 10 HPBWs was determined. This surface was plotted in Fig. 12 (solid curve) together with the corresponding Petzval surface (dashed curve). The scan plane patterns are plotted in Fig. 13 for

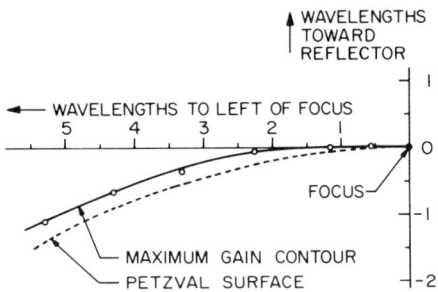

FIG. 12. Maximum gain contour of paraboloid illuminated by laterally defocused TE_{11} feed for an *H*-plane scan where $D = 25$ wavelengths, $F/D = 0.4$, and $a/\lambda = 0.5$.

[37] E. M. Kennaugh and R. H. Ott, Antenna Lab. Rep. 1223-16, Contract AF33(616)-8039, Ohio State Univ. Res. Foundation, Columbus, Ohio (August 31, 1963).

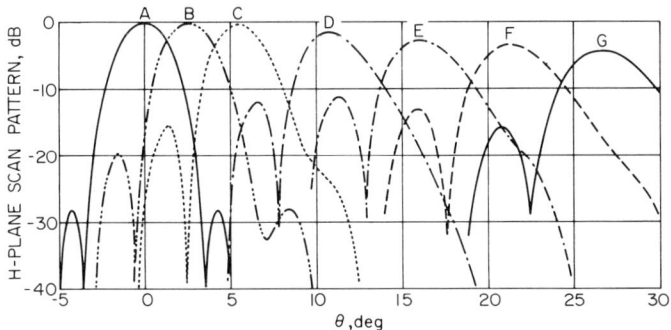

FIG. 13. H-plane pattern of paraboloid illuminated by laterally defocused TE_{11} feed.

focused conditions (A), one-HPBW scan (B), two-HPBW scan (C), four-HPBW scan (D), six-HPBW scan (E), eight-HPBW scan (F), and ten-HPBW scan (G). With increasing scan the peak gain drops, the main beam broadens, the sidelobe on the axis side of the main beam (coma lobe) rises, and the sidelobe on the other side of the main beam broadens and merges into the main beam.

The following scan characteristics apply to more general feed/reflector configurations designed to achieve maximum scan gain[38]:

(1) Unless the F/D is very large or spillover is excessive, a higher scan gain is achieved when the axis of a directional feed is parallel to the axis of the reflector than when the feed is directed toward the reflector vertex.

(2) The contour of maximum scan gain is a function of illumination taper and reflector F/D. In general, larger F/D values (greater than 0.5) tend to have a maximum-gain contour close to the focal plane, while the smaller F/D values tend to have a maximum-gain contour closer to the Petzval surface.

(3) Maximum-gain contours are plotted in Fig. 14 for a 34-wavelength paraboloid with 10-, 15-, and 20-dB illumination edge tapers for the class of feed functions of the form $\cos^n \theta$. The axial and lateral components are divided by the F/D value. Three solid maximum-gain curves corresponding to F/D values of 0.433 (60° edge angle), 0.604 (45° edge angle), and 0.687 (40° edge angle) are plotted for each edge taper. Superimposed on each figure are dashed curves to indicate the scan angle in HPBWs. Normalizing the coordinate axes by the F/D value and the wavelength makes the maximum-gain contours relatively insensitive to F/D and frequency. Consequently, Fig. 14 can be used to select the feed position for maximum scan gain over a wide range of scan angle and reflector shape by simple interpolation.

[38] W. V. T. Rusch and A. C. Ludwig, *IEEE Trans. Antennas Propagat.* **AP-21**, 141 (1973).

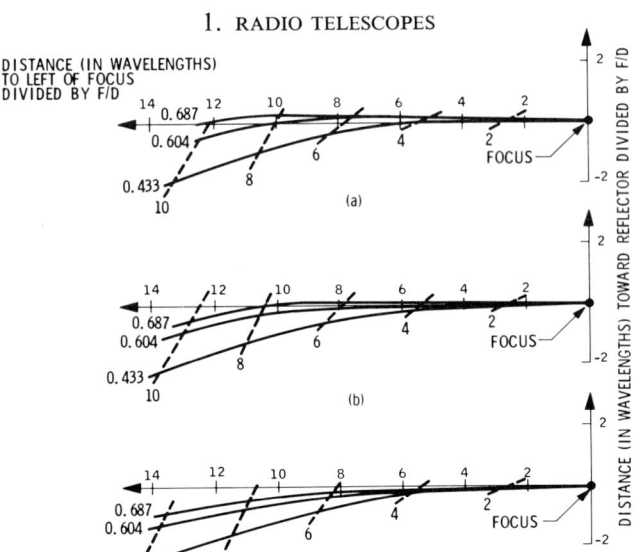

FIG. 14. Generalized maximum field loci for edge tapers of (a) 10, (b) 15, and (c) 20 dB. The solid curve represents the maximum field locus for a constant F/D and the dashed curve indicates the scan angle in HPBWs.

1.3.3. Analysis of Multireflector Systems

1.3.3.1. Classical Dual-Reflector Antenna Systems. In current microwave practice most dual-reflector systems are designed after the Cassegrain optical telescope. Commonly used subreflectors are pure hyperboloids, degenerate hyperboloids (splash plates), and modified hyperboloids. The basic properties of the system may be derived using the principles of ray optics: the subreflector is placed so that its foci coincide with the focus of the paraboloidal reflector and the phase center of the primary feed (Fig. 15). A spherical wave from the primary feed will be transformed by the subreflector into a spherical wave emerging from the focus of the paraboloid; this wave will then be reflected by the paraboloid and radiated into space. Every ray emerging from the point-source feed and undergoing reflection at the two surfaces will travel an equal distance to a plane in front of the antenna and perpendicular to its axis. Ray tracing indicates no spillover beyond the edge of the paraboloid and, consequently, a low effective noise temperature since radiations will not be received from the ground.

In 1663, nine years prior to the invention of the Cassegrain telescope, James Gregory proposed a two-reflector telescope using a concave ellipsoidal subreflector placed beyond the focus of the paraboloid so that its two foci coincide with the paraboloid focus and the primary feed phase center (Fig. 16).

1.3. ANALYSIS OF PARABOLOIDAL-REFLECTOR SYSTEMS

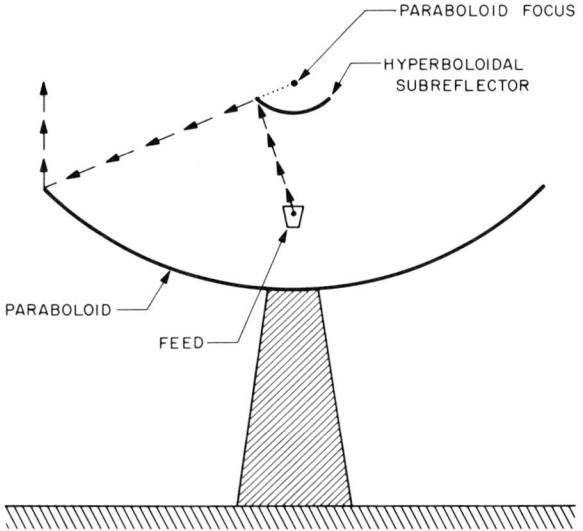

FIG. 15. Geometry of Cassegrain antenna.

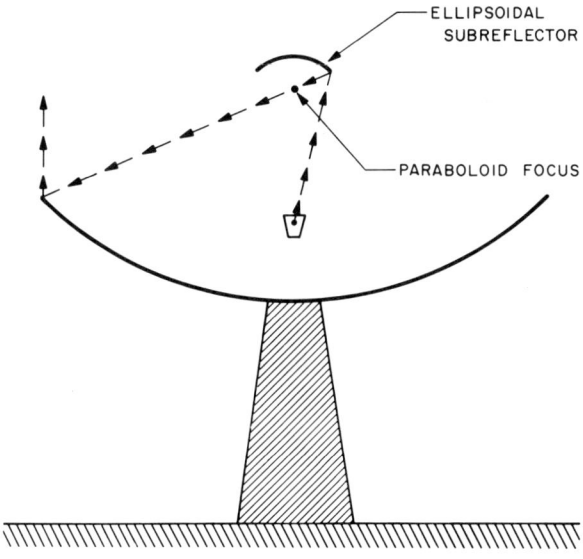

FIG. 16. Geometry of Gregorian antenna.

Every ray emerging from the primary feed and reflected by the ellipsoid will converge toward the paraboloid focus and then diverge toward the paraboloid. The ray behavior is similar to the Cassegrain system except for "inversion" and longer ray paths. The Gregorian system clearly suffers from the disadvantage of considerably greater overall length than an equivalent Cassegrain.

1.3.3.1.1. HYPERBOLOIDAL SUBREFLECTOR. The technique of physical optics can be used to determine the fields scattered from a hyperboloidal subreflector, similar to the procedures outlined in Section 1.3.2. for a paraboloidal reflector. The feed-reflector geometry is that of Fig. 1, except that the reflector is convex, instead of concave, and the distance from the hyperboloid's external focus to the surface is $\rho = -ep/(1 - e \cos \psi)$, where $p = c(1 - 1/e^2)$ and $e = c/a$. The two vector components of the far-zone scattered field are[39,40]

$$E_\theta = j\left(\frac{kep}{2}\right) \sin \phi \, \frac{e^{-jkR}}{R} \int_0^\Psi \frac{\exp[-jk\rho(\cos\theta \cos\psi + 1)]}{(1 - e\cos\psi)^2}$$

$$\times \{A_1(\psi) \cos\theta(e - \cos\psi)[J_0(\beta) - J_2(\beta)]$$

$$+ C_1(\psi) \cos\theta(e \cos\psi - 1)[J_0(\beta) + J_2(\beta)]$$

$$- 2j \sin\theta \sin\psi J_1(\beta) A_1(\psi)\} \sin\psi \, d\psi \qquad (1.3.19a)$$

$$E_\phi = j\left(\frac{kep}{2}\right) \cos \phi \, \frac{e^{-jkR}}{R} \int_0^\Psi \frac{\exp[-jk\rho(\cos\theta \cos\psi + 1)]}{(1 - e\cos\psi)^2}$$

$$\times \{A_1(\psi)(e - \cos\psi)[J_0(\beta) + J_2(\beta)] + C_1(\psi)(e \cos\psi - 1)$$

$$\times [J_0(\beta) - J_2(\beta)]\} \sin\psi \, d\psi, \qquad (1.3.19b)$$

where $\beta = k\rho \sin\theta \sin\psi$.

Geometrical techniques are also available for subreflector analysis. The classical geometrical optics formulation for the field scattered from a hyperboloid is available in the work of Rusch and Potter.[41] For highly tapered feeds with low illumination of the subreflector rim, the geometrical optics results may be sufficient for most purposes. The geometrical theory of diffraction (GTD) provides subreflector analysis with a useful supplement to the optical ray.[G9] The singularities at shadow boundaries and reflection boundaries of the Keller version of GTD can be eliminated with transition functions.[G12] A typical scattered field pattern from a 25-wavelength hyperboloid is shown in Fig. 17. Both the amplitude and phase characteristics are pre-

---

[39] W. V. T. Rusch, *IEEE Trans. Antennas Propagat.* **AP-11**, 414 (1963).
[40] W. V. T. Rusch, *IEEE Trans. Antennas Propagat.* **AP-14**, 266 (1966).
[41] W. V. T. Rusch and P. D. Potter, "Analysis of Reflector Antennas," pp. 30–32. Academic Press, New York, 1970.

1.3. ANALYSIS OF PARABOLOIDAL-REFLECTOR SYSTEMS

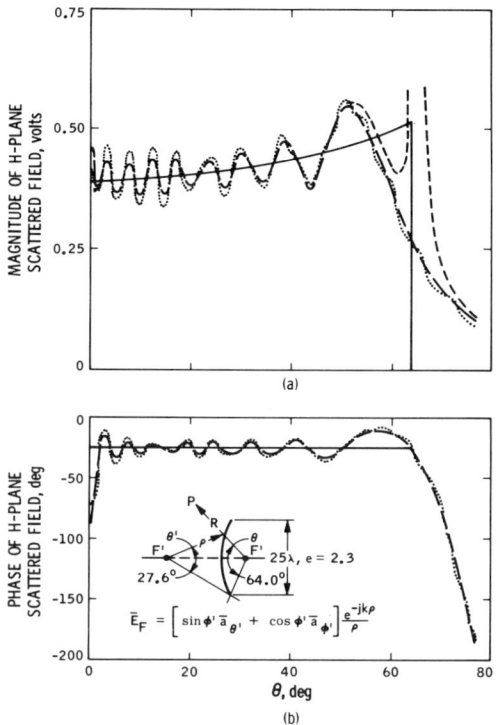

FIG. 17. Comparison of GO (—), GTD (Keller: - - -; Kouyoumjian: – – –), and PO (· · ·) fields scattered from a hyperboloid using amplitude (a) and phase (b) characteristics.

sented for geometrical optics, physical optics, GTD (Keller), and GTD (Kouyoumjian). The advantage of the GTD results is a 100fold or better saving in computational time, while, at the same time, achieving a more accurate representation of the scattered field because the edge currents are modeled more accurately.

1.3.3.1.2. ELLIPSOIDAL SUBREFLECTOR. The analytical techniques of the previous section apply equally well to ellipsoidal subreflectors in Gregorian systems. Rusch[42] indicates a surprising similarity between the scattered fields of an ellipsoid and an equivalent hyperboloid (equal diameter; equal extremum ray).

1.3.3.1.3. NONOPTICAL SUBREFLECTORS. The classical hyperboloidal/ ellipsoidal subreflector contours are sometimes modified to achieve improved rf performance. These modifications may be empirically or analytically

[42] W. V. T. Rusch, *Proc. IEEE* **51**, 630 (1963).

derived.[43] Such scattering from these modified shapes may also be analyzed using the techniques of the previous sections.[44]

Such a modified hyperboloid is shown in Fig. 18. This configuration is a

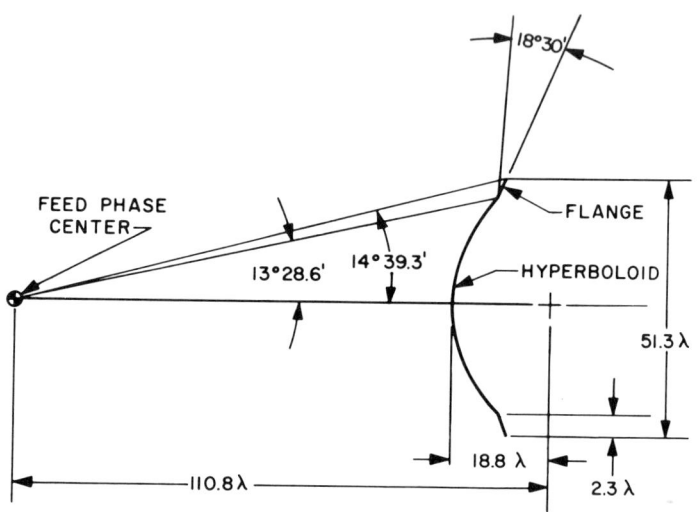

FIG. 18. Subreflector configuration for scattered field patterns.

51.3-wavelength diameter Cassegrainian subreflector consisting of a hyperboloid with a conical flange extension.[45] The amplitude and phase patterns of the feedhorn used to illuminate the subreflector were measured on a 16-GHz antenna range, and these patterns were used as tabular input for the field calculations. Although the nature of the flange extension violates the assumptions of physical optics, this technique was used to determine the scattered field. The computed and experimental scattered-field amplitude and phase patterns (E-plane) are plotted in Fig. 19. Close agreement between the calculated and experimental patterns is evident, except for discrepancies near 0°, due to unavoidable feedhorn blockage of the experimental pattern, and at wide angles where the reduced amplitude is more significantly affected by "parasitic" errors caused by ground reflections, inexact feedhorn pattern determination, etc. Furthermore, the field equations based on the geometrical

[43] P. D. Potter, *IEEE Trans. Antennas Propagat.* **AP-15**, 727 (1967).
[44] W. V. T. Rusch, Tech. Rep. 32-434, Jet Propulsion Lab., Pasadena, California (May 27, 1963).
[45] A. C. Ludwig and W. V. T. Rusch, Tech. Rep. No. 32-1190, Jet Propulsion Lab., Pasadena, California (November, 1967).

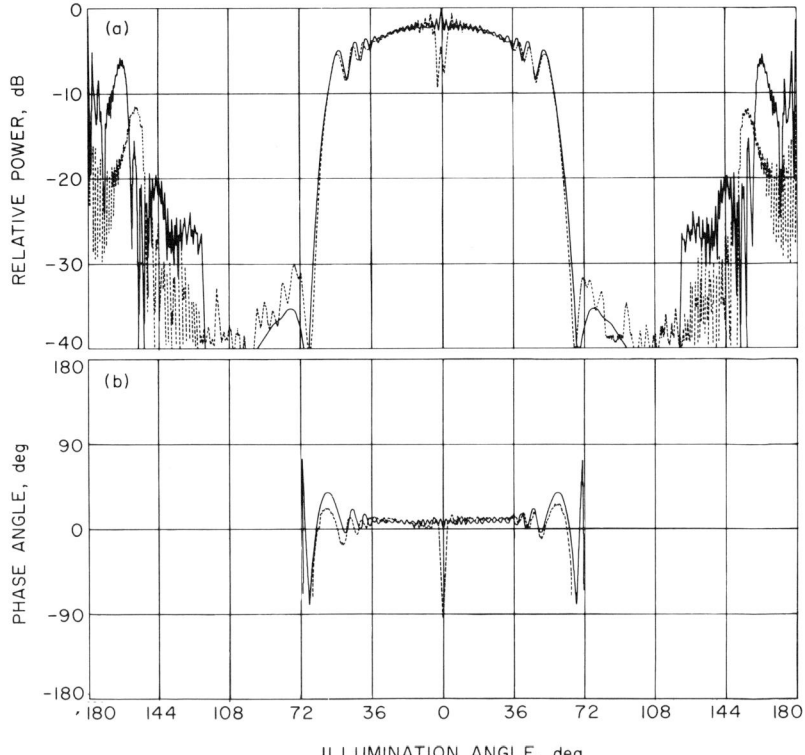

FIG. 19. Comparison of computed (solid line) and experimental (dotted line) scattered field patterns: (a) amplitude patterns and (b) phase patterns.

optics approximation for the current distribution have questionable validity when extrapolated into the rear hemisphere.

1.3.3.1.4. DUAL-REFLECTOR SCATTERING. Radiation from dual-reflector systems may be calculated by applying physical optics (PO) to both reflectors, where the fields scattered from the subreflector are used to illuminate the main reflector[39, 46-48] Under conditions of complete axial symmetry, a two-dimensional numerical integration must be carried out. When the reflectors lack axial symmetry, or the feed is laterally defocused, the azimuthal integrations must also be evaluated numerically for each reflector. Thus the double PO technique for a symmetric antenna is very time consuming, while

[46] A. C. Ludwig, Tech. Rep. 32-1430, Jet Propulsion Lab., Pasadena, California (February 15, 1970).
[47] H. Zucker and W. H. Ierley, *Bell Syst. Tech. J.* **47**, 897 (1968).
[48] W. V. T. Rusch and H. L. Strachman, *1968 NEREM Record*, Boston p. 20 (1968).

the absence of symmetry virtually makes the technique prohibitively time consuming. An attractive compromise to the double PO technique is the application of GTD or other stationary-phase techniques to determine the fields scattered from the subreflector, and PO to determine the fields from the main reflector.[F3,F4,F6] It is not possible to use GTD to evaluate scattering from the main reflector in the direction of the main beam, because the GTD expressions become singular in focused directions.

1.3.3.1.5. EQUIVALENT PARABOLOID. An alternative compromise to the dual-reflector analysis problem is the application of geometrical optics to the subreflector and physical optics to the main reflector.[49-51] As shown in Fig. 20 the combination of main dish and subdish may then be considered

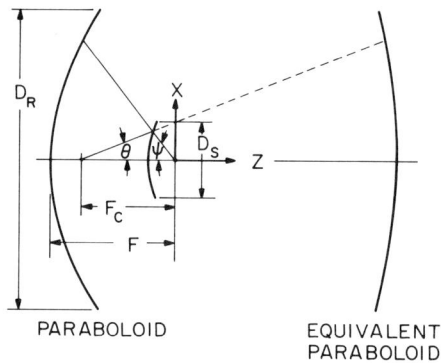

FIG. 20. Geometry of equivalent paraboloid.

as being replaced by an equivalent focusing surface at a certain distance from the real primary focus of a Cassegrain or Gregorian system. This surface is defined as the locus of intersection of incoming rays parallel to the antenna axis with the extension of the corresponding rays converging toward the real focal point. This equivalent reflecting surface has a paraboloidal contour with a focal length equal to the distance from its vertex to the real focal point. As a result, this surface could be employed as a reflecting dish which would focus an incoming plane wave toward the real focal point in exactly the same manner as does the combination of main dish and subdish. (Actually, the plane wave would have to be incident from the opposite direction.) Thus the analysis–design problem reduces to the analysis–design

---

[49] R. T. Jones, *J. Opt. Soc. Amer.* **44**, 630 (1954).

[50] P. W. Hannam, *IRE Trans. Antennas Propagat.* **AP-9**, 140 (1961).

[51] R. E. Collin and F. J. Zucker, "Antenna Theory," pp. 51–55. McGraw-Hill, New York, 1969.

of a prime-focus system with the same feed and a paraboloidal reflector with the same diameter but a much larger focal length given by

$$F^{equiv} = F\left(\frac{e+1}{e-1}\right). \tag{1.3.20}$$

Because of the large equivalent F/D ratio, the equivalent space-attenuation in Cassegrainian systems is considerably less.

The advantage of the equivalent paraboloid is its simplicity. It fails, however, to take into account the additional illumination taper produced across the paraboloid by the diffraction field of the subreflector. Thus predicted aperture efficiencies will be higher than actual, typical differences are generally $< 10\%$.[52]

1.3.3.2. Shaped Dual-Reflector Systems. Geometrical optics techniques have been developed[53-57] to determine the necessary dual-reflector combination which satisfies the boundary conditions imposed by (1) an assumed primary-feed field distribution, and (2) a desired field distribution in the aperture of the modified paraboloid (Fig. 21). A system of coupled equations must be solved numerically. These equations are based on the principles of geometrical optics:

(a) At each reflector the incident and reflected rays and the surface normal are coplanar, and the angle of incidence equals the angle of reflection (Snell's law). Hence,

$$\frac{1}{\rho}\frac{d\rho}{d\theta} = \tan\left(\frac{\theta + \beta}{2}\right) \tag{1.3.21}$$

$$\frac{dy}{dx} = -\tan\frac{\beta}{2}. \tag{1.3.22}$$

(b) Energy flow along each differential tube of rays remains constant, even when the tube undergoes reflection (conservation of energy). Hence,

$$I(x)x\,dx = F(\theta)\sin\theta\,d\theta, \tag{1.3.23}$$

where $F(\theta)$ is the known axially symmetric angular distribution of the power from the primary feed and $I(x)$ the desired power distribution in the aperture.

(c) Ray directions are normal to the constant-phase surfaces, and this condition is maintained after reflections (theorem of Malus).

[52] W. C. Wong, *IEEE Trans. Antennas Propagat.* **AP-21**, 335 (1973).
[53] V. Galindo and W. J. Welch, Consolidated Quart. Progr. Rep., No. 11, Elec. Res. Lab., Univ. of California, Berkeley, California (November, 1963).
[54] V. Galindo, *IEEE Trans. Antennas Propagat.* **AP-12**, 403 (1964).
[55] W. F. Williams, *Microwave J.* **8**, No. 7, 79 (1965).
[56] K. A. Green, *IEEE Trans. Antennas Propagat.* **AP-11**, 589 (1963).
[57] S. P. Morgan, *IEEE Trans. Antennas Propagat.* **AP-12**, 685 (1964).

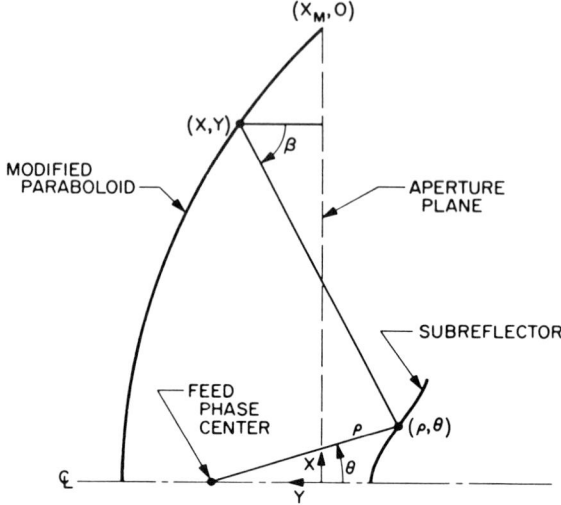

FIG. 21. Geometry of dual-shaped reflector antenna system.

Rearranging these three simultaneous differential equations in the unknowns ρ, θ, and y yields a form suitable for numerical integration:

$$\frac{d\rho}{dx} = \rho \frac{d\theta}{dx} \tan \frac{\theta + \beta}{2}, \tag{1.3.24}$$

$$\frac{d\theta}{dx} = \frac{xI(x)}{\sin \theta \, F(\theta)}, \tag{1.3.25}$$

$$\frac{dy}{dx} = -\tan \frac{\beta}{2}. \tag{1.3.26}$$

Because the procedure is based on geometrical optics, valid solutions are limited to surfaces with large radii of curvature. Furthermore, diffraction effects such as spillover are not included.

A high-efficiency dual-reflector system generally requires that (1) a high percentage of the feed energy be intercepted by the reflectors (i.e., reduction of spillover), and (2) the field in the aperture of the main reflector be distributed as uniformly as possible. Ordinarily these two effects work against each other, i.e., reduction of spillover requires tapering the field distribution and a uniform aperture distribution generally involves substantial spillover. Consequently, optimum performance generally involves a compromise which has limited efficiencies of conventional systems to about 55 to 60%.

The shaped dual-reflector concept permits the apparent contradiction between the two requirements for high efficiency to be overcome with the following rationale: a feed is selected with a high taper at the edge of the

1.3. ANALYSIS OF PARABOLOIDAL-REFLECTOR SYSTEMS

subdish to minimize forward spillover. The subdish shape is designed to distribute the highly tapered feedhorn energy uniformly over the aperture of the main reflector. The classical hyperboloidal subreflector is transformed into an empirical contour which has a smaller radius of curvature than a hyperboloid in the central section to deflect more of the rays to the outer part of the main reflector. In this manner, there is little spillover and, at the same time, nearly a uniform aperture distribution. The main reflector must then be slightly reshaped from its original paraboloidal contour to produce a constant-phase aperture distribution. Because of complete axial symmetry, these systems can be used with orthogonal linear or circular polarization, and can be made compatible with a monopulse requirement.

The shaped subreflector (modified hyperboloid) resulting from a typical calculation[58] is illustrated in Fig. 22. The central region of the modified

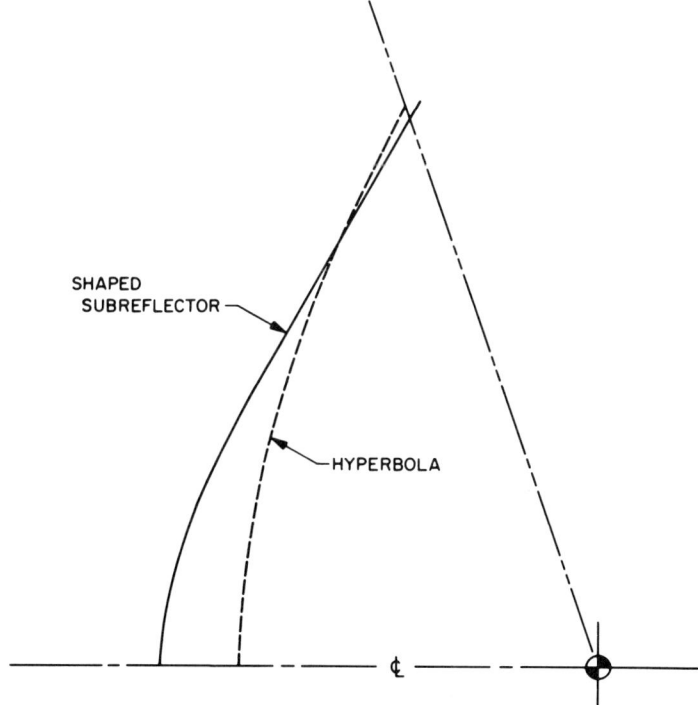

FIG. 22. Shaped subreflector.

[58] A. C. Ludwig, SPS No. 37-35, Vol. IV, p. 266, Jet Propulsion Lab., Pasadena, California (October, 1965).

contour has a considerably smaller radius of curvature than the hyperboloid.

After the two reflector profiles have been determined from geometrical optics as outlined above, a complete diffraction theory analysis of the system can be carried out using the techniques of previous sections to determine overall system performance, including such diffractive effects as noise temperature due to spillover.[59]

Maximum aperture efficiency is generally achieved with uniform illumination, i.e., setting $I(x)$ equal to a constant in Eq. (1.3.23). Other distribution functions may, however, be used if different radiation characteristics such as ultra low sidelobes are desired.[60]

1.3.3.3. Analysis of a Three-Reflector Feed System. It may be shown that an electromagnetic field $\bar{E}(\rho, \theta, \phi)$ in a region outside a sphere (which encloses all sources) of radius $\rho_0 > 0$ may be expressed as a superposition of vector spherical waves

$$\bar{E}(\rho, \theta, \phi) = -\sum_m \sum_n a_{e,o,m,n} \bar{m}_{e,o,m,n} + b_{e,o,m,n} \bar{n}_{e,o,m,n},$$
(1.3.27a)

$$\bar{H}(\rho, \theta, \phi) = (k/j\omega\mu) \sum_m \sum_n a_{e,o,m,n} \bar{n}_{o,e,m,n} + b_{e,o,m,n} \bar{m}_{o,e,m,n},$$
(1.3.27b)

where the well-known spherical waves $\bar{m}_{e,o,m,n}$ and $\bar{n}_{e,o,m,n}$ are defined by Ludwig.[61] If the tangential $\bar{E}(\rho_1, \theta, \phi)$ is known on a sphere of radius ρ_1 (where $0 \leq \rho_0 \leq \rho_1 \leq \infty$), then the coefficients $a_{e,o,m,n}$ and $b_{e,o,m,n}$ can be determined uniquely. For example, determining the tangential components E_θ and E_ϕ at great distances from the sources will permit the complete field to be known for all values of ρ subject to $\rho_0 \leq \rho \leq \rho_1$. Consequently, this technique provides a useful analytical tool for transforming far-field data into near-field data. A unique application of this technique is described next.

A dual-frequency microwave feed system has been developed for the NASA Jet Propulsion Laboratory's (JPL) 64-m antenna in Goldstone, California.[62] This system is capable of simultaneous low-noise reception at S- and X-bands and high-power transmission at S-band. To fulfill this requirement, a particularly attractive approach, the reflex-feed system, has been implemented by Potter at JPL. A cross-sectional view of the reflex-feed system geometry is shown to scale in Fig. 23. The system is comprised of four basic components: the S-band feedhorn, an ellipsoidal reflector, a planar dichroic reflector, and

---

[59] T. Kitsuregawa and M. Mizusawa, 1968 *G-AP Int. Symp. Digest, Boston* 391 (1968).
[60] F. Rouffy, Design of Dual-Reflector (Cassegrainian type) Antennas, p. 88, U.R.S.I. 1968 Fall Meeting Digest, pp. 10–12, Boston, Massachusetts (September, 1968).
[61] A. C. Ludwig, *IEEE Trans. Antennas Propagat.* **AP-19**, 214 (1971).
[62] P. D. Potter, JPL Tech. Rep. 32–1526, Vol. VIII, p. 53, Jet Propulsion Lab., Pasadena, California (April 15, 1972).

1.3. ANALYSIS OF PARABOLOIDAL-REFLECTOR SYSTEMS

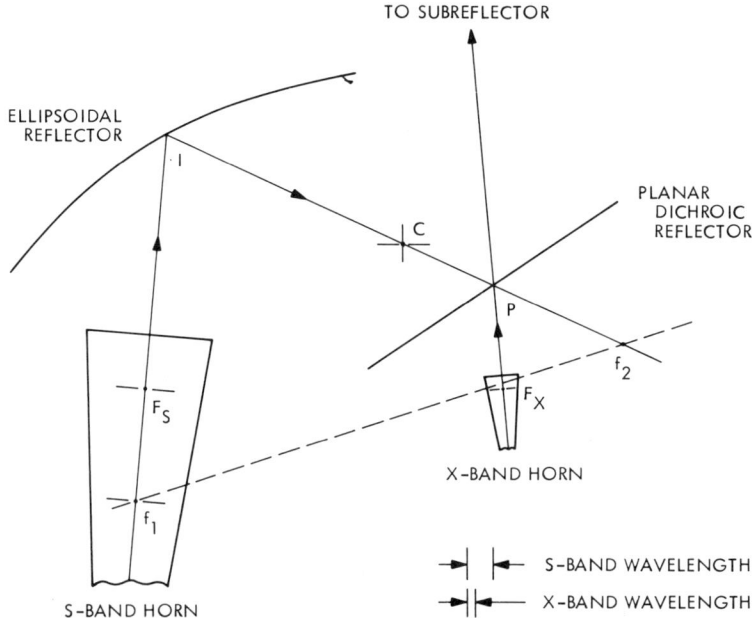

FIG. 23. Geometry of reflex-feed system.

the X-band feedhorn. By reciprocity, the operation of the reflex feed is the same in the receiving mode as in the transmitting mode; for simplicity, Fig. 23 shows only the transmitting mode. For S-band operation, from a geometrical optics standpoint, radiated energy from one of the ellipsoid foci f_1 is focused to the point f_2. As shown in Fig. 23, however, the system is not large compared to a wavelength. Because of this consideration and the fact that the S-band feedhorn does not represent a point source, the radiated energy from the elliposoid is actually found to focus to a small region centered at the point C. This energy is then redirected by the planar reflector to the antenna subreflector. By the principle of images, this redirected radiation appears to emanate from the point F_x, which is the far-field phase center of the X-band feedhorn and also coincides with one of the subreflector foci. To permit simultaneous X-band operation, the central region of the planar reflector is perforated with an array of X-band slots, thereby making the reflector essentially transparent to X-band but reflective to S-band.

Design of the reflex feed system has been carried out with simultaneous 1 : 7 scale model testing and physical optics scattering programs. Sample calculated and measured field patterns are compared in Fig. 24. Because each reflection is essentially in the near-zone of the previous horn or scatterer, it is necessary to use far-field–near-field transformations with spherical waves.

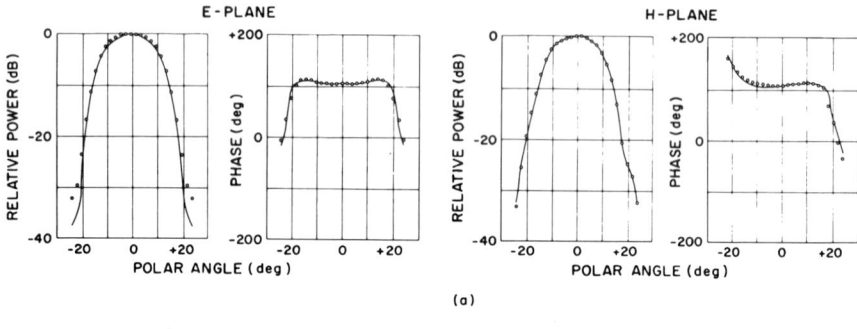

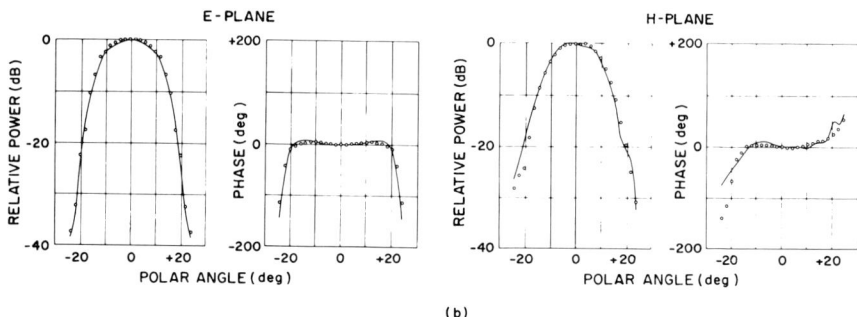

FIG. 24. Computed (circles) and measured (solid curves) patterns of reflex-feed system; (a) ellipsoidal reflector and (b) ellipsoidal and flat-plate reflectors.

For example:

(1) the far-field radiation pattern of the S-band horn was determined experimentally;
(2) this field was transformed into its proper value (including radial field components) at the surface of the ellipsoid;
(3) using geometrical optics for the currents on the front of the ellipsoid, the far-field scattered amplitude and phase patterns have been determined and are plotted in Fig. 24;
(4) this field is then transformed from the far-field to the vicinity of the surface of the planar reflector;
(5) geometrical optics then yields the currents on the planar reflector, from which can be determined the final scattered field, also plotted in Fig. 24.

1.3. ANALYSIS OF PARABOLOIDAL-REFLECTOR SYSTEMS

REFERENCES

A Dyadic Green's Function

A1 J. A. Stratton, "Electromagnetic Theory," p. 460. McGraw-Hill, New York, 1941.
A2 S. Silver, "Microwave Antenna Theory and Design," p. 144. McGraw-Hill, New York, 1949.
A3 C. J. Bouwkamp, *Phys. Soc. London Rep. Progr. Phys.* **17**, 56 (1954).
A4 R. E. Collin and F. J. Zucker, "Antenna Theory." McGraw-Hill, New York, 1969.
A5 W. V. T. Rusch and P. D. Potter, "Analysis of Reflector Antennas," p. 42. Academic Press, New York, 1970.
A6 C-T. Tai, "Dyadic Green's Functions in Electromagnetic Theory." Intext Educ. Publ., Scranton, Pennsylvania, 1971.

B Geometrical and Physical Optics

B1 M. Born and E. Wolf, "Principles of Optics," Chapter 3. Pergamon, Oxford, 1959.
B2 P. Beckmann, "The Depolarization of Electromagnetic Waves," Chapter 3. Golem Press, Boulder, Colorado, 1968.
B3 M. I. Sancer, *Radio Sci.* **3**, 141 (1968).
B4 J. J. Bowman, T. B. A. Senior, and P. L. E. Uslenghi, "Electromagnetic and Acoustic Scattering by Simple Shapes," p. 29. North-Holland Publ., Amsterdam, 1969.
B5 W. V. T. Rusch and P. D. Potter, "Analysis of Reflector Antennas," p. 46. Academic Press, New York, 1970.

C Method of Moments

C1 A. W. Maue, *Z. Phys.* **126**, 601 (1949).
C2 M. G. Andreasen, *IEEE Trans. Antennas Propagat.* **AP-12**, 746 (1964).
C3 M. G. Andreasen, *IEEE Trans. Antennas Propagat.* **AP-13**, 303 (1965).
C4 F. K. Oshiro and K. M. Mitzner, *1967 IEEE Int. Antennas Propagat. Symp. Digest, Ann Arbor, Michigan* 257 (1967).
C5 R. F. Harrington, "Field Computation by Moment Methods." Macmillan, New York, 1968.
C6 J. R. Mautz and R. F. Harrington, *Appl. Sci. Res.* **20**, 405 (1969).
C7 F. K. Oshiro, K. M. Mitzner, and S. S. Locus, Tech. Rep. AFAL-TR-70-21, Part II, Northrup Corp. (April 1970).
C8 P. L. E. Uslenghi, *Alta Freq.* **39**, 709 (1970).
C9 W. V. T. Rusch, JPL Tech. Memo. 33-478, Pasadena, California (May 15, 1971).

D Boundary-Value Solutions

D1 R. W. P. King and T. T. Wu, "The Scattering and Diffraction of Waves." Harvard Univ. Press, Cambridge, Massachusetts, 1959.
D2 R. F. Harrington, "Time-Harmonic Electromagnetic Fields," Chapters 5 and 6. McGraw-Hill, New York, 1961.
D3 A. Sommerfeld, *Math Ann.* **47**, 317 (1896).
D4 W. Pauli, *Phys. Rev.* **54**, 924 (1938).
D5 E. Pinney, *J. Math Phys.* **25**, 49 (1946).
D6 E. Pinney, *J. Math Phys.* **26**, 42 (1947).
D7 C. W. Horton and F. C. Karal, Jr., *J. Appl. Phys.* **22**, 575 (1951).
D8 I. N. Korbanskiy, *Radio Eng. Electron. Phys.* **13**, 1460 (1968).
D9 S. E. Stone, Univ. of Michigan Radiat. Lab. Rep. No. 8525-3-T, Ann Arbor, Michigan (ASTIA Doc. No. AD 818382).

E Numerical Techniques

E1 W. Romberg, *Kon. Korsk. Videnskab, Forhandl.* **28**, 30 (1955).
E2 C. C. Allen, *IRE Trans. Antennas Propagat. Spec. Suppl.* **AP-7**, 5387 (1959).
E3 J. H. Richmond, *IRE Trans. Antennas Propag.* **AP-9**, 358 (1961).
E4 J. W. Cooley and J. W. Tukey, *Math. Comput.* **19**, 297 (1965).
E5 A. C. Ludwig (ed.), Tech. Rep. No. 32-979, Jet Propulsion Lab., Pasadena, California (April 1967).
E6 W. V. T. Rusch and H. L. Strachman, *1968 NEREM Record*, Boston, p. 20 (1968).
E7 A. C. Ludwig, *IEEE Trans. Antennas Propagat.* **AP-16**, 767, (1968).
E8 J. B. Davies, *Alta Freq.* **38**, 277 (1969).
E9 E. K. Miller and G. J. Burke, *IEEE Trans. Antennas Propagat.* **AP-17**, 669 (1969).

F Stationary-Phase Evaluation of Radiation Integrals

F1 P. M. Morse and H. Feshbach, "Methods of Theoretical Physics," p. 437. McGraw-Hill, New York, 1953.
F2 A. F. Kay, Rep. No. 3, TRG Inc., East Boston, Massachusetts (November 1962).
F3 C. C. Allen, Final Rep. on the Study of Gain-to-Noise-Temperature Improvement for Cassegrain Antennas, General Electric Co., Schenectady, New York (1967).
F4 C. C. Allen, *1968 G-AP Int. Symp. Digest*, Boston, p. 419 (1968).
F5 W. H. Ierley and H. Zucker, *Bell Syst. Tech. J.* **49**, 431 (1970).
F6 C. C. Allen, *1971 G-AP Int. Symp. Digest*, Los Angeles, p. 212 (1971).

G Geometrical Theory of Diffraction

G1 J. B. Keller, *J. Appl. Phys.* **28**, 426, 570 (1957).
G2 J. B. Keller, A Geometrical Theory of Diffraction, *Proc. Symp. Appl. Math.*, Vol. 8, pp. 27–57. McGraw-Hill, New York, 1958.
G3 J. B. Keller, *J. Opt. Soc. Amer.* **52**, 116 (1962).
G4 P. Ia. Ufimtsev, *Sov. Phys.—Tech. Phys.* **2**, 1708 (1957).
G5 P. Ia. Ufimtsev, *Sov. Phys.—Tech. Phys.* **3**, 2386 (1958).
G6 P. Ia. Ufimtsev, Method of Edge Waves in the Physical Theory of Diffraction (English transl.), Doc. ID No. FTD-HC-23-259-71 (1971).
G7 B. E. Kinber, *Radio Electron.* **6**, 481 (1961).
G8 R. G. Kouyoumjian, *Proc. IEEE* **53**, 864 (1965).
G9 W. V. T. Rusch, JPL Tech. Rep. 32-1113, Jet Propulsion Lab., Pasadena, California (June 1967).
G10 M. I. Sancer, TRW Rep. 08051-6008-RO-OO (November 20, 1968).
G11 P. A. J. Ratnasiri, R. G. Kouyoumjian, and P. H. Pathak, Ohio State Univ. Electrosci. Lab., Scientific Rep. No. 4 (March 23, 1970).
G12 P. H. Pathak and R. G. Kouyoumjian, The Dyadic Diffraction Coefficient for a Perfectly Conducting Edge Structure, URSI Digest, pp. 64–65, UCLA, Los Angeles, California (September 21–23, 1971).
G13 G. L. James and V. Kerdemelidis, *IEEE Trans. Antennas Propagat.* **AP-21**, 14 (1973).

H Reciprocity

H1 M. K. Hu, *1958 IRE Nat. Convent. Rec. New York* **8**, 128 (1958).
H2 J. H. Richmond, Reciprocity Theorems and Plane Surface Waves, Eng. Exp. Station, College of Eng., Ohio State Univ., Vol. XXVIII, No. 4 (July 1959).
H3 R. F. Harrington, "Time-Harmonic Electromagnetic Fields," pp. 116–120. McGraw-Hill, New York, 1961.
H4 J. H. Richmond, *IRE Trans. Antennas Propagat.* **AP-9**, 515 (1961).
H5 R. E. Collin and F. J. Zucker, "Antenna Theory," p. 24. McGraw-Hill, New York, 1969.
H6 J. R. Pace, *IEEE Trans. Antennas Propagat.* **AP-17**, 285 (1969).
H7 P. Wood, *Electron. Lett.* **6**, 326 (1970).
H8 P. Wood, *IEEE Trans. Antennas Propagat.* **AP-14**, 191 (1971).

1.4. Feed Systems for Paraboloidal Reflectors*

1.4.1. Introduction

The purpose of this chapter is to acquaint the radio astronomer with various types of feeds for parabolic reflectors. Such knowledge is valuable in evaluating proposals for equipment and in judging the value of various commercial feeds available. As the parabolic radio telescope is a versatile instrument operable over a very large range of wavelengths, it is inevitable that the original feeds supplied will be changed as the interest of the observer and the necessary experiments change.

As indicated in Chapter 1.3 the design of feed equipment for a focal-fed or Cassegrain system to achieve high antenna efficiency (collecting area) is a careful procedure with attention paid to many details. Such a design effort can only be performed in a suitably equipped antenna laboratory where automatic pattern, polarization, and phase-measuring equipment[1] is available. The cost of such a development is justified for the basic and long-term feed equipment of the radio telescope. Frequently, however, a feed that radiates a major part of its energy onto the parabolic surface and is placed at its focus will result in a usable though less efficient and higher noise system. Such feeds may be available (or may be wavelength-scaled from existing designs) from commercial firms at moderate cost.

1.4.2. Basic Feed Types

A large variety of feeds have been developed for parabolic reflectors. Many of these have been for complex radar applications and are not required for use in radio astronomy. The usual requirements for radio astronomy are the generation of a pencil beam with a high efficiency (collecting area), polarization purity, and low-noise reception. The last requirement implies (1) the reception of the energy reflected from the parabolic reflector and no other such as that reflected from the warm earth, and (2) a well-matched and low-loss transmission line for connection to the energy detector. Typical feed types useful for radio astronomy are shown in Fig. 1.

---

[1] J. S. Hollis *et al.*, "Antenna Measurements." Scientific-Atlanta, Atlanta, Georgia, 1970.

---

* Chapter 1.4 is by **John Ruze**.

1.4. FEED SYSTEMS FOR PARABOLOIDAL REFLECTORS

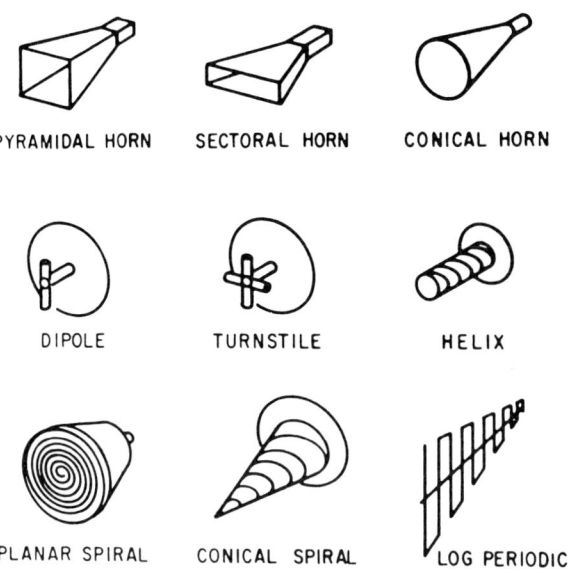

FIG. 1. Typical reflector feeds. [R. C. Hansen (ed.), "Microwave Scanning Antennas," p. 148, Fig. 17. Academic Press, New York, 1964].

1.4.3. Horn Feeds

The most common microwave feed is the horn radiator. Basically, it is nothing more than a flared section of waveguide. Due to the gradual transition from the impedance of the fundamental mode in the waveguide to that of free space the feed can be readily matched over a wide wavelength band. With the appropriate dimensions, it can be used either as a focal-point feed or as a Cassegrain feed.

As the boundary conditions on the metal walls require a zero electric field tangential component, the field distribution in the horn aperture is cosinusoidal in the H-plane and uniform in the E-plane. This different field distribution in the two principal planes requires that the horn aperture be made rectangular since equal beamwidths are usually desired to illuminate the circular reflector.

The horn aperture should be proportioned so that the reflector edge illumination is about 10 dB down for maximum collecting area and about 14 dB down for an optimum collecting-area/antenna–noise–temperature design. For focal-point feeds empirical formulas are used because stray currents on the horn sides disturb the theoretical calculations. The formulas usually used for the horn pattern 10-dB beamwidth are[2,3]

$$E(10 \text{ dB}) = 88\lambda/B \quad (\text{deg}) \quad \text{for} \quad B/\lambda < 2.5, \qquad (1.4.1a)$$

$$H(10 \text{ dB}) = 31 + 79\lambda/A \quad (\text{deg}) \quad \text{for} \quad A/\lambda < 3.0, \qquad (1.4.1b)$$

where B and A are the horn aperture dimensions in the E- and H-planes, respectively. The angle used is the total angle to the reflector rim at the prime focus. With the addition of space attenuation[4] this should yield a satisfactory low-noise design.

For the Cassegrain focus the theoretical formulas are applicable and we have for the 14-dB beamwidth

$$E(14 \text{ dB}) = 95\lambda/B \quad (\text{deg}), \qquad (1.4.2a)$$

$$H(14 \text{ dB}) = 133\lambda/A \quad (\text{deg}), \qquad (1.4.2b)$$

where now the angle used is the total angle subtended by the subreflector at the Cassegrain focus. As this angle is substantially smaller than the prime focus angle (roughly by the system magnification), Cassegrain feeds are considerably larger.

The required length of the horn is determined by the allowed phase variation across the horn aperture. This is the distance of the aperture chord from the spherical phase-front arc originating at the horn throat. Setting this separation to $\lambda/16$, we obtain as the horn slant length

$$r = 2B^2/\lambda. \qquad (1.4.3)$$

We note that Cassegrain horn feeds for systems of high magnification can be quite long especially at long wavelengths. This may preclude the use of the Cassegrain focus at these wavelengths. An additional limitation on the long-wavelength limit of Cassegrain systems is the requirement that the subreflector should be in the far field of the horn feed, so that a spherical wave is developed. This sets the maximum wavelength

$$\lambda_{\max} < B^2/s, \qquad (1.4.4)$$

where s is the feed–subreflector spacing.

The horn length [Eq. (1.4.3)] can be significantly reduced by incorporating a lens in the horn aperture.[5] Here the lens performs the wave-front collimation

---

[2] S. Silver, "Microwave Antenna Theory and Design," Sect. 10. McGraw-Hill, New York, 1949.

[3] H. Jasik, "Antenna Engineering Handbook," Sect. 10. McGraw-Hill, New York, 1961.

[4] J. V. Evans and T. Hagfors (eds.), "Radar Astronomy," Chapter 8. McGraw-Hill, New York, 1968.

[5] H. Jasik, "Antenna Engineering Handbook," Sect. 14. McGraw-Hill, New York, 1961.

1.4. FEED SYSTEMS FOR PARABOLOIDAL REFLECTORS

and the horn sides act as a guide for the electromagnetic energy. In such designs the lens surfaces should be matched by a quarter-wave layer. The increased noise temperature due to the lossy dielectric should be investigated. Although a lens-compensated horn reduces the horn length, the radiator may be heavier than desired.

The literature also describes a so-called "optimum horn." This horn is optimum in the sense that it provides maximum horn gain for a given slant length. It is not "optimum" as a feed for reflector antennas where we generally require smaller aperture–phase variations than provided by this horn.

The rectangular pyramidal horn described previously is a simple device that can be well matched over a comparatively wide frequency band and provides essentially axially symmetric patterns and high collecting area. For focal-fed systems, it can be readily constructed. It suffers, however, from the disadvantage that it can receive only one polarization although it can be rotated and provides excellent polarization discrimination (35–40 dB).

To provide both orthogonal polarizations, the horn aperture must be axially symmetric; that is, square or round. To provide the two outputs of orthogonal polarization either two orthogonal probes are inserted or the horn is fed with a set of orthogonal waveguides. Such devices are referred to as "orthomode transducers." If oppositely sensed circular polarization (CP) is desired, a "quarter-wave plate" is inserted in the horn, or for wider bandwidths the outputs are connected to a 90° hybrid junction. Axial ratios better than 1.05 or port isolation better than 30 dB are obtainable.

The requirement that for dual polarization the horn be square or round causes two problems: (1) The horn radiation patterns are now unequal in the two principal planes [Eq. (1.4.1)–(1.4.2)] causing the antenna gain function or collecting area to be elliptical in cross section. This may cause distortion in mapping radio sources and degrades the off-axis axial ratio; (2) As the horn beamwidths are unequal in the principal planes an optimum beamwidth cannot be chosen causing a reduction in antenna collecting area of up to 20%.

Various stratagems are used to obviate these difficulties. Fortunately for focal-fed systems a simple conical horn of about 0.9 wavelength yields essentially equal beamwidth. The pattern data[7] indicates that such a radiator would be suitable for reflectors with a f/D of 0.3 to 0.4.

Other useful horns, primarily for Cassegrain systems, are the *diagonal* or *Love horn*.[8] Here the patterns are equalized by exciting the electric vector across the diagonal of the square aperture. This radiator, however, has higher cross-polarized energy which will cause off-axis antenna cross-

[6] A. F. Harvey, "Microwave Engineering," p. 115. Academic Press, New York, 1963.

[7] H. Jasik, "Antenna Engineering Handbook," Fig. 10-15. McGraw-Hill, New York, 1961.

[8] A. W. Love, *Microwave J.* 5, No. 3, 117 (1962).

polarization response; similar to the inherent reflector cross polarization or Condon lobes.[9,10] Other methods of compensating for the unequal beamwidths are by subdividing the horn aperture by means of septums which decrease the H-plane beamwidth or by means of internal fins which increase the E-plane beamwidth as the E-field does not penetrate into the finned area.[11] For focal-fed systems pattern control can also be exercised by the use of external fins[12] or flanges. Flanges and chokes located at, or set back from, the aperture are also useful in reducing the spurious and back radiation of horn antennas[13] with a resultant decrease in antenna-system temperature. These designs are largely empirical requiring cut-and-try modifications and corresponding pattern measurement.

Recent interest in high-efficiency, low-noise systems has spurred the development of new types of feeds. An ideal feed for radio astronomy applications would have very low sidelobes (that is, a high beam-efficiency), axially symmetric radiation patterns, a well-defined phase center, a good VSWR, low internal losses, and a frequency bandwidth adequate for the experiment to be performed. A feed embodying many of these characteristics is the *corrugated* horn, shown in Fig. 2a with its radiation patterns, Fig. 2b.[14] This is a conical horn with its inner surface corrugated with circular ridges approximately a quarter-wave deep. The ridged surface reduces to zero the amplitude of the normal component of the electric vector thereby reducing the edge discontinuity of the E-plane uniform illumination of the conventional TE_{11} excited horn. Removal of this discontinuity also reduces the far-out sidelobes of the horn radiation pattern by eliminating the stray currents which creep around to the outside surface of a conventional horn as a result of the high E-plane edge illumination. The corrugated horn may be thought of as a multimode horn generating a tapered spherical wave resolvable into many higher-order modes instead of the single fundamental mode of a conventional horn. The horn has a well-defined phase center at the vertex of the cone. Bandwidths as high as an octave have been reported.[15] To date, the corrugated horn has been developed for focal-fed systems or Cassegrain systems of low magnification. This restriction is due to the large flare angle usually required to keep the various modes in phase. A lens-compensated

---

[9] S. Silver, "Microwave Antenna Theory and Design," Sect. 12.4. McGraw-Hill, New York, 1949.

[10] J. D. Kraus, "Antennas," p. 471. McGraw-Hill, New York, 1950.

[11] H. Jasik, "Antenna Engineering Handbook," Fig. 17-24. McGraw-Hill, New York, 1961.

[12] J. J. Epis, *Microwave J.* **4**, No. 5, 84 (1961).

[13] A. H. Lagrone and G. F. Roberts, *IEEE Trans. Antennas Propagat.* **AP-14**, 102 (1966).

[14] H. D. Minnett and B. M. Thomas, *IEEE Trans. Antennas Propagat.* **AP-14**, 654 (1966).

[15] R. E. Lawrie and L. Peters, Jr., *IEEE Trans. Antennas Propagat.* **AP-14**, 605 (1966).

(a)

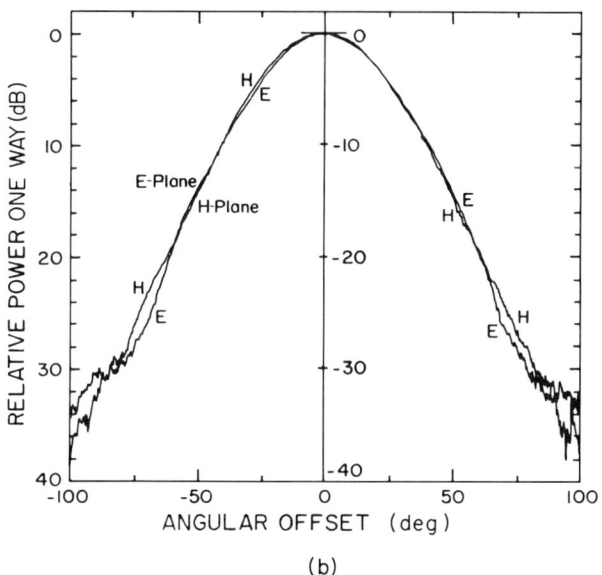

(b)

FIG. 2. (a) Corrugated horn; (b) its radiation patterns in the HE_{11} mode with $\lambda = 11.3$ cm and $2a/\lambda = 1$. [H. D. Minnett and B. M. Thomas, *IEEE Trans. Antennas Propagat.* **AP-14**, 656 (1966).]

corrugated horn can also be designed for Cassegrain systems of high magnification.[16] The term "scalar" feed denoting independence of vector or directional effects is a commercial designation of a corrugated horn.

A highly efficient *multimode* horn has been developed for the use of satellite-communications earth-stations.[17] Here a taper or a step discontinuity at the horn throat is used to generate a set of higher-order modes, a section of waveguide, beyond cutoff for all but the desired modes, is then introduced followed by the normal horn flare. Care must be taken that all modes arrive in the proper phase relation at the horn aperture, and this limits the bandwidth to about 10%. These horns may be either conical or square and have usually been designed for Cassegrain geometries. The design of the mode generation and suppression section is largely empirical and involves considerable effort. The design has been highly successful and with a dual-shaped reflector has achieved overall aperture efficiencies as high as 74%.[18] An alternate method of generating the higher-order modes has recently been suggested.[19] Here mode generation is accomplished by changes in the flare angle at discrete points along the length of the horn. This technique has the promise of wider-band operation as the mode generation is made in sections far from cutoff where the relative mode phase velocity is almost that of free space. Unfortunately such horns are rather long.

The basic horn radiator is a broadband device with good impedance characteristics. An impedance mismatch occurs at the horn aperture–free space discontinuity and at the throat waveguide–flair junction. These two discontinuities may be separated by many wavelengths creating an interference phenomena between these two reflection coefficients as a function of wavelength. For large horn apertures (Cassegrain systems), the aperture mismatch is small and may be compensated, where necessary, by an iris-matching device slightly inside the aperture. The throat discontinuity can be reduced by a gradual flare or by a ridge transition.

The *ridge* horn shown in Fig. 3 (doubly ridged for dual polarization)[20,21] provides a gradual transition from the connector input to the horn aperture. Horns of this type are available to cover a 2:1 bandwidth with a VSWR less than 1.5:1. For radio astronomy it is generally desirable to design for constant collecting area or maximum antenna gain over the frequency band instead of constant antenna beamwidth. This is done by designing the feed radiator for constant beamwidth over the frequency band. In any broadband

---

[16] P. J. B. Clarricoats, and P. K. Saha, *Proc. Inst. Elec. Eng.* **118**, 1167 (1971).
[17] P. D. Potter, *Microwave J.* **6**, No. 6, 71 (1963).
[18] R. Lindsey, *Aerosp. Technol.* 28-30 (November 6, 1967).
[19] S. B. Cohn, *Microwave J.* **13**, No. 10, 41 (1970).
[20] J. K. Shimuzu, *IRE Trans. Antennas Propagat.* **AP-9**, 223 (1961).
[21] K. L. Walton and Sundberg, *Microwave J.* **7**, No. 3, 96 (1964).

1.4. FEED SYSTEMS FOR PARABOLOIDAL REFLECTORS

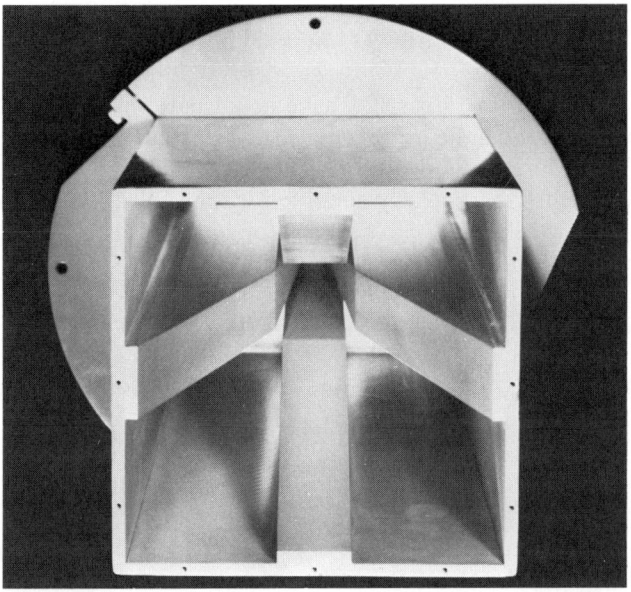

FIG. 3. Broadband ridge horn. (Courtesy of Scientific Atlanta Inc., Atlanta, Ga.)

design, however, compromises in aperture efficiency are generally necessary to obtain the convenience of wideband operation.

The horn aperture is generally sealed against the weather by means of a dielectric sheet or window. Provisions are frequently made to maintain a small positive pressure of dry air or nitrogen to prevent the entry of moisture. The dielectric window may be positioned in the horn to correct partially the horn–aperture mismatch. The dielectric sheet, being a lossy material, introduces a noise temperature due to the material emission equal to

$$T = 293[(2\pi d/\lambda)e \tan \delta] \quad °K \qquad (1.4.5)$$

where d is the membrane thickness, e the membrane dielectric constant, and $\tan \delta$ the membrane loss tangent (dissipation factor).

1.4.4. Dipole Feeds

At wavelengths longer than about $\frac{1}{2}$m a horn feed becomes too large and bulky even for focal-fed systems. A common and simple feed is the half-wave dipole supported a quarter-wave above a ground screen a half-wave on a side. The dipole is fed with a coaxial cable through one of its supports and the dipole length adjusted for optimum match. Two cross dipoles can be used for dual polarization. In the latter case the two outputs can be connected by means of a commercial 3-dB, 90° hybrid to yield both senses of circular polarization.

A simple single dipole does not have equal patterns in the two principal planes. Two dipoles can be arranged in parallel, spaced a half-wave apart on a wavelength-square ground plane, to yield an axially symmetric pattern.[22] This arrangement is particularly suitable for long focal-length dishes, $f/D = 0.35-0.45$. The two elements can be fed through a 3-dB 180° hybrid to provide a pencil and a split beam for greater pointing accuracy. The other orthogonal polarization can be obtained with another dipole pair.

The bandwidth of the dipole antenna is only about 10%, but multifrequency operation can be achieved by providing another longer wavelength dipole pair at a larger spacing. In addition the central area is available for placement of a microwave horn radiator.

Figure 4 illustrates a commercially available design. Crossed dipoles are recessed in a cylindrical cavity which serves to equate the E- and H-plane beamwidths and provides an improved front to back ratio. For the design shown the cavity diameter is 0.762 m (30 in.) for an operating frequency of 215 to 260 MHz with a VSWR less than 1.5:1.

FIG. 4. Dual-polarized crossed-sleeve dipoles. (Courtesy of Radiation Systems Inc., McLean, Va.)

[22] W. N. Christiansen and J. A. Hogbom, "Radio Telescopes," p. 68. Cambridge Univ. Press, London and New York, 1969.

1.4.5. Loop Feed

A useful and simple linearly polarized feed has been suggested[23] for radio astronomy use at long wavelengths. This is simply a square loop of wire or rods (alternately a circular loop may be used) placed an eighth-to a quarter-wave above a ground plane and fed with a half-wave coaxial balun. The circumference of the square or circular loop is one wavelength and the E- and H-plane beamwidths are reasonably equal. Operation is for prime focus only and bandwidth is only a few percent which may be altered by providing a tuning condenser across the loop terminals. The focal point is unobstructed and available for a shorter wavelength radiator.

1.4.6. Log-Periodic Dipole Array

The log-periodic feed is basically a number of radiating elements, usually dipoles, of different lengths or resonant wavelengths disposed along a transmission feeder. The unit is fed at the short or high-frequency end. At the various wavelengths of interest a particular element or elements are automatically excited due to their resonant property. The longer elements act as reflectors so that the array radiates in the direction of the short end. On this basis very broadband feeds can be designed that cover up to a decade of wavelength range. The input VSWR can be maintained below 2:1. The patterns are reasonably constant over the frequency band and can be made essentially axially symmetric by arraying two log-periodic arrays in the H-plane to narrow the H-plane beamwidth.

The difficulty with the log-periodic antenna is that the phase center changes with wavelength, progressing toward the base as the wavelength increases. A compromise position near the high-frequency end is usually selected as the position of the parabolic reflector focus. It is this phase-center variation that usually limits the available operating bandwidth as a reflector feed. Naturally, provisions can be incorporated to provide a focusing adjustment as a function of wavelength.

Due to its comparatively wide beamwidth the structure is only suitable for focal-fed parabolas. Naturally, compromises in aperture efficiency, polarization purity, and impedance match are necessary to cover the extremely wide bandwidth.

Figure 5 is a photograph of a dual-polarized log-periodic array which provides equal E- and H-plane beamwidths. The dipole elements and the interconnecting transmission line is printed on a dielectric substrate. The aperture efficiency ranges from 30 to 45% over the frequency band.

---

[23] A. T. Moffet, *IEEE Trans. Antennas Propagat.* **12**, 132 (1964).

Fig. 5. Dual-polarized log-periodic array. (Courtesy of Radiation Systems Inc., McLean, Va.)

1.4.7. Helix Feed

A circular helix with a circumference of about one wavelength and a pitch angle of about 15° is suitable as a broadband circularly polarized radiator. The helix normally is periphery-fed against a ground plane 1–1.5-design wavelengths in diameter. Maximum radiation occurs along the axis of the helix and over the frequency band of $\pm 25\%$ of the design frequency the patterns, axial ratio, and impedance are reasonably constant. The input impedance is essentially resistive, of about 100 to 150 ohm, and an input

1.4. FEED SYSTEMS FOR PARABOLOIDAL REFLECTORS

transformer is generally required to match a 50-ohm feeder. Design details are available in the literature.[24,25]

The directivity of the helix increases with its length or number of turns. A single 6-turn helix is suitable as a prime-focus feed for long focal length dishes ($f/D = 0.45$–0.6). The E- and H-plane patterns are reasonably equal, the axial ratio about 1.15 and the phase center is about a half-wavelength above the ground plane.[26,27] The polarization purity or the axial ratio can be improved to about 1.03 by making the cross section elliptical with an eccentricity of about 0.9 and using an integral or half-integral number of turns.[28]

For reflectors with more conventional f/D ratios (0.3–0.4) only a 2- or 3-turn helix is required as the total angle seen from the prime focus varies from 160 to 130°. Unfortunately the characteristics of a helix with a few turns deteriorate approaching in the limit of a single turn that of an unterminated loop. The characteristics can be greatly improved and made acceptable for measurements not requiring high polarization purity by placing the helix in a conical cavity.[29,30] Here a 2.5-turn helix is placed in a conical cavity with a base of 0.4 wavelengths, an aperture of 0.75 wavelengths, and a height of 0.4 wavelengths. This yields equal E- and H-plane beamwidths and an axial ratio of 1.25. Frequently, in the case of helices with a small number of turns the input impedance is stabilized by inserting a 150–200-ohm resistor at quarter-wavelength from the open end. Although this introduces a lossy element, thermal noise considerations are not of prime importance at the long wavelengths where the helical feed finds its principal application.

A single helix does not possess sufficient directivity for use at the Cassegrain focus. For systems of moderate magnification (3–4) a square broadside array of four helices[31] may be used. For higher magnifications a single helix may be used to excite the circularly polarized TE_{11} mode in a cylindrical waveguide connected to a conical horn.

The sense of circular polarization may be reversed by winding the helix in the opposite direction. Two such contrawound helices placed side by side with a spacing of about a wavelength will then provide both senses of polarization.

[24] H. Jasik, "Antenna Engineering Handbook," Sect. 7. McGraw-Hill, New York, 1961.

[25] J. D. Kraus, "Antennas," p. 173. McGraw-Hill, New York, 1950.

[26] S. Adachi and Y. Mushiake, Directive Loop Antennas, Sci. Rep. RITU, B-(Elec. Comm. Japan), Vol. 9, No. 2, pp. 105–112 (1957).

[27] S. Sander and D. K. Cheng, *1958 IRE Nat. Convent. Rec., New York* **1**, 152 (1958).

[28] J. Y. Wong, and S. C. Loh, *IRE Trans. Antennas Propagat.* **AP-7**, 46 (1959).

[29] G. Svennerus, *Proc. Int. Congr. UHF Circuits Antennas, Paris*, 1957 724 (1957).

[30] A. Bystrom and Bernsten, *IRE Trans. Antennas Propagat.* **AP-4**, 53 (1956).

[31] H. Jasik, "Antenna Engineering Handbook," Fig. 7-10. McGraw-Hill, New York, 1961.

1.4.8. Conical Spiral

The conical spiral[32] shown in Fig. 1 appears as a helix with a tapered diameter. Actually its design is based on the same principal as the log-periodic dipole array wherein a scaling factor is successively applied to the radiating structure to produce an identical structure wavelength-scaled. Such structures are theoretically bandwidth-limited only by the truncations at the ends. Practically, size limits the low-frequency end and the physical construction of the apex region limits the high-frequency response. Typical conical spirals are available commercially over the frequency band of 100 MHz to 10 GHz and individually cover frequency ratios of 5 to 20:1.

The feed receives circularly polarized radiation in the direction of the apex. The antenna is constructed by winding a metallic tape of decreasing width on a conical form in the shape of a logarithmic spiral. Another tape is displaced by 180° to provide a balanced structure. The structure is fed by a small coaxial cable bonded to one of the tapes with its center conductor connected to the other tape at the apex forming an "infinite" balun. Generally no matching is attempted; however, the VSWR is typically below 2.5:1 on a 50-ohm cable. The E- and H-plane beamwidths are reasonably equal and suitable as feeds for focal-fed parabolas. Increasing the spiral pitch angle increases the beamwidths. The feed phase center is located at the point where the average circumstance is a wavelength and therefore its position varies over the frequency band. A focusing adjustment is necessary to prevent degradation of performance; generally, however, the conical spiral is set in a fixed compromise position.

With any feeds of such extreme bandwidth performance is expected to be compromised. For radio-astronomy applications the disadvantages are principally the low collecting area (30%) and poor axial ratio (3.0:1). Variations and erratic performance may occur in portions of the frequency band. Therefore, pattern, impedance, and axial-ratio data should be directly measured at the frequencies of particular interest. Nevertheless such broadband feeds may find use in certain experiments where observations are made over a swept-frequency band.

The phase of the energy received by a conical spiral or a helix can be altered by mechanical rotation about its axis; the phase change being equal to the angle of rotation. By progressively rotating an array of such elements the direction of reception can be changed as determined by the emerging uniform phase front. This principle has been applied to the design of line source feeds for cylindrical parabolic reflectors.[33]

---

[32] J. D. Dyson, *IRE Trans. Antennas Propagat.* **AP-7**, 329 (1959).
[33] G. W. Swenson, Jr. and Y. T. Lo, *IRE Trans. Antennas Propagat.* **AP-9**, 9 (1961).

1.4.9. Flat Spirals

The flat spiral[34,35] is essentially a conical spiral compressed on a plane sheet. Unfortunately the radiation now becomes bidirectional. To provide the omnidirectional radiation required for a reflector feed, the flat spiral is mounted in a cavity. This expedient reduces the large bandwidth potentially available to about 2 or 4:1 depending on the polarization purity required.

The feed radiates a circularly polarized beam of essentially equal E- and H-plane beamwidth suitable for use as a prime focus feed for typical parabolic reflectors. The axial ratio is below 1.5:1. The spiral diameter is one wavelength and the cavity depth a quarter of the low-frequency wavelength. The cavity is usually mounted in a larger ground plane for reflector feed application. Cavity-backed spiral antennas are available commercially, usually being of printed circuit construction. The cavity-backed spiral has also been used as an exciter for conical horns. This provides the possibility of use in Cassegrain systems of low magnification.

1.4.10. Tertiary Reflector Systems

As discussed in Section 1.4.3 the horn feeds become too long when used as Cassegrain feeds in systems of high magnification and at long wavelengths. Outside of their bulk such horns may not fit into the space provided between the Cassegrain focus and the reflector vertex. Although the horn can be significantly shortened by the use of a dielectric lens, this involves considerable engineering effort and still results in a heavy structure. In such cases, it is preferable to use the prime focus or if this is not possible to decrease the Cassegrain magnification.

If neither alternative is feasible, it is still possible to use the Cassegrain focus with a tertiary reflector. This is another parabolic reflector which illuminates the Cassegrain or secondary reflector which in turn illuminates the main or primary reflector. The size of the tertiary must naturally be sufficient to illuminate the secondary with the desired illumination taper, about 14 dB. Such systems are inherently of low efficiency because in addition to the normal spillover energy of the primary and secondary reflectors there is spill over the rim of the tertiary from its own feed. Calculations indicate that a 40 % aperture efficiency is possible although actual systems have yielded lower figures.

The tertiary reflector must be carefully designed to yield high beam efficiency; that is, as large an amount of the input energy must be concentrated onto the secondary as is concentrated onto the tertiary reflector. To do this

---

[34] R. Bower and J. J. Wolfe, *1960 IRE Int. Convent. Rec. New York* **1**, 84 (1960).
[35] J. D. Dyson, *IRE Trans. Antennas Propagat.* **AP-7**, 181 (1959).

TABLE I. Characteristics of Various Types of Antenna Feeds

Feed type	Polarization and frequency	Polarization purity	Efficiency	Bandwidth	Focus	Construction	Remarks
Rectangular horn	Linear >500 Mc	High	High	High with ridge	Prime or Cass.	Simple for prime focus	Rarely used for Cass. system due to single polarization
Diagonal horn	Dual >1000 Mc	High	Moderate	Moderate	Prime or Cass.	Precise transition required	
Multimode horn	Dual >1000 Mc	Moderate	High	Low	Cass.	Complex	
Corrugated horn	Dual >1000 Mc	High	High	High	Prime or Cass.	Corrugation fabrication	Requires further development
Ridged square horn	Dual >500 Mc	Moderate	Moderate	High	Prime	Ridge fabrication	Unequal E- and H-beamwidths
Dipole	Linear 100–3000 Mc	High	Moderate	Low	Prime	Simple	Unequal E- and H-beamwidths

Type	Polarization / Frequency				Focus	Complexity	Remarks
Two parallel dipoles	Linear 100–1000 Mc	High	Moderate to low	Low	Prime	Simple	Focal point unobstructed
Loop	Linear 100–500 Mc	Moderate	Moderate	Low 100–500 MHz	Prime	Simple	Focal point unobstructed
Turnstile	Dual 100–1000 MHz	Moderate	Moderate	Low	Prime	Moderate	Unequal E- and H-beamwidths
Log-periodic dipole array	Dual 100–5000 MHz	Low	Low	Extreme	Prime	Moderate	Phase center changes with wavelength
Helix	Circular 50–3000 MHz	Moderate to high	Moderate	Moderate	Prime or Cass.	Simple	
Conical spiral	Circular 100–10,000 MHz	Low	Low	Extreme	Prime	Simple	Phase center wavelength-dependent
Flat spiral	Circular 300–10,000 MHz	Low	Low	High	Prime	Printed circuit construction	
Tertiary reflector	Plane or circular	Moderate	Low	Moderate	Cass.	Spill-over consideration	Limited to where prime focus cannot be used

the spillover energy of the tertiary must be minimized. This is done by using relatively deep dishes for the tertiary ($f/D = 0.25$ or smaller) and using a feed for the tertiary which has low side- and backlobes. A Clavin feed[36] has been used for this purpose as it is a rear feed with no feed support blockage.

Figure 6 shows a tertiary reflector system used on the 36.6-m (120-ft)

FIG. 6. Feed system for the 120-ft Cassegrain Haystack antenna. (Courtesy of MIT, Haystack Observatory.)

Cassegrain Haystack antenna. For observation at 180-cm wavelength, a 2.13-m (7-ft) diameter paraboloidal reflector with a Clavin feed is used in the on-axis position. For observation at shorter wavelengths the large reflector is retracted, exposing a scaled-down tertiary reflector feed for 3.8-cm wavelength and a pair of horn feeds for 2-cm wavelength. The on-axis 2-cm feed can be rotated to change the position angle of the plane of polarization. It will be observed that the 18-cm wavelength tertiary is not at the same focal

[36] A. Chlavin (sic), *IRE Trans. Antennas Propagat.* **AP-2**, 113 (1954).

point as the higher-frequency antennas. This displacement is permissable as the depth of focus for this high magnification system is about 1.5 m (5 ft) at this wavelength. The higher-frequency antennas in the focal plane do not all occupy the axial position so that their directions of reception are displaced by a small but calculable amount.

1.4.11. Summary of Feed Types

Each feed type has advantages and limitations as to operating frequency, polarization capability, efficiency, bandwidth, and the type of telescope on which it can be used. Table I gives a summary of this information on the various types of feeds.

SPECIAL JOURNAL ISSUES AND OTHER REFERENCES OF INTEREST

1. Radio astronomy, *Proc. IRE* **46**, No. 1 (1958).
2. Radio astronomy, *Trans. IEEE Antennas Propagat.* **AP-9**, No. 1 (1961).
3. Radio astronomy, *Proc. Inst. Radio Eng. Aust.* **24**, No. 2 (1963).
4. Radio and radar astronomy, *Trans. IEEE Antennas Propagat.* **AP-12**, No. 7 (1964).
5. Millimeter waves and beyond, *Proc. IRE* **54**, No. 4 (1966).
6. Millimeter wave antennas and propagation, *Trans. IEEE Antennas Propagat.* **AP-18**, No. 4 (1970).
7. Cumulative index, *IEEE Trans. Antennas Propagat.* **AP-17**, No. 6 (1952–1968), **AP-22**, No. 5 (1969–1973).

1.5. Antenna Calibration*

1.5.1. Introduction

Direct antenna calibrations of radio telescopes, such as the measurement of the polar pattern on a turntable, are usually not possible. Present-day radio telescopes are too large for these methods, which are generally employed in the testing of antennas for radio communication purposes. For a turntable measurement the antenna must be in the far field. The far field of an antenna starts at a distance d given by

$$d = 2a^2/\lambda, \qquad (1.5.1)$$

where a is the aperture and λ the wavelength. In a case of a 100-m radio telescope at $\lambda = 10$ cm the far field begins at the distance $d = 2 \times 10^5$ m, i.e., 200 km. Scale down and model measurements are sometimes in fact used. To achieve meaningful calibration by scaling, both a far-field measurement of the polar pattern of the model and near-field probing of the aperture field in the model and of the full antenna need to be combined. Consequently very specialized and intricate measurements need to be made before a successful antenna calibration is available. Some smaller radio telescopes which are fully steerable may be calibrated by placing a transmitter on a distant tower. The far-field requirement usually means that high towers at substantial distances are necessary. As a consequence the elevation angles are low, which immediately results in ground reflection problems and turbulent air-refraction jitter of the beam. Balloon-borne transmitters have been used in some cases to test large antennas. This method requires a high degree of wind stability and skilled surveying of balloon's position from the ground. Aircraft and helicopters flying a transmitter have been employed in certain cases with varying degrees of success. To measure the exact position of a flying aircraft relative to the antenna requires extensive instrumentation and would in most cases appear to be too elaborate for a radio astronomy establishment. It is therefore logical to use for calibration of large antennas the abundantly occurring natural radio sources. Each radio astronomy observatory facing the problem of calibrating a radio telescope would use the source data available from surveys of other observatories. Sources will be chosen to be calibrators for a number of different reasons. A grid of calibrators covering

---

* Chapter 1.5 is by R. Wielebinski.

1.5. ANTENNA CALIBRATION

the whole sky is desirable so that later other sources can be related to these calibrators or antenna performance can be checked in minimum time.

In this section the criteria for the selection of calibration sources will be described. Lists of calibrators, both for positional and for flux scale calibration, will be given. A discussion of present-day accuracy of the flux scale will be undertaken, particularly in view of the discrepancies between observatories at frequencies of 408 MHz and below. The techniques of calibration will be described, particularly as used in the tests of the 100-m fully steerable, paraboloidal radio telescope at Effelsberg. A description of the aims of a complete pointing theory will be discussed, which is an attempt to obtain the ultimate in positional accuracy in a radio telescope.

1.5.2. Natural Radio Sources

Ideally a radio source to be used for calibration purposes should fulfill a number of requirements:

(a) Accurate position of the source must be known, preferably an optical identification.

(b) The source should be clear of other confusing sources, such as other strong point sources or extended galactic objects.

(c) The source should have small angular size or, if extended, the intensity distribution must be well known.

(d) The source flux must be known and must not vary with time.

(e) The frequency spectrum should be straight on a logarithmic plot (i.e., a power law) to allow interpolation between frequencies.

(f) The polarization characteristics of the source must be known. Some highly polarized calibrators are required for the determination of the antenna polarization characteristic.

The strongest radio sources such as Cassiopea A or Cygnus A are useful calibrators for smaller antennas where beam size is not comparable with source diameter. In addition, Cassiopea A has a secular decrease of flux of about 1.1% per year. The radio source Taurus A is particularly useful as a calibrator for higher frequencies due to its flat spectrum. The planets, wandering across the sky with the range $-25° < \delta < 25°$, are used for positional telescope calibrations, again at the highest frequencies. For tests of a large radio relescope, a large number of calibrating sources are required, distributed all over the sky. Ideally all the requirements set out in points (a)–(f) should be met.

By a strange trick of nature there are few if any sources which combine the properties of an exact position, small angular dimension, a power-law spectrum, and source-flux invariability. Sources with power-law spectra are often extended. Sources with small angular dimension do not, in general, have straight spectra. In addition some sources of well-known position and with small angular extent required for antenna positional calibration are variable.

Thus in assembling calibrating sources the following classes of sources are chosen:

I. a class of sources with well-known flux densities, where possible, related to an absolute flux determination,
II. sources with straight spectra,
III. sources with reliable flux densities,
IV. sources of extremely small angular size,
V. sources with very precisely known position, and
VI. strong sources useful for sidelobe messurements.

A discussion of the determination of flux scale must follow before lists of calibration sources may be presented.

1.5.3. Determination of the Flux Scale

The measurement of a source that can subsequently be used for antenna calibrations is a difficult problem, which in the past was not often carefully considered. There are two separate problems which must be solved in obtaining a flux scale. The first problem is that of radiometric intensity calibration, discussed elsewhere in this volume or by Findlay.[1] In view of the published reports of calibration procedures used, there appears to be no reason why at frequencies below 10 GHz a 1% accuracy in antenna-temperature determination cannot be reached when sufficient care is taken in the calibration procedures. Thus the second and, at present, major problem of flux-scale determination is the derivation of antenna gain. The basic relation between antenna gain (in fact directivity), antenna temperature, and flux of a point source is given by

$$S = \frac{8\pi k T_a}{D\lambda^2} = \frac{2k T_a}{A_e}, \qquad (1.5.2)$$

where T_a is the antenna temperature in degrees Kelvin D the antenna directivity, A_e the effective area of the antenna in meters squared, k Boltzmann's constant $= 1.4 \times 10^{-23}$ mks units, and λ the wavelength in meters. Formulas (1.5.2) hold for a narrow bandwidth across which the intensity varies only slowly. The fact that a linearly polarized antenna abstracts only $\frac{1}{2}$ of the incident energy from a randomly polarized radiation is also included. To

---

[1] J. W. Findlay, Absolute intensity calibrations in radio astonomy, *Ann. Rev. Astron. Astrophys.* No. 14, 77 (1966).

calibrate the source flux S, the antenna directivity D (or effective area A_e) must be determined with a degree of precision comparable to the precision of antenna temperature determination. Two methods have been used for absolute source flux determination. Both depend on the availability of antennas of known gain.

One of the antenna types with a highly precisely determinable gain is a pyramidal horn. The gain (or rather directivity) is usually calculated from formulas based on the theoretical work of Schelkunoff[2] and the modification introduced by Slayton[3]. The assumption used for the computation of the "Schelkunoff gain" is that the horn aperture is excited by the field due to the rectangular waveguide TE_{10} mode expanded in an infinite horn. This assumption does not take into account the fact that due to the E edge, a diffraction field exists, that due to the sudden truncation of the waveguide reflected waves can be set up, or that due to machining imperfections higher modes may be excited in the horn. Experimental investigations have been published indicating that for particular horns the theoretical Schelkunoff gain was measured to be within 2.5%[4] and 0.6%[5], respectively. Theoretical methods of computation of the E-plane horn pattern can now be used[6,7] which in turn by integration of the polar pattern can give horn directivity. In a series of measurements and computations[8] it was demonstrated that large pyramidal horn antennas (for example $6.1\lambda \times 4.5\lambda$ aperture, 10λ long) can be expected to agree better than 2.5% with the computed Schelkunoff gain. On the other hand small horns ($2\lambda \times 1.5\lambda$ aperture, 4λ long) will deviate as much as 10% from the computed gain. A useful indication for exact antenna performance is the probing of the horn aperture field. For large horns the aperture field closely resembles the expected expansion of the waveguide mode with some interference due to the radiating edge. This interference is, of course, much stronger in small horns resulting in lowering of the gain.

The direct method of absolute source calibration is to take an antenna of known gain and calibrate the antenna temperatures of the strongest sources. For example, a horn with a gain of 30 dB will for Cassiopea A produce an antenna temperature of 2.75°K at 21-cm wavelength. This can be measured with great care to 1% accuracy. A summary of Cas A measurements is

[2] S. A. Schelkunoff, "Electromagnetic Waves," p. 360, Van Nostrand-Reinhold, Princeton, New Jersey, 1943.

[3] W. T. Slayton, Naval Res. Lab., Rep. No. 4433, Washington (1954).

[4] E. V. Jull and E. P. Deloli, *IEEE Trans. Antennas Propagat.* **AP-12**, 439 (1964).

[5] T. S. Chu and R. A. Semplak, *Bell. Syst. Tech. J.* **44**, 527 (1965).

[6] P. M. Russo, R. C. Ruddock, and L. Peters, Jr., *IEEE Trans. Antennas Propagat.* **AP-13**, 219 (1956).

[7] J. S. Yu, R. C. Ruddock, and J. Peters, Jr., *IEEE Trans. Antennas Propagat.* **AP-14**, 138 (1966).

[8] R. Wielebinski, *Proc. Astron. Soc. Aust.* **1**, No. 2, 62 (1967).

found in the work of Findlay[1] and Baars et al.[9] Large horns such as those used by Findlay exist only in a few observatiories. They can be used to determine the absolute flux of only a few of the strongest radio sources.

The next step in the evaluation of a flux scale is to determine the relative flux of a number of "secondary" calibrators. These are sources of medium intensity which are in turn used to calibrate other sources. One of the disadvantages of this direct method of flux scale determination is the large flux difference between primary calibrators and the actual sources being calibrated. Also at frequencies below 1.4 GHz it becomes impracticable to build large enough horns. Thus different standard gain antennas and also a different method is preferred.

A different method of absolute source calibration which also avoids the difficulty of the large ratios between primary standards and actual sources was developed by Little[10] and extended recently by Wyllie.[11] In this method a small standard gain antenna such as a dipole is used in conjunction with a large radio telescope like a Mills cross. A simultaneous measurement can be made of the total power received from a radio source during transit and of the interferometric source intensity, the interferometer being made up of the radio telescope and the standard gain antenna. Since the effective area of the interferometer is twice the geometrical mean of the two interferometer components, the effective aperture of the radio telescope can disappear in the algebraic manipulations. At a frequency of 85 MHz, Little[10] used a dipole as the standard. Near-field probing can determine the current distribution in the dipole from which the polar pattern can be computed and hence by integration the directivity. The same method was used by Wyllie[11] except that an array of eight dipoles was used as the standard gain antenna. It would be interesting to repeat these calibrations where dipole and horn standard-antennas were used at one frequency.

1.5.4. The Calibrating Sources

There are many source catalogs at frequencies from 10 MHz up to 86 GHz. The review of these original catalogs, particularly in papers which investigate the spectra of radio sources, provide the basis for compilation of a calibration sources list. In the process of comparison of source catalogs, correction factors are invariably used to allow for discrepancies in calibrating techniques. A recent work of this nature by Kellermann et al.[12] covered sources at 38, 178, 750, 1400, 2695, and 5000 MHz. There is, however, evidence

[9] J. W. M. Baars, P. G. Mezger, and H. J. Wendker, *Astrophys. J.* **142**, 122 (1965).
[10] A. G. Little, *Aust. J. Phys.* **11**, 70 (1958).
[11] D. V. Wyllie, *Mon. Not. Roy. Astron. Soc.* **142**, 229 (1969).

1.5. ANTENNA CALIBRATION

that the flux scale of Kellermann et al.[12] is in error as much as 20% at the lowest frequency. The effect at 1.4 GHz is probably less than 2%, being negligible at higher frequencies. In the southern skies the Parkes sources catalog has been edited by Ekers.[13] The frequencies of 408, 1410, and 2650 MHz are covered. An extension of this catalog to 5009 MHz has been made by Shimmins et al.[14] There is evidence again that at 408-MHz flux value discrepancies of up to 25% are common, when compared with the Molonglo calibration sources (Wyllie[11]). At the frequency of 10.6 GHz the three catalogs of Docherty et al.[15], Zimmermann[16] and Pauliny-Toth et al.[17] are available. Recent measurements by Kellermann and Pauliny-Toth[18] extend the source catalogs to 32 and 86 GHz. There are also special types of sources which are useful for particular calibrating experiments. Line emission of OH at 1.6 GHz gives sources with an intense, pure circular polarization or a high degree of linear polarization. Sources of H_2O emission are also useful calibrators at 22 GHz since they have negligible angular size and very high flux. A catalog of H_2O sources has recently been published by Sullivan.[19] However, the H_2O sources and to lesser degree the OH sources are known to have time-varying flux.

Three tables with sources have been compiled. In Table I the strongest radio sources are listed. These, according to the classification set out in the first paragraph of this section are all of class VI, i.e., useful for sidelobe measurements and for calibration of smaller telescopes; some are of class I where absolute flux has been determined. The main list of calibration sources is given in Table II; some 90 sources observable from Bonn are listed. The selection follows the criteria described and the class of the source is included in the table. Some sources satisfy two criteria such as class IV and class V—an extremely small angular size source and a source of extremely precisely known position. Most source positions are known to better than 10 arc sec. Some sources observed with interferometers are better than 2 arc sec. Optical identifications which give an even better position are listed separately by class V_0. The final list given, Table III, lists the planets. The exact planet diameter and position must be obtained for the observing data from Astronomical Ephemeris. The values of planet mean disk temperatures are steadily being improved and Table III should be used as an approximate guide only.

[12] K. I. Kellerman, I. I. K. Pauliny-Toth, and P. Williams, *Astrophys. J.* **157**, 1 (1969).
[13] J. A. Ekers (ed.), *Aust. J. Phys. Astrophys. Suppl.* No. 7 (1969).
[14] A. J. Shimmins, R. N. Manchester, and B. J. Harris, *Aust. J. Phys. Astrophys. Suppl.* No. 8. (1969).
[15] L. H. Doherty, J. M. Macleod, and C. R. Purton, *Astrophys. J.* **74**, 827 (1969).
[16] P. Zimmermann, *Beitr. Radioaston.* **1**, No. 6, 161 (1970).
[17] I. I. K. Pauliny-Toth and K. I. Kellerman, *Astron. J.* **78**, 828 (1973).
[18] K. I. Kellermann and I. I. K. Pauliny-Toth, *Astrophys. Lett.* **8**, 153 (1971).
[19] W. T. Sullivan, III, *Astrophys. J. Suppl.* No. 222 **35**, 393 (1973).

TABLE I. The Strongest Radio Sources

Source	Class	Right Ascension (1950)	Declination (1950)	1 (85 MHz)	2 (150 MHz)	3 (408 MHz)	4 (1.4 GHz)	5 (2.7 GHz)	6 (5 GHz)	7 (10.7 GHz)	Size (arcmin)	Polarization (%)	Remarks
0320-37	VI	03h20m42s	−37° 25′	950	475	177	—	94	—	—	20	10	Fornax A
0518-45	VI	05 18 24	−45 50	570	343	117	52	32	—	—	5	3	Pictor A
0539-01	VI	05 33 11	−01 56	—	—	37	51	56	—	—		<2	
3C144	VI	05 31 31	+21 59	1550	1400	1150	930	790	680	600	5	<1	Taurus A
3C145	VI	05 32 51	−05 25	69	42	177	281	51	—	—	4.5		Orion nebula
0916-12	VI	09 15 43	−11 53	580	340	110	36	25	13	—	5	<2	Hydra A
3C273	VI	12 26 34	+02 19	167	150	49	45	42	45	45		2.5	Variable
3C274	VI, I	12 28 17	+12 40	1800	1015	600	214	120	72	37	4.7	<0.1	Virgo A
3C348	VI	16 48 42	+05 04	890	420	145	45	22.5	12	5	3.2	5	Hercules A
3C353	VI	17 17 56	−00 56	400	265	130	55	35	22	12.5	4	6	
3C405	VI	19 57 45	+40 36	14,000	10,000	4700	1600	785	370	200	2.3		Cygnus A
2152-69	VI	21 52 58	−69 56	253	114	60	26	16	10	—	4	4	
3C461	VI, I	23 21 11	+58 33	20,800	13,100	6050	2400	1450	920	520	4		Cassiopea A
2356-61	VI	23 56 29	−61 12	296	66	49	19	11	—	—	4	5	

S (flux units 10^{-26} W m^{-2} sec^{-1})

TABLE II. Calibration Sources

Source	Right Ascension (1950)	Declination (1950)	38 (MHz)	178 (MHz)	750 (MHz)	1400 (MHz)	2695 (MHz)	5000 (MHz)	10,700 (MHz)	Size (arcsec)	Class	Remarks
3C2	00 03 48.8	−00 21 06	44	15.0	5.7	3.5	2.25	1.41	0.83	<6	III	
0023-26	00 23 18.9	−26 18 51				7.1	5.62	3.76		<15 × <35	III	
3C27	00 52 45.6	68 06 06	59	26.5	10.7	7.3	4.26	2.48	1.15	31 × 22	III	
3C32	01 05 48.0	−16 20 21					2.09	1.08		<10	III	
3C38	01 17 59.7	−15 36 00					2.74	1.42		Double 8	III	
3C43	01 27 15.1	23 22 53	39	11.6	4.1	2.7	1.68	1.09	0.65	<3	II	
3C47	01 33 40.2	20 42 04	125	26.4	6.7	3.6	2.06	1.10	0.60	Double 60	II, V	
3C48	01 34 49.8	32 54 20	61	55.0	24.4	15.3	8.97	5.37	2.62	<1	III	
NRAO 91	02 02 07.4	14 59 51					3.00	2.27	2.58	<3	IV	
3C62	02 13 11.5	−13 13 21					2.82	1.60		Double 5	III	
3C71	02 40 07.1	−00 13 31	33	16.1	7.3	4.9	2.99	1.90	1.00	9	III	
3C75	02 55 04.9	05 50 41	92	25.8	9.3	6.2	3.61	2.36	1.31	Double 168	II	
3C78	03 05 49.1	03 55 14	52	17.8	9.6	7.1	5.05	3.40	2.31	<20	III	
CTA 21	03 16 09.1	16 17 40					4.95	2.86		<0.4	IV	
3C84	03 16 29.6	41 19 52	360	62.6	20.8	12.7	(9.9)	(18.0)	(~50)	<0.01	IV	Complex below 3 GHz
3C89	03 31 41.8	−01 21 10	102	20.2	5.1	2.7	1.42	0.81	0.30	13	II	
NRAO150	03 55 45.3	50 49 21					~4	(7.40)	(~10)	<3	IV	Complex below 1.4 GHz

S (flux units 10^{-26} W m^{-2} sec^{-1})

Table II (continued)

Source	Right Ascension (1950)	Declination (1950)	\multicolumn{7}{c}{S (flux units 10^{-26} W m^{-2} sec^{-1})}	Size (arcsec)	Class	Remarks
			38 (MHz)	178 (MHz)	750 (MHz)	1400 (MHz)	2695 (MHz)	5000 (MHz)	10,700 (MHz)			
0403-13	04 03 13.9	−13 16 19				3.2	2.85	2.32	1.94	<15 × <25	IV	
3C119	04 29 07.9	41 32 09	18	15.7	11.5	8.3	5.36	3.42	1.99	<2	III	
3C120	04 30 31.6	05 15 00					~7	(5.09)	(~50)	<3	IV	
3C123	04 33 55.2	29 34 14	577	189.0	72.3	45.9	27.2	16.32	8.23	9 × 29	III	
3C130	04 48 54.3	51 59 42	50	15.5	4.3	2.6	1.58	0.89	0.47	Complex ~150	II	
3C133	04 59 54.2	25 12 12	58	22.3	8.5	5.7	3.38	2.16	1.28	13 × <16	III	
3C138	05 18 16.5	16 35 27	13	22.2	12.0	9.2	5.99	4.16	2.30	<2	III	
0521-36	05 21 13.2	−36 30 19				14.7	11.28	9.37	6.7	14 × <18	III	
3C147	05 38 43.5	49 49 42	18	60.5	32.6	22.2	12.98	8.18	4.14	<1	III	
3C161	06 24 43.0	−05 51 14					11.06	6.73	3.31	<3	III	
0704-23	07 04 27.3	−23 06 59				3.0	2.34	1.45		<15 × <30		
3C175	07 10 15.6	11 51 21	81	17.6	4.5	2.5	1.23	0.66	0.30	Double 45	II	
3C180	07 24 33.5	−01 58 44	49	15.1	4.5	2.6	1.59	0.94	0.58		II	
3C195	08 06 30.6	−10 19 03					2.52	1.58	0.78		III	
3C196	08 09 59.4	48 22 07	166	68.2	22.8	13.9	7.66	4.36	1.95	Double 5	III	
3C197.1	08 18 01.1	47 12 11	24	8.1	3.0	1.9	1.16	0.86	0.30	<8 × 12	II, v	
DA 251	08 31 04.4	55 44 42					~7		(2.77)	<2	IV	
0859-25	08 59 36.8	−25 43 30				4.6	2.94	1.75	0.82	36 × 35	III	

Name	RA	Dec	S₁	S₂	S₃	S₄	S₅	S₆	S₇	Structure	Class	Notes
3C218	09 15 41.3	−11 53 05					23.3	13.78	6.98	16	III	
4C39.25	09 23 55.3	39 15 24					4.54	7.57	(~11)	<1	IV	
3C223	09 36 50.5	36 08 05	44	14.7	5.3	3.1	2.06	1.29	0.72	Double 300	II, V	
3C227	09 45 08.5	07 39 19	124	30.4	11.6	6.8	4.16	2.60	1.51	Triple 180	II	
3C231	09 51 43.8	69 54 59	23	14.6	10.2	8.4	5.59	3.94	2.35	35 × 20	III	
3C245	10 40 06.0	12 19 15	40	14.4	4.9	3.2	2.09	1.39	1.03	4	III	
1055+01	10 55 55.3	01 50 03				3.5	3.5	3.07	(2.86)	<15 × <30	IV	
3C249.1	11 00 25.0	77 15 11	39	12.5	3.5	2.3	1.40	0.78	0.40	Double 19	II	
1127−14	11 27 35.7	−14 32 55				6.0	6.8	7.25	(5.46)	<15 × <30	IV	
3C263	11 37 10.8	66 04 23	46	15.2	4.9	2.9	1.73	1.04	0.57	Double 47	II	
3C268.1	11 57 46.4	73 17 27	51	21.4	9.6	6.6	4.00	2.62	1.36	Double 42	III	
3C270	12 16 50.1	06 06 09	122	51.8	24.9	18.1	12.65	8.32	5.00	Double 51	II	
3C273	12 26 33.3	02 19 42	140	62.8	45.3	(45.0)	(41.8)	(44.9)	(~45)	Double 20	IV, V₀	
3C274	12 28 17.7	12 40 00	3570	1050	352	214	118	72.1	37.9	Complex	I	Virgo A
3C278	12 52 00.1	−12 17 07					4.45	2.84	1.43	120	III	
3C279	12 53 35.9	−05 31 08				(~12)	2.83	(15.34)	(~12)	<1	IV, V₀	
3C280	12 54 41.6	47 36 02	62	23.7	7.7	4.9	2.83	1.53	0.77	Double 10	III	
3C286	13 28 49.6	30 45 58	32	24.0	18.4	14.4	10.26	7.48	4.68	<5	III	Polarized 10%
1345+12	13 45 06.2	12 32 20				4.9	3.71	2.89	2.30	<15 × <18	IV	
3C295	14 09 33.5	52 26 13	94	83.5	35.3	22.7	11.83	6.53	2.76	Double 4	III	
3C298	14 16 38.8	06 42 20	73	47.5	11.5	5.7	2.71	1.46	1.03	<5	III	
3C303	14 41 23.7	52 14 19	37	11.2	3.9	2.4	1.55	0.94	0.65	25 × <18	II	
1453−10	14 53 12.2	−10 56 52				3.6	2.43	1.40	0.82	<15 × 30	III	
3C309.1	14 58 56.6	71 52 11	37	22.7	11.0	7.9	5.30	3.76	2.58	<2	III	
1510−08	15 10 08.9	−08 54 51				2.9	3.1	3.25	2.58	<15 × <30	IV	

Table II (continued)

Source	Right Ascension (1950)	Declination (1955)	38 (MHz)	178 (MHz)	750 (MHz)	1400 (MHz)	2695 (MHz)	5000 (MHz)	10,700 (MHz)	Size (arcsec)	Class	Remarks
3C317	15 14 17.1	07 12 18	165	49.0	11.3	5.4	2.11	0.87	0.30	<20	III	
3C327.1	16 02 12.9	01 25 59	86	23.6	7.4	4.1	2.10	1.10	0.56	Double 7	III	
CTD 93	16 07 09.3	26 49 18					2.89	1.67	0.67	<15 × <15	IV	
3C343	16 34 01.1	62 51 41	<8	12.4	7.6	4.9	2.68	1.49	0.70	<3	III	
3C343.1	16 37 55.3	62 40 34	8	11.5	7.6	4.1	2.23	1.20	0.50	<3	III	
3C345	16 41 17.6	39 54 11	23	10.8	7.6	6.6	(5.3)	(5.5)	(~11)	<3	IV	
3C348	16 48 40.8	05 04 36	1690	351	83.7	44.5	22.4	11.89	5.24	Double 84	I	
3C349	16 58 05.0	47 07 32	50	13.3	4.8	3.2	1.88	1.14	0.58	Complex 78	II	
3C351	17 04 05.0	60 48 44	40	13.7	5.0	3.2	2.03	1.21	0.59	Double 60	II, v	
3C353	17 17 55.6	−00 55 53	713	236	84.5	54.9	34.8	21.5	11.50	Double 150	I	
3C358	17 27 40.7	21 27 11					10.1	7.29	4.70	162	III	
1730-13	17 30 13.5	−13 02 46				5.1	4.9	4.0	(~4)	<2	IV	
3C380	18 28 13.5	48 42 41	211	59.4	22.2	14.8	10.1	7.5	(~4.8)	2	III	Variable
3C390.3	18 45 52.8	79 42 47	120	47.5	16.8	11.6	6.76	4.48	2.56	Double 204	III	
1938-15	19 38 24.6	−15 31 35				6.7	3.95	2.29	1.11	<15 × <25	III	
3C401	19 39 38.6	60 34 30		20.9	7.8	4.7	2.76	1.37	0.67	<8 × 16	III	
3C409	20 12 18.0	23 25 42	280	76.6	25.1	13.9	6.44	3.12	1.30	<15 × 22	III	
3C410	20 18 04.2	29 32 41	96	34.6	15.5	10.1	6.00	3.79	2.07	Double 6	III	
3C418	20 37 07.4	51 08 36		13.1	7.0	5.3	4.71	(3.81)	(~3)	<2	IV	
2104-25	21 04 24.9	−25 39 06				9.5	6.40	2.45		Double 198	III	

3C430	21 17 02.4	60 35 34	126	33.7	7.5	4.69	3.32	1.69	Double 57	III
3C433	21 21 30.6	24 51 18	187	56.2	11.5	6.49	3.74	1.83	11 × 52	III
2145+06	21 45 36.1	06 43 40			2.8	(3.5)	(4.71)	(~2.5)	<3	IV
2203-18	22 03 25.7	−18 50 17			6.0	5.1	4.05	3.14	<18 × <43	IV
3C441	22 03 49.2	29 14 49	41	12.6	2.6	1.51	0.92	0.43	29	II
3C444	22 11 43.0	−17 16 49				4.47	2.29	0.94	Double 90	III
3C446	22 23 11.0	−05 12 18				(~4)	(4.07)	(~5.3)	5	IV	Polarized 5%
CTA 102	22 30 07.8	11 28 23				4.93	3.63	2.80	<0.1	III
3C454	22 51 29.5	15 52 54	27	11.6	2.1	1.21	0.79	0.42	<2	IV
3C459	23 14 02.3	03 48 55	53	25.6	4.0	2.28	1.36	0.62	12	III

TABLE III. Planets—Size and Mean Disk Temperature

Planet	Size (arc sec) 1971 semidiameter[a]		T_b (°K)						
	Max	Min	408 MHz	1.4 GHz	2.7 GHz	5 GHz	10 GHz	15 GHz	24 GHz
Mercury	6.02	2.32			380	380	370	350	330
Venus	31.33	4.95	450	400	700	700	600	530	470
Mars	2.92	1.76	700	640	190	190	180	170	170
Jupiter	20.72	14.36		230	770		145	145	145
Saturn	9.15	7.33	700	2250	190	160	140	130	130
Uranus	1.96	1.77		260	180	180	170	150	140
Neptune	1.25	1.17		200	120	130	140	140	120

[a] Note: The exact planet diameter for an observing period must be obtained from *Astronomical Ephemeris*. Variations particularly for nearer planets are enormous.

1.5.5. Mechanical Measurements

Electrical calibrations of a radio telescope must be preceded by mechanical measurements. Two separate sets of measurements are required, namely:

(1) surface measurements, and
(2) axis-geometry measurements.

The surface measurements are necessary to obtain the required (paraboloidal) surface of a radio telescope. The precision with which the measurements and adjustments have to be made depends on the frequency range of the telescope. Axial-geometry measurements are also connected with the telescope frequency range, since they determine the pointing precision of the telescope.

1.5.5.1. Surface Measurements. The need for developing methods of accurate surface measurements was recognized very early, since to increase the frequency range of the telescope a reliable method of measurement is necessary. Three methods have been widely employed, with varying degrees of success. The simplest method available from the classical surveying techniques is that of tape and theodolite. It is interesting to note that this method is being used for the surface measurement of the 100-m Effelsberg radio telescope.[20] It is a development of a method used by Puttock et al.[21] for surface measurement of the Parkes dish. This method is shown diagramatically in Fig. 1. A

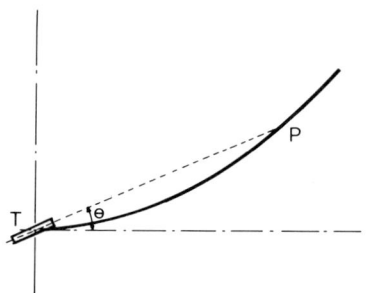

FIG. 1. A diagram showing tape and theodolite measurements of an antenna surface. The tape is used to measure the distance TP, and the theodolite to measure the angle θ.

set of markers are glued on to the surface panels through holes drilled in a tape appropriate to each ring of markers. The tape has to be made of a material which equals the expansion of the telescope surface. A special theodolite (Kern, Aarau type DKM3 is used for the Effelsberg telescope) can resolve 1 arc sec angles, and at a distance of some 60 m can distinguish 1 mm deviation of marker from cross hair. For smaller telescopes this method could provide even better surface panel setting.

[20] O. Hachenberg, B. H. Grahl, and R. Wielebinski, *Proc. IEEE* **61**, 1288 (1973).
[21] M. J. Puttock and H. C. Minett, *IEEE Proc.* (*London*) **113**, 1723 (1966).

A different method available for surface setting uses pentaprisms. This method is shown diagramatically in Fig. 2. A pentaprism carefully machined

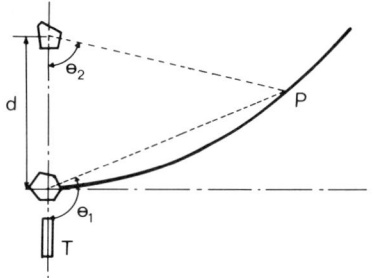

FIG. 2. A diagram showing the arrangement of pentaprisms for antenna surface measurements. (See text for further discussion of this technique.)

reflects light at an angle θ_1. A second pentaprism placed accurately a distance d above the first reflects light through angle θ_2. Ideally this system allows absolute distance measurement if accurate pentaprisms are held rigidly at a fixed distance apart. This method was applied by Zeiss[22] to a 34-m precision dish at Werthoven, Germany. A set of pentaprisms was necessary to allow measurement of a number of rings (a set of two is required for each ring) giving setting accuracies of 0.3 mm. Further methods for surface setting such as photogrammetic, modulated light beam, or laser beam have been used at various times.

The measurements of the Haystack antenna and radome were described by Meeks and Ruze.[23] A good description of comparative merits of the various methods is given by Slater.[24] In particular, the difficult task of determining the surface deviation during tipping can be elegantly handled by a laser which, however, cannot give good absolute positioning. Great care and patience is necessary to obtain precise setting of telescope surface, which in turn achieves maximal aperture efficiency of the antenna.

1.5.5.2. Axis Geometry Measurement. The telescope must be related to the fixed geodetic coordinates as a point on the Earth sphere. In the first instance it is necessary to determine the direction of the beam. A strong source of large diameter like the sun should allow the determination of this, usually in the first attempt. In the 100-m telescope at 11-cm wavelength the beam is 5 arc min wide and in the first instance a deviation of some 3 arc min was observed. As the mechanical adjustments proceeded it was possible to

[22] C. Kühne, "Design and Construction of Large Steerable Array," *Inst. Elec. Eng. (London) Conf.* **21**, 187 (1966).
[23] M. L. Meeks and J. Ruze, *IEEE Trans. Antennas Propagat.* **AP-19**, 733 (1971).
[24] R. H. Slayter, *IEEE Proc. (London)* **118**, 1691 (1971).

point the telescope to any source. There is at such a stage of antenna calibration an uncertainty about the accuracy of the structure, the absolute accuracy of the position encoders, and the repeatability of settings. Thus numerous measurements following a source from transit to the telescope horizon are useful in determining the pointing of the telescope. At this stage the correction for the atmosphere must be considered since it is large compared to the beam of a large radio telescope. When the radio telescope becomes astronomically steerable a useful test is to follow a radio source at the 3-dB edge. Stability of the source intensity is a good indication of the accuracy of the telescope drives.

1.5.6. Pointing Theory

Assuming that the mechanical telescope adjustments have been made to a stage of perfection which cannot be bettered and if a computer is available to steer the telescope, the final corrections can be ascertained from measurements. In a transit telescope such as a Mills cross, only one dimension correction must be considered. In a fully steerable paraboloidal reflector two dimensions must be considered. A "pointing theory" has been laid down by Stumpff.[25] In this the starting assumption is that the position of the source is known at the time of observation. From the geodetic position and transit time of the source one can compute the position in azimuth and zenith angle. A number of corrections must then be considered, for example, refraction, which is zenith-angle dependent. Collimation errors can be included as well as beam deviation due to the elastic banding of the telescope structure. In the end, after measurements of a grid of sources at various elevation angles have been carried out, a correction set can be determined for the telescope. This would be included in the computer drive program to allow exact pointing in astronomical observations. A detailed description of the tests of the 100-m MPIfR telescope is available as an internal report.[26]

ACKNOWLEDGMENTS

For the compilation of the table of calibrating sources I wish to express my gratitude to Dr. I. I. K. Pauliny-Toth. Dr. W. B. McAdam provided the planetary data. Helpful discussions about telescope tests took place with Prof. O. Hachenberg and Dr. B. Grahl.

ADDENDUM†

At wavelengths of a few centimeters and shorter it becomes difficult to use calibration standards established at longer wavelengths. These standard sources become too weak, too large in angular size, or have curved spectra at millimeter wavelengths. One exception

[25] P. Stumpff, *Kleinheubacher Ber.* **15**, 431 (1972).
[26] I. I. K. Pauliny-Toth and W. Altenhoff (eds.).

† September 1975

1.5. ANTENNA CALIBRATION

is the compact, ionized-hydrogen source, DR-21. Dent[27] has singled out this source as the best flux-density standard for high frequencies and has given the following relationship for flux S_v (Jy) at frequencies $v \geq 7$ GHz:

$$S_v = 26.8 - 5.6 \log v \quad \text{(GHz)}.$$

Klein[28] has confirmed this relationship by comparision with Cas A at 21.9 GHz, and measurements around 90 GHz are consistent[28] with this formula. DR-21 is a double source[29] with component separation of 21 arc sec: the comparatively weak northern component is smaller than 7 arc sec in angular diameter and the southern component appears elliptical 14 by 22 arc sec in angular size. The position[29] of the southern component, epoch 1950.0, is $\alpha = 20^h\ 37^m\ 14.3^s$ and $\delta = 42°08'55''$. Although this source is located in a complex region of the galactic plane, Dent found the background sufficiently linear within 18 arc min of the source to ensure that symmetrical off-source measurements introduce negligible error.

Since Chapter 1.5 was written, more accurate values for the absolute flux densities of many sources have been reported by Baars and Hartsuijker[30] and by Ross and Seaquist.[31]

---

[27] W. A. Dent, *Astrophys. J.* **177**, 93 (1972).
[28] M. J. Klein, *Astron. J.* **79**, 139 (1974).
[29] C. G. Wynn-Williams, *Mon. Not. R. Astron. Soc.* **151**, 397 (1971).
[30] J. W. M. Baars and A. P. Hartsuijker, *Astron. Astrophys.* **17**, 172 (1972).
[31] H. N. Ross and E. R. Seaquist, *Mon. Not. R. Astron. Soc.*, **170**, 115 (1975).

1.6. Practical Problems of Antenna Arrays*

1.6.1. Introduction

Practical problems and techniques of array construction are discussed in this chapter. Emphasis is on antenna arrays in the HF and VHF frequency bands, roughly from 3 to 300 MHZ.

The need for a moderate-sized array sometimes arises unexpectedly. In such a case the following section could serve as a reference for rapid array construction. The subsequent sections discuss principles that help in understanding practical problems, and the last section is a compilation of practical techniques.

1.6.2. A 16-Dipole Array

An array having a gain of up to 25 dB can be very quickly and inexpensively constructed. As an example, consider the construction of a 16-dipole array at 38 MHz. The gain of a half-wave dipole is 2.15 dB in free space and 7.45 dB when placed one-quarter wavelength over a reflecting ground-plane. The ground-reflected wave nearly doubles the field pattern which quadruples the power gain.

The gain of the 16-dipole array should be approximately 16 times that of a single dipole, or about 20 dB, provided the element spacing is large enough to prevent mutual interference. When a spacing is used that allows a physical area equal to the effective area of the element, only a small amount of mutual coupling is present, and we have what is termed a *filled array* (see Section 1.2.2). The effective area of a 7.25-dB gain dipole is 0.422 square wavelengths which is 26 m² at 38 MHz. Effective area A_e and gain G, are related by

$$G = 4\pi A_e/\lambda^2, \qquad (1.6.1)$$

where λ is the wavelength.

The procedure in constructing the array is first to install the posts, then the dipoles, and then the feeder lines. Space the posts on a square grid about $(26)^{1/2}$ m apart. Use 10 × 10 cm (4 × 4 in.) timbers 2.7 m (9 ft) long, buried (0.76–0.9 m) ($2\frac{1}{2}$ to 3 ft) in the ground and backfilled with dirt or preferably sand. Place the dipoles in a plane about $\frac{1}{4}\lambda$, or slightly less, above ground.

---

* Chapter 1.6 is by J. C. James.

1.6. PRACTICAL PROBLEMS OF ANTENNA ARRAYS

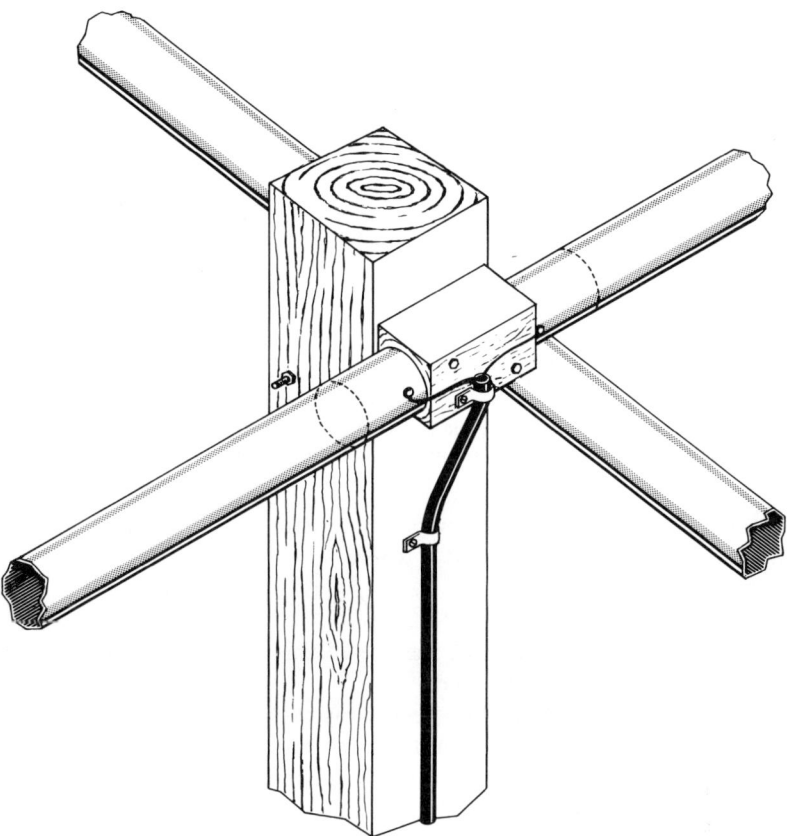

FIG. 1. Diagram showing the arrangement of orthogonal dipoles on a post. (See text for discussion.)

Use 3.8-cm (1.5-in.) diameter, 0.127-cm (0.05-in.) wall thickness aluminum tubing press-fitted onto wooden mounts as shown in Fig. 1, so that the tip-to-tip length is 3.6 m (142 in). This produces an impedance of 75 to 80 ohm.

Standard 1-cm (0.4-in.) flexible coaxial feeder lines may be attached directly to the dipoles without a balun as shown in Fig. 1. Figure 2a shows two possible feeder schemes. Note that commercially available cables are used and that each junction is impedance matched. When 25-ohm cable cannot be located, it may be simulated by two 50-ohm cables in parallel.

The Smith chart can be very useful in solving the matching problem. In some cases quarter-wavelengths of line are useful for matching. When the

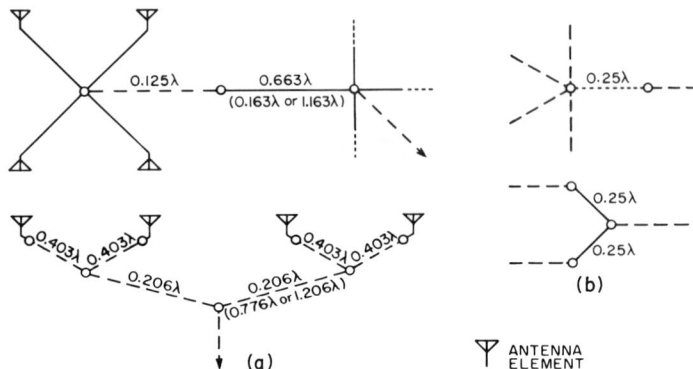

Fig. 2. The two diagrams in (a) show alternative arrangements of coaxial feeder lines. In (b) two examples of matched feeder junctions are shown. The solid line represents a 75-ohm line, the dashed line a 50-ohm line, and the dotted line a 25-ohm line.

quarter-wave section has a line or characteristic impedance Z_0, it matches an impedance Z_1, to impedance Z_2, where

$$Z_1 Z_2 = Z_0^2.$$

Figure 2b shows two examples of the use of this formula. Lumped constant matching circuits also may be used.

The array described above has been constructed and works well even in rain. The tubing and wooden mount are mechanically rigid, but the dipoles could be constructed of wires separated by insulators.

The length of the dipoles should be trimmed so that the dipole impedance is resistive. To determine the proper length, mount one dipole that is $\lambda/2$ tip to tip. Then measure the admittance of this dipole as 1-in. increments are cut from each end. Use the length for which the reactive term of the admittance is zero, then cut all other dipoles the same length. When the dipoles will be spaced closer than 0.6λ, this length determination should be repeated on one dipole when surrounded by other excited or resistance-terminated dipoles.

With no added phasing cables the array is sensitive to signals arriving from the zenith. The computation of the required length of phasing cables for any beam pointing is a straightforward geometrical problem. Phasing cables of proper length or phase must be added to the line from each dipole or each group of dipoles so that the time for each signal component to travel from a wavefront in space to the receiver is the same for all dipoles. When a phasing cable is one wavelength or longer, it may be reduced in length by an integral number of wavelengths; however, such a procedure will limit the useful bandwidth of the array. Several phasing cables of lengths $\frac{1}{2}$, $\frac{1}{4}$ and $\frac{1}{8}\lambda$

should be available, and each phasing completed to an accuracy of $\pm \frac{1}{16}\lambda$ by adding up to three cables in series. Certain feeder arrangements, such as the second one shown in Fig. 2a, would allow four dipoles to be phased simultaneously. This method could be employed when the radio source moved perpendicular to the row of dipoles.

Cross dipoles may be added to the posts as shown in Fig. 1. This results in another linear array on the same ground and is necessary to receive all the energy in the arriving signal, and to measure the polarization angle of the signal. The outputs of the two arrays may be combined in a hybrid to produce right- and left-circular polarization. Figure 3 shows how this can be done using a rat-race hybrid made of coaxial cable.

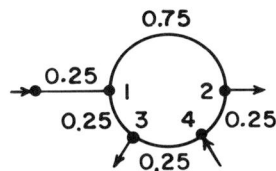

FIG. 3. Diagram of a rat-race hybrid made of coaxial cables connected to produce right- and left-circular outputs at ports 2 and 3 from orthogonal linear-polarized inputs at ports 1 and 4. The four lines that constitute the circle have characteristic impedance $\sqrt{2}Z_0$, and all other lines have impedance Z_0.

1.6.3. Array Patterns

The normalized antenna pattern for a single row of evenly spaced elements having equal currents is given by[1]

$$P_1 = \frac{1}{N_1}\left[\frac{\sin N_1 (\pi d_1 \cos A \sin Z + S_1)}{\sin (\pi d_1 \cos A \sin Z + S_1)}\right]^2, \quad (1.6.2)$$

where d_1 is element spacing in wavelengths and S_1 is one-half the phasing added to each successive element to force the axis of the beam to point in direction (A_0, Z_0). Here A and Z are azimuth and zenith angles, and S is given by

$$S_1 = -\pi d_1 \cos A_0 \sin Z_0 \quad \text{rad}. \quad (1.6.3)$$

A similar expression for the orthogonal column of N_2 elements with spacing d_2 and successive element phasing $2S_2$ is

$$P_2 = \frac{1}{N_2}\left[\frac{\sin N_2 (\pi d_2 \sin A \sin Z + S_2)}{\sin (\pi d_2 \sin A \sin Z + S_2)}\right]^2. \quad (1.6.4)$$

The total number of elements in a rectangular array is $N = N_1 N_2$.

[1] J. D. Kraus, "Antennas," McGraw-Hill, New York, 1950.

The complete expression for the pattern of a rectangular array is

$$P = P_1 P_2 G, \qquad (1.6.5)$$

where G, the pattern of a single element, is weakly dependent on S_1 and S_2 except for large scan angles or when the element spacing is less than about $A_e^{1/2}$. Except in these cases, the array gain P is the product of a variable array factor $P_1 P_2$ and a fixed element factor G. The element gain may be represented by a large transparent shell above the element so that the distance from the element to a point on the shell is proportional to the element gain in that direction. The array factor then moves the array's narrow beam around inside the shell, with the maximum normalized gain of the array always being limited by the shell.

The beamwidth B of the array factor can be quickly estimated from the relation

$$B = 60/D \quad \text{deg}, \qquad (1.6.6)$$

where D is the aperture dimension in wavelengths. This can be remembered as twice the angle that produces an out-of-phase condition between the two elements at the aperture extremes, with $180/\pi$ rounded to 60.

1.6.4. Grating Lobes

When an antenna is phased to point in a certain direction (A_0, Z_0), the array factor $P_1 P_2$ has its maximum value of N for signals received from that direction. When the separation of elements is large, the array factor may also have a maximum value in other undesired directions. These other maxima are called *grating* or *diffraction lobes*.

Note from Eq. (1.6.2) and (1.6.4) that major lobes occur when the denominators are zero, and that when Eq. (1.6.3) is substituted for S_1 the condition for $P_1 = N_1$ is that

$$\pi d_1 (\cos A \sin Z - \cos A_0 \sin Z_0) = n\pi, \qquad (1.6.7)$$

where n is an integer. For any $n \neq 0$, this is the condition for a grating lobe. It is often desirable to eliminate grating lobes for all desired values of (A_0, Z_0). This can be done by choosing d_1 small, by choosing the element type so that G is small in the grating lobe directions, or by spacing the elements nonuniformly so that a more restrictive relation than Eq. (1.6.7) applies.

When $A = A_0 = 0$, the separation of the first grating lobe from the main lobe is $Z - Z_0$. From Eq. (1.6.7) the following relation is derived from which $Z - Z_0$ may be computed:

$$d = (\sin Z - \sin Z_0)^{-1}. \qquad (1.6.8)$$

1.6. PRACTICAL PROBLEMS OF ANTENNA ARRAYS

Here d is the separation of rows in the $A_0 = 0$ direction that produces a grating lobe at Z when the array is phased to Z_0. Suppose there are no grating lobes for $Z_0 = 0$, and that it is desired to find the smallest Z_0 for which the first grating lobe does appear. In this case the lobe would first appear at the opposite horizon from the direction of Z_0 increase, and the limiting condition becomes

$$Z_0 = \arcsin\left(\frac{1}{d} - 1\right).$$

This defines the limiting value of Z_0, when $A_0 = 0$, for no grating lobes.

In our 16-dipole array, with a 0.65-λ spacing, no grating lobes appear until Z_0 increases to be 32.6°. No grating lobes exist for any zenith angle for this array when the azimuth angle is 45°. A physical explanation for this is that when $A = A_0 = 45°$, the effective separation of rows is $d/\sqrt{2}$.

The spacing d of the elements then is critical. When d is too large, grating lobes appear. When d is too small, mutual coupling among the elements becomes a problem and adjustable impedance-matching devices may have to be used at each element. The proper matching adjustment may vary with scan angle. When d is small enough to create this situation, the effective area of each element is reduced and the element beamwidth is larger.

When elements are spaced a distance equal to the square root of their effective area, grating lobes are normally not a serious problem. In this case the separation of the main lobe from the first grating lobe is roughly equal to d^{-1} rads by Eq. (1.6.8), but this is just the beamwidth of the element pattern. (In Eq. (1.6.6) replace D by d to obtain the approximate element beamwidth.) Consequently a spacing approximately equal to the square root of the effective area of a single element alone is a rule-of-thumb compromise that avoids serious problems due to mutual coupling and due to grating lobes.

Lo[2] has shown that on a statistical basis a random spacing of elements in a large array has several advantages. These include (1) the elimination of grating lobes, (2) reduced sidelobe levels, (3) the reduction of mutual coupling problems, and (4) wide bandwidth. The gain of a randomly spaced array will be proportional to the number of elements, and these can be spread over a large area to obtain the desired angular resolution.

Ruze[3] had earlier reached similar conclusions, and pointed out that as the average spacing increases with a given number of elements, the main beam

---

[2] Y. T. Lo, *IEEE Trans. Antennas Propagat.* **AP-12**, 257 (1964).

[3] J. Ruze, Private communication. See Final Report on Investigation of Wide Angle Scanning VHF Antenna Arrays, Prepared for Air Force Cambridge Res. Lab., Contract No. AF19 (604)-6191 (Jan. 22, 1961).

narrows but the gain remains constant. Because the gain integral must remain constant, there must be an increase in total sidelobe level. This increase may appear as additional full strength diffraction lobes, or as smaller but more numerous sidelobes. There should be no difficulty in solving the phasing problem with random spacing provided that a computer is employed to control the phasing.

1.6.5. Considerations When Building an Array

1.6.5.1. Cost. The cost of an array to meet a given set of performance specifications can vary considerably, and will depend on many factors including the amount of preplanning, the experience of the designer, and how much money is available.

In order to decide among several designs of large phased arrays, each must be designed in detail and realistic cost estimates obtained. Afterwards, further study and development can further reduce the cost. A cost reduction of just one dollar per element in a 50,000-element array is a saving of $50,000, so the point of diminishing returns only comes after much study.

Part of the design of a large array should include a small pilot antenna to test as many aspects of the proposed large one as possible. A large array should never be constructed without having first tested a small part of it.

1.6.5.2. Site Selection. Careful selection of a site has long range benefits. A site should be selected so that the sources to be studied by a broadside array are as near the zenith as possible. An array to study astronomical sources near the ecliptic, for example, should ideally be located near the equator. A few degrees difference in latitude will have an appreciable effect on grating-lobe and mutual-coupling problems.

Other characteristics of a good site are flat ground without large rocks and trees, the absence of radio interference, a climate that permits many days per year of outside working conditions, and available electrical power. A ground surface flat within 0.05 wavelengths is normally desired, and ground excavation to flatten a rough site is expensive. The site should not be closer than one or two miles to a busy highway, or to an industrial plant. The presence of mountains with elevations of 2 to 5° or more on the horizon may eliminate some radio interference unless relay stations are on top of the mountains. The antenna can be designed so that hurricanes and rain do not interfere with its performance but an arid region reduces the mud and grass-mowing problems and results in less corrosion. An investigation should be made of the history of flooding by interrogating long-time residents of the area.

1.6.5.3. Phasing Techniques and Feeder Networks. Many different types of phasing schemes have been used, ranging from sliding contacts on two-wire lines to fast computer-controlled diode switches that add or remove lengths of coaxial line.

1.6. PRACTICAL PROBLEMS OF ANTENNA ARRAYS

The principle of an antenna array is very simple. Each element collects radio energy in its vicinity and this energy flows along a line of suitable length to a central collecting point where all desired received components are added in phase. The total energy received by the array from a point source should be equal to the power received by one dipole multiplied by the number of dipoles, provided that the feed cables have negligible loss and that all elements receive equal powers. Actually, the loss in the lines will be great unless their cross sections are physically large in diameter, and, of course, expensive. The larger the array the larger this lossy-feeder problem becomes. The phase stability of the feed cables is also a problem and this increases with the size of the array.

Some saving in total amount of cable required may be realized by using a corporate[4] feed system, and when possible, by decreasing the lengths of cables by $n\lambda$, where n is an integer. As soon as this is done, however, the antenna only works properly at the one frequency determined by the wavelength λ. At nearby frequencies the beam of such an array points in slightly different directions. This effect is called *frequency sensitivity* and becomes worse as the array size increases. Ideally, the lengths of feed cable are such that the signal delay time going from a wave front in space to the central collecting point is the same for each element. For narrow-band antennas very little frequency sensitivity is introduced when the path lengths from some elements are decreased by $n\lambda$, provided n is small. Sections of transmission line provide trouble-free phasing increments; however, lumped constant phase shifters also have been used. For example, Mills *et al.*[5] used pi-network phase shifters. When the line lengths are not decreased by $n\lambda$ the antenna is not frequency sensitive. Such an array could more properly be called a time-delay array rather than a phased array. The longest lengths of cable would be required after some signal combining, and thus very few long lengths are necessary.

Another problem that must be considered as arrays become larger and larger is the method of phasing. For a small array the phasing can be done manually because there are not many cables to adjust, and because the beam is large so that continuous tracking is not required. A 38-MHz, 500-kW, 1024-dipole solar radar array[6] at El Campo, Texas, was manually phased on many occasions in less than 10 man hours. For this array four phasing cables of RG-8A/U with Type N connectors were stored at each dipole, and 0 to 4 of them were connected in series to realize any phase within $\pm 1/32\lambda$. The beam was 6.5° in the East-West direction so that no phasing was required throughout a solar radar experiment. Had the antenna been larger, a beam

---

[4] W. H. Kummer, *in* "Microwave Scanning Antennas" (R. C. Hansen, ed.), Vol. 3, p. 6. Academic Press, New York, 1966.

[5] B. Y. Mills, A. G. Little, K. V. Sheridan, and O. B. Slee, *Proc. IRE* **46**, 67 (1958).

[6] J. C. James, *IEEE Trans. Antennas Propagat.* **AP-12**, 876 (1964).

tracking operation throughout the time of one experiment would have been required. Also, more dipoles would have required phasing, so that automatic phasing would have been necessary.

Very large transmission lines will be required in a large array unless amplifiers are used in the array to drive smaller lines. The amplifiers should be very linear to avoid intermodulation especially for broadband systems,[7] and they should have low noise figures. In some cases it is desirable to have a receiver that introduces much less noise than cosmic noise. An example occurred during a study[6] of the sun when adjacent signal bands were compared for signal integration. More frequency sensitivity could be tolerated when the noise figure was decreased.

To obtain a high signal-to-noise ratio the noise figure of the field receiver should be several decibels less than the lowest cosmic noise level to be studied. In the 20- to 100-MHz range, field receivers with noise figures less than 4 dB are desirable and are available. The overall noise figure of an array employing field amplifiers is the average of the noise figures of each of the field receivers, weighted according to the gain of each amplifier circuit. The line loss between the antenna element and the first amplifier should be kept low because this in effect increases the noise figure of the amplifier by the amount of the line loss.

If a dipole array has a separate amplifier and phase shifter at each dipole, the cost of the elements will be much less than the cost of the electronics. Conceptually there are several possible ways to decrease the relative cost of the field electronics. These include (1) manual phasing of small groups of dipoles that feed the first amplifier; (2) using a higher gain element than the dipole; (3) waiting until the cost of the integrated amplifier decreases; and (4) automatically phasing a small group of dipoles with switches prior to the first amplifier. Several types of low-loss switches including diode switches and reed switches may be used. The total power required to operate the phasing switches could be large regardless of the type of switch used. For example, if $\frac{1}{4}$ W/element is required for the switches, a 40,000-element array would require 10 kW. More diode-switch current is required if the array is also used for transmitting.

Amplifiers are available now that are reliable and stable in gain and phase shift. Such an amplifier can be placed near the antenna element in the field and operated for a long period of time with little maintenance. Integrated circuit amplifiers show promise because all those manufactured in a given run are alike in gain and phase shift. The price of a suitable integrated amplifier could drop to $5 or less per unit provided the production capacity were large enough. The production should be measured in the tens of millions

---

[7] T. W. Clarke, H. S. Murdoch, and M. I. Large, *Proc. IRE (Aust.)* **30**, 236 (1969). Also see accompanying paper by M. I. Large and R. H. Frater, p. 227.

1.6. PRACTICAL PROBLEMS OF ANTENNA ARRAYS

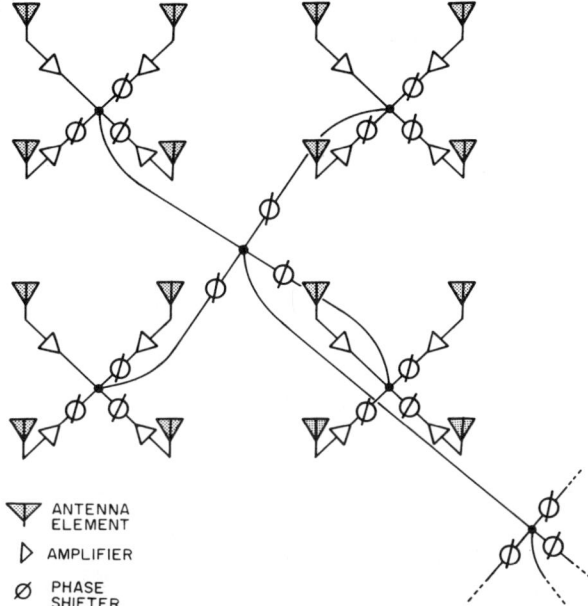

FIG. 4. One of the many possible ways in which array elements may be interconnected.

rather than in the tens of thousands which is the number likely to be required for an antenna array.

Figure 4 shows one of the many possible ways that an array may be interconnected. The first amplifier should have sufficient gain to overcome the subsequent losses in lines, matching networks, and phasing networks up to the second amplifier, plus about 10 dB. Too much gain in the first amplifier invites calibration and stability problems.

In addition to field amplification to permit the use of higher-loss cable, frequency mixing may also be performed in the field for the same purpose, because lines and cables have less loss at the lower frequencies. This procedure is used in the Molonglo[7] radio telescope.

The phasing increment need not be any smaller than error tolerances that affect phasing accuracy throughout the array. Such tolerances are found in element spacings, line lengths, amplifier phase shifts, etc. The gain loss due to uniformly distributed phase errors over a range of $\pm I$, or due to the use of phasing increments of minimum length $2I$, is computed by summing N unit vectors uniformly distributed over an angle $2I$, where N is large and is the number of elements in the array. The power-gain factor is

$$\left(\frac{1}{2I}\int_{-I}^{I}\cos x\, dx\right)^2 = \left(\frac{\sin I}{I}\right)^2. \tag{1.6.9}$$

The power not accumulated at the antenna terminal is reradiated. Some of the power from sidelobe directions that should cancel at the array summing point does not because of phase errors, and this increases the sidelobe level. When the source is a uniform sky, the power lost from the main beam due to phase errors is equal the power gained from the sidelobes. Barton and Ward[8] and Cheston and Frank[9] present formulas and plots for the effect of phase quantization on sidelobe levels.

When the sidelobe level is not of great importance, a smallest phasing increment of $\frac{1}{8}\lambda$ is sufficient. From (1.6.9) the gain loss is 0.22 dB for a $\frac{1}{8}$-λ phasing increment and 0.056 dB for a $\frac{1}{16}$-λ increment. The effect of a $\frac{1}{8}$-λ phasing increment on main-beam pointing accuracy is insignificant.

1.6.5.4. Choice of Element. An element should satisfy the bandwidth, polarization, and sky-coverage requirements of the array to be built, but of equal importance are simplicity, ruggedness, and low cost. Fabrication and installation expense should not be overlooked, and the element should be able to withstand likely storms in the area.

A large, general-purpose receiving array should be broadbanded (perhaps 10–100 MHz), should cover much of the celestial hemisphere without grating lobes (perhaps a coverage of all zenith angles up to 60°), and should receive two orthogonal polarizations separately and simultaneously. There is, as yet, no known element that will allow these specifications to be easily met. Some modification of a log spiral will meet the bandwidth requirements. Erickson[10] uses 720 teepee-shaped conical spiral elements in a Clark Lake T-array that is capable of operating between 15 and 125 MHz.

A very good element for a filled array would have a small effective area and large beamwidth; it would be useful over a wide frequency range, but have an effective area that was not a function of frequency. Such an element should be realizable but has not yet been developed. Until such an element is developed, compromise solutions must be used.

One way to compromise is to permit grating lobes, but always choose the receiving frequency so that no lobe falls on a strong source; or use a receiving bandwidth so large that the effective "broadband lobe" is not strong in any given direction. Grating lobes can be tolerated in many radio astronomical studies. Another way to compromise is to construct several arrays to cover the desired frequency range or to cover the desired area of the sky.

---

[8] D. K. Barton and H. R. Ward, "Handbook of Radar Measurement," p. 194. Prentice-Hall, Englewood Cliffs, New Jersey, 1969.

[9] T. C. Cheston and J. Frank, in "Radar Handbook" (M. I. Skolnik, ed.), pp. 11–36. McGraw-Hill, New York, 1970.

[10] W. C. Erickson and J. R. Fisher, *Radio Sci.* **9**, 387 (1974). Also see J. R. Fisher, Design Tests of the Fully Steerable, Wideband, Decametric Array at the Clark Lake Radio Astronomy Observatory, PhD. Thesis, Univ. of Maryland (1972).

1.6. PRACTICAL PROBLEMS OF ANTENNA ARRAYS

To provide a large angular coverage we could use an element with a small beamwidth compared to the celestial hemisphere, and provide some means of mechanically rotating the element. An advantage of this method is that fewer elements, fewer field amplifiers, and a less complex feeder arrangement are required. The major disadvantage is that the complex rotatable elements are likely to be more expensive than a larger number of smaller gain elements, amplifiers and phase shifters. Another disadvantage is that steering to directions outside the element beamwidth will be slow.

Mechanical positioning often looks good until tried, so before the final design decision is made, a full-scale model of one element should be made and tested. Elements that may be useful in such arrays are dipole subarrays, crossed yagis, crossed backfires, helixes,[1] paraboloids,[11] corner reflectors, log spirals, and log periodics.

Much expense and complication arises when mechanical motion of a large number of elements is required. Mechanical motion must be avoided in an inexpensive array unless it is to be performed manually. It may be cheaper to build several arrays, or to have available several sets of elements with fixed orientations in different directions, than to build one large array with one set of elements that require mechanical positioning.

A half-wave dipole constructed of thin-walled aluminum tubing is simple, inexpensive, and rugged. Dipoles with length-to-diameter ratios of about 100 are usable in arrays with bandwidths of 15% of the center frequency. One deficiency of the dipole is that the gain varies considerably with zenith angle in the plane of the dipole. For example, for a North–South dipole, the gain at a 45° zenith angle in the north or south direction is 4–7 dB less than in the zenith direction. This problem can be alleviated by tilting the dipole or by choosing an antenna site so that the zenith angles of the sources to be studied are near zero.

Fat dipoles have larger bandwidths,[12] but are not as simple to construct and install. The UTR-1 array of the Academy of Sciences of the Ukranian SSR[13] operates over the frequency range 10–25 MHz and uses dipoles having a total length of 8 m, a diameter of 1.8 m, and a height above gound of 3.5 m. These fat dipoles are constructed similarly to those shown in Fig. 5.

Helixes and log-spiral antennas may be phased by rotating the element about the longitudinal axis,[14] or by selecting the appropriate winding of a multiple-wound element. The latter method is used by Erickson and Fisher[10]

---

[11] R. N. Bracewell and G. Swarup, *IRE Trans. Antennas Propagat.* **AP-9**, 22 (1961).

[12] E. A. Wolff, "Antenna Analysis," p. 56, Wiley, New York, 1966.

[13] Yu. M. Bruk, N. Yu. Goncharov, A. V. Men, L. G. Sodin, and N. K. Sharykin, *Izv VUZ, Radiofiz.* **10**, 608 (1967) [English trans.: *Radio Phys. Quantum Electron.* **10**, 331 (1967)].

[14] G. W. Swenson, Jr. and Y. T. Lo, *IRE Trans. Antennas Propagat.* **AP-9**, 9 (1961).

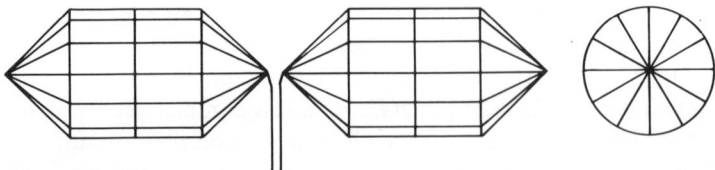

FIG. 5. Diagram showing the structure of fat dipoles, designed to give increased bandwidth.

in a Clark Lake array. A log-spiral element similar to that developed by Erickson is sketched in Fig. 6.

Other elements that have been used in arrays are helixes,[1] full-wave dipoles[15,16] folded dipoles,[5,17] half-wave dipoles,[6] full-wave dipoles backed by a corner reflector,[18] paraboloids,[11] rhombics, log periodics,[7,19] loops, turnstiles, and colinear sections of coaxial cable.[20,21] The latter element is used in the large, inexpensive, 34-MHz Cocoa cross at Clark Lake.

1.6.5.5. Ground Screen. A wire mesh is sometimes placed on the ground under an array to guarantee constant ground reflectivity for varying conditions of soil moisture. Such a ground screen should be avoided when possible, because it is expensive and interferes with array construction and maintenance. A ground screen whose purpose is to provide a constant gain in a very large array may be eliminated, because the gain of the array can be calibrated daily with one of several radio sources. When field amplifiers are used, daily calibrations will likely be made anyway.

Some elements including the log spiral and log periodic, require no ground screen. When a ground screen must be used it should preferably be placed 2 or 3 in. beneath the ground surface and should be a conductor not readily corroded by the soil. Possibilities are copper, aluminum, nonmagnetic stainless steel, and heavily galvanized steel. When wire strands are used there should be two sets running in orthogonal directions unless a single linear polarization is received.

1.6.6. Practical Suggestions

Practical problems, when not properly solved, can lead to the greatest deficiencies of an array. These include problems associated with contraction

---

[15] C. H. Constain, J. D. Lacey, and R. S. Roger, *IEEE Trans. Antennas Propagat.* **AP-17**, 162 (1969).

[16] W. C. Erickson, *IEEE Trans. Antennas Propagat.* **AP-13**, 422 (1965).

[17] J. A. Galt, C. R. Purton, and P. A. G. Scheuer, *Publ. Dom. Obs.* **25**, 295 (1967).

[18] C. H. Constain and F. G. Smith, *Mon. Not. Roy. Astron. Soc.* **121**, 405 (1960).

[19] H. T. Howard, *IEEE Trans. Antennas Propagat.* **AP-13**, 365 (1965).

[20] S. D. Shawhan and W. C. Cronyn, *Bull. Amer. Astron. Soc.* **5**, 286 (1973).

[21] G. R. Ochs, Nat. Bur. of Std. Rep. 8772, Boulder, Colorado (March 5, 1965).

1.6. PRACTICAL PROBLEMS OF ANTENNA ARRAYS

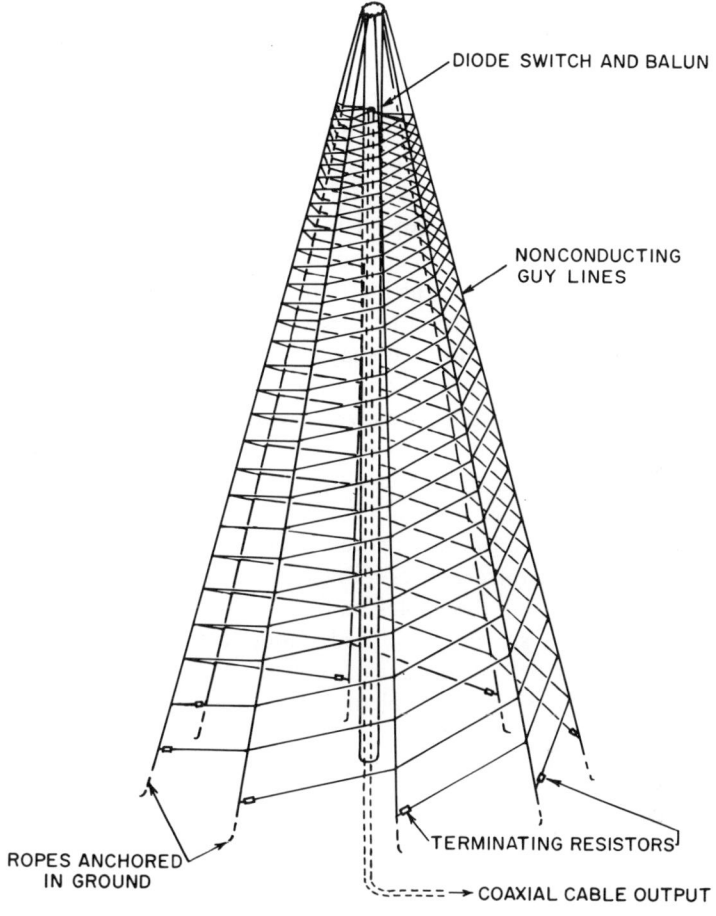

Fig. 6. Diagram of a log-spiral antenna element.

and expansion of lines, lubrication, corrosion, construction procedures, power-line noise, radio interference, grass mowing, maintenance, etc.

1.6.6.1. Design and Construction Procedure. The design process should include many iterations of conceptual design, detailed design, pricing, and testing. The construction sequence for the final design should be well planned in advance. Trench digging and cable laying should be completed early. Trenches should be straight and well marked on prints. A sufficient number of power outlets should be provided for test instruments and tools. Field amplifiers and coaxial cables may be waterproofed and buried for greater temperature stability; however, Erickson[22] and others have found that well-designed amplifiers in boxes above the ground are sufficiently stable and accessible.

1.6.6.2. Ground Cover. In some regions grass in the array may be a problem. It may be controlled with herbicides, including the growth-inhibitor maleic hydrazide, and a low-silhouette lawn mower that produces no ignition noise. Maleic hydrazide should be applied in a water spray to the grass in the spring after the grass is growing well. This slows the growth of the plants and can decrease the number of required mowings throughout the summer by a factor of three or more. A herbicide that prevents the germination of seed and another that kills only broad-leaf plants may be mixed with this spray. Later a herbicide that kills all plants may be applied to selected areas. Herbicides were available at a minimum cost of about $7.50/1000 m^2/yr ($30/acre/yr) in 1970 that would kill all plants. An elevated ground screen could help solve the grass problem but is very expensive. A ground cover of gravel is also expensive.

One solution to the grass problem would be to kill all plants in the antenna array with an application of one of several herbicides every one to five years, but this results in mud during certain seasons. Grass mowing can be a problem when the mower cuts the cables and the ground screen. To prevent this the cables and screen should be buried under the ground. Such problems should be considered before the array is constructed.

In addition to the mowing problem, grass creates a fire hazard in some localities at certain times of the year. In temperate regions where there is sufficient rainfall, *Ophiopogon Japonica*, commonly called "monkey grass," will provide a ground cover that requires no mowing and is evergreen.

Dry lake beds, or playas, of the southwestern United States, are in many respects ideal sites for astronomical arrays. No ground cover is required, plants cannot grow, there are no insects or spider webs, corrosion is minimum, insulators stay dry, and the land is flat and low without rocks. The greatest hazard is occasional flooding.

1.6.6.3. Gain Measurements. The gain of a dipole above ground will depend on the reflection coefficient at the ground surface. The phase lag on reflection at normal incidence will usually be between 165 and 180° depending on the dielectric constant and conductivity of the surface. The magnitude of the reflection coefficient may range between 0.3 and 1.0 at UHF. Reed and Russell[23] derive expressions for the complex reflection coefficient R, as a function of wavelength, grazing angle, surface conductivity C (in mhos/m) and

[22] W. C. Erickson, Clark Lake Radio Observatory, Borrego Springs, California, Private communication, 1974.

[23] H. R. Reed and C. M. Russell, "Ultra-High Frequency Propagation," 2nd ed. p. 88, Boston, Tech. Publ., Lexington, Massachusetts, 1964.

1.6. PRACTICAL PROBLEMS OF ANTENNA ARRAYS

relative surface permittivity or dielectric constant E. For horizontal polarization and normal incidence the formula is

$$R = \frac{1 - (E - j\,60\lambda C)^{1/2}}{1 + (E - j\,60\lambda C)^{1/2}}.$$

The desirable characteristic of the ground surface is a consistent reflectivity from dry to wet conditions. Low reflectivity is of secondary importance, but leads to larger element beamwidth. Most soils have relative permittivities between 3 and 30 and conductivities between 10^{-1} and 10^{-5} mhos/m. Both E and C will change with soil moisture, but the relative change in C will be larger. At higher frequencies E is more likely to dominate $60\lambda C$ so that the change in reflectivity as moisture changes will not be large.

The power gain of a $\frac{1}{2}$-wavelength dipole one-quarter wavelength above very dry ground at El Campo, Texas, was measured to be 2–3 dB less than the gain of a similar dipole above a very good ground screen. When the ground was wet this decrease in gain was only 0–2 dB. These measurements were made at a frequency of 75 MHz using a transmitting element borne aloft by a weather balloon.

Erickson[22] found only a 1-dB difference in gain between an element with and without a ground screen and rather independent of ground moisture. For these measurements he used dipoles at various decametric wavelengths and $\frac{1}{8}$ wavelength above the Clark Lake playa.

One effective test arrangement for studying the properties of small groups of antenna elements requires weather balloons filled with hydrogen or helium. A test transmitter on a helicopter or on a tall steel tower produces problems because reflections near the radiating antenna produce a complex radiation pattern that make gain measurements very difficult. The radiation pattern of a balloon-borne dipole is easily calculated and measured. The only difficulty with the balloon tests is that a slight amount of wind may cause the balloon to be unstable. This can often be solved by performing the experiments within two hours after dawn. The balloons, gas, and guy lines are inexpensive. The generator may be a small, inexpensive, crystal-controlled oscillator which is stable in frequency and amplitude.

When the far zone of the antenna under test is more than a few hundred feet away in the zenith direction, the balloon will be difficult to control. In this case an aircraft or astronomical radio source should be used as the test source.

Radio sources may be used to measure gain, pointing accuracy, and beamwidth of large arrays. The positions and flux densities of the more intense astronomical sources are known with sufficient accuracy for most calibration purposes. The reported flux density is normally that for both polarizations, unless otherwise stated, which means that a single antenna is only sensitive to one-half the flux unless the radiation is polarized.

The sky background radiation from the galaxy may be used to measure roughly the antenna losses, but should not be used to measure the directivity of an antenna because of the relationship between directive gain and beamwidth. The antenna temperature of any lossless antenna is that of the sky within the beam regardless of the gain and beamwidth of the antenna, so the output power from a small antenna is the same as that from a large antenna, when both are looking at uniform sky noise.

The output of a lossless antenna into a matched load is the product of flux density from the radio source, assumed to have a small angular dimension, and the effective area of the antenna. The effective area of a large array can be essentially as large as the physical area when the elements are matched and spaced for this purpose. In this case and when there are no grating lobes, a space wave arriving from the phased direction encounters an impedance at the array which is equal to the free-space impedance so there is no reflection. Waves from other directions may be partially or totally reflected.

1.6.6.4. Cables and Lines. Two-wire lines are simple and easy to construct and have low loss when good insulators are used.[24] The phase stability depends on the change in physical length with temperature, and this can be minimized by initially installing the lines under high tension. The strain of the conductors at low temperature can largely balance the termal expansion at high temperatures. Copper-clad steel wires are often used to construct two-wire lines. Openwire lines are much cheaper than coaxial lines for a given attenuation and work well in a dry climate.

Flexible, solid-dielectric, braided outer conductor, coaxial cables are commercially available and are convenient to use. The phase stability of solid polyethylene cables is -100 to -250 ppm/°C. Examples are the 50-ohm RG-213/U and the 75-ohm RG-11A/U cables having overall diameters of about 1.04 cm (0.41 in.). These cables have stable characteristics for many years especially when buried in the ground. In direct sunlight the outer covering will begin to deteriorate slowly in 5 to 10 yr. These two cables have the noncontaminating outer covering, but some similar cables, the RG-11/U for example, are covered with a plastic that may leach a lossy film though the braid over a period of years. Even this additional loss may be negligible. The propagation factor of these cables is about 0.66 due to the dielectric constant of polyethylene. This velocity factor increases with temperature and more than overcomes the increase in phase length of a given cable due to the thermal expansion of the copper.

[24] E. A. Laport, "Radio Antenna Engineering," Chapter 4. McGraw-Hill, New York, 1952.

1.6. PRACTICAL PROBLEMS OF ANTENNA ARRAYS

Other varieties of commercially available coaxial cable include braided flexible cable with a foamed dielectric, semiflexible cable with solid outer conductors and spiral or foamed dielectrics, and rigid lines with rigid outer and center conductors and mostly air dielectric. The latter type in the standard 7.9-cm ($3\frac{1}{8}$ in.), 15.6-cm ($6\frac{1}{8}$-in.), and 23.5-cm ($9\frac{1}{4}$-in.) copper lines have low loss, can carry high power, and are expensive.

Coaxial cables with foam dielectric are available with propagation factors of 0.80 for polyethylene and 0.91 for polystyrene. Common impedances for coaxial cables are 50, 75, and 95 ohm. Some manufacturers have 25- and 125-ohm cables. Good references for cable characteristics and availabilities are the catalogs of some cable manufacturers; also see radio and antenna handbooks.[25]

Large rigid lines may be custom fabricated by the antenna builder to suit his special needs, one of which might be economy. Aluminum extrusions of almost any specified cross-section can be purchased from some extrusion plants for only about 10% more than the cost of the raw metal. An aluminum line was specially fabricated and used with success in the National Bureau of Standards (NBS) 50-MHz array[21] at Jicamarca, Peru. A feed line made by forming an inverted trough from galvanized steel sheets and covering on the bottom with chicken wire was used by Mills et al.[5] Aluminum irrigation tubing is used in an array of log-periodic monopoles near Platteville, Colorado.[26] A 12.7-cm (5-in.) inverted trough line was fabricated from sheet aluminum and used extensively in a 38-MHz array[6] at El Campo, Texas. The loss of this trough line was about 0.5 dB/1000 ft when the humidity was less than 80%. The loss above 80% depended on the design and age of the spacer insulator. One rod-and-sleeve insulator was used that became lossy when the humidity was high after a few years because of corrosion products deposited between the rod and sleeve by capillary action. At 95% humidity, a line with these dirty rod-and-sleeve insulators had a loss of 2 dB/304.8 m (2dB/1000 ft), whereas an old line with new polyethylene rod insulators had a loss of 1 dB/304.8 m (1 dB/1000 ft). This trough line was easy to service, but an extruded 25.4-cm (10-in.) diameter, cylindrical line could have been installed at less cost, because of the proximity to an extrusion plant, and the cylindrical line not subject to increased loss at high humidity.

A conductive lubricant should be used on mating surfaces of lines and radiating elements to prevent corrosion, assure good electrical contact, and permit smooth relative motion of sliding joints such as occurs during thermal expansion and contraction. Commercially available antisieze compounds

---

[25] K. S. Packard and R. V. Lowman, in "Antenna Engineering Handbook" (H. Jasik, ed.), Chapter 30. McGraw-Hill, New York, 1961.
[26] P. W. Arnold, *IEEE Trans. Antennas Propagat.* **AP-19**, 584 (1971).

make good conductive lubricants, but equally good is a mixture of low-vapor-pressure grease and finely ground metal. The layer of lubricant should be thin so that individual metal particles contact both surfaces.

The absorptive losses in a large coaxial line may be investigated quite easily before installation by using a $\lambda/2$ section of line shorted at both ends. This produces a resonant cavity which may be excited through a small, loosely coupled loop at one end. The standing-wave amplitude may then be sampled, by means of another loosely coupled probe so that the resonant frequency f and the half-power width Δf of the resonant curve can be measured. The line loss α in decibels per λ may than be computed using the relation

$$\alpha = 27.3 \, \Delta f/f.$$

Semirigid coaxial cables with a solid outer conductor, a plastic outer covering, and a foamed plastic dielectric were used in a 75-MHz, 128-element array at El Campo, Texas. The phase stability of the cables was about -20 ppm/°C, which is very low because the phase change due to the copper and the dielectric are almost equal but opposite in sign. Cable can be purchased from some manufacturers that has the metallic and dielectric phase changes with temperature nearly balanced to produce a line with a temperature coefficient as low as 5 ppm/°C.

Shielded, multiconductor telephone cables make excellent control lines, and may be buried in the same trench with the rf lines. These telephone cables can supply amplifier power and various types of switching signals.

Connectors for the coaxial cables in the feeder network can cost almost as much as the cables themselves. This problem is alleviated when satisfactory permanent connections are made without connectors. When connectors are used, the early part of the summing network where many connections and short lengths of cable are required should consist of less expensive cables and connectors than the latter part of the feeder network.

1.6.6.5. A 75-MHz Dipole Array. An array of 128 half-wave dipoles spaced one-quarter wavelength above ground was constructed at El Campo, Texas, in order to test an electronic phasing scheme. The array consisted of 8 bays of 16 dipoles each. Within each bay the dipoles were spaced 0.63λ, and the bays were spaced uniformly on an east-west line with an overall aperture length of about 800 ft. Each dipole fed a field amplifier through a length of RG-11A/U cable without a balun. The output of each amplifier was phase shifted before being combined with other outputs and fed to the central receiver.

The phase shifter consisted of lumped constant components printed on circuit boards. There were seven $\lambda/16$ and one $\lambda/2$ phase shifters at each dipole that were selectable by a matrix driven by four flip flops. Any phase shift from 0 to $15/16\lambda$ in steps of $\lambda/16$ could be chosen by feeding a chain of 0 to 15 pulses to the flip flops.

The pointing accuracy of the array was excellent, and the gain changed by no more than 1 dB over a period of 1 yr. This gain change included that of the amplifiers plus that of the dipoles from wet to dry ground conditions. The effect of the $\lambda/16$ phasing increment on gain could not be measured; a $\lambda/8$ increment would have given the same results. Neither the gains nor phase shifts of the amplifiers were adjusted over the 1-yr test period. Cassiopeia and Cygnus were the principle test sources. The gain of the array was as expected when a 7-8-dB gain was assigned to each dipole. No ground screen was used except that of the buried combiner-network, power cables, and control cables. All cables, including the RG-11 A/U, were buried directly without conduits; the amplifiers were in boxes above ground.

1.6.7. Miscellaneous Notes

1.6.7.1. Balun. A balun is not required when the impedance looking along the outside of the cable from the antenna junction is much greater than the impedance looking into the antenna from that point. This condition is usually met for a half-wave dipole above ground.

1.6.7.2. Relative Costs of Elements and Electronics. The cost of one amplifier and phase shifter may be about 10 times the cost of one dipole element. Consequently, it may be cheaper to have higher-gain elements and fewer amplifiers. In this case the array may consist of several sets of high-gain elements, each set oriented to look in different directions, with a switching system to connect the proper elements to the amplifier.

The technique of having several arrays of elements intermingled on a given piece of land, a single set of amplifiers, and a single phasing network and feeder-line system reduces the overall cost. Erickson and Fisher[10] use a modification of this principle for the purpose of simplifying the phasing in a Clark Lake array. They essentially have 8 elements at a given location, 8 log spiral windings on a single cone. Each winding is separated by 45° in phase angle from the adjacent ones, and a switch selects the proper one to provide the correct phasing.

1.6.7.3. Height above Ground. The height of half-wave, full-wave, and folded dipoles above the ground will affect their impedance and the angular coverage. In one of the El Campo arrays[3,6] the height of the half-wave dipoles was set to 0.18λ to better match a 50-ohm feed cable. A 10-MHz array[17] at Penticton, British Columbia, uses three-wire folded dipoles spaced $1/8\lambda$ above a reflecting screen, and realizes a dipole resonant impedance of 365 ohm. Another 22-MHz array[15] at Penticton uses full-wave dipoles spaced $\lambda/8$ above the reflecting screen. The dipoles were constructed from two loops of wire, and have a broad bandwidth and an impedance of 2000 ohm.

1.6.7.4. Dipole Mounts. All-weather insulators may be turned from wood and then treated with wax. A large number of dipole mounts such as shown in Fig. 1 were turned oversize, treated in paraffin, and then forced inside aluminum tubing. A small amount of wood was shaved off as the tubing went on to produce a snug fit. Pop rivets were then used to attach the cable and to lock the tubing in place. The wood was impregnated by being heated in molten paraffin to a temperature above 100°C and then being removed from the paraffin after the temperature dropped below 100°C. A wax with a melting temperature just below 100°C should be chosen.

1.6.7.5. Power-line Noise. Power-line radio noise can be caused by broken or dirty insulators, or by loose hardware or broken ground lines on the poles. The static-generating spark does not necessarily occur on the power line itself, rather it may be the result of induced voltages. A more common type of noise results when the normal leakage path through the insulators to ground is interrupted by thin insulators such as corrosion on metal surfaces.

The common hanging type insulator is a potential source of trouble when the air is dry. The dirt and corrosion that collect over a period of time between the metallic supporting hardware of adjacent sections of the insulator string can create an added resistance to the leakage current when the air is dry. The capacitor thus formed by the supporting hardware, with the corrosion as dielectric, is easily charged to its breakdown voltage by the leakage current. These capacitors alternately charge and breakdown. The spark generates radio noise. In some cases the capacitor acts as a diode so that noise pulses appear for one polarity of the 60-cycle voltage and not the other. The problem may be cured by shorting each of these "capacitors" within at least one mile of the antenna with a copper braid firmly attached to the insulator supports. Underground power lines also solve this line-noise problem.

1.6.7.6. Materials. All metallic hardware exposed to the weather should be "noncorrosive." Stainless steels, brass, bronze, copper, and aluminum will have longer lives than steel with a thin cadmium or zinc coat. Hot-dipped galvanized bolts are good for some purposes. Many plastics deteriorate in the sun, especially those with no pigment.

Wooden posts should be pressure treated with creosote or pentachlorophenol, and the latter is much less irritating to the skin. Before installing many treated posts, a few samples should be tested by a chemical laboratory to ensure that the preservative completely penetrated the wood. Posts that are pressure treated will last more than 20 yr. Of course, the design of an array will be influenced by the sources and prices of materials that the designer finds.

2. ATMOSPHERIC EFFECTS

2.1. The Ionosphere*

2.1.1. Introduction

Since Marconi successfully transmitted the first messages across the Atlantic, the ionosphere has been the subject of interest and extensive research because of its value to long-distance radio communication. A major advance in the method of monitoring the ionosphere took place with the invention in 1926 by Breit and Tuve of the pulsed swept-frequency sounder which was also the first radar. Through a network of such sounders a relatively detailed set of empirical laws was established governing the regular behavior of the ionosphere below the height of maximum electron density.

The space age brought about many changes in relation to the ionosphere in a number of ways. The relative importance of the ionosphere as a communication medium declined as a result of the advent of communication satellites. At the same time the availability of space vehicles, both rockets and satellites, placed new tools in the hands of the ionospheric scientist which now enable him to measure many parameters not previously accessible to him.

In parallel with the development of space vehicle-based exploration techniques came a number of new powerful radio methods which vastly expanded the range of parameters available to ground based observers. The most important of these by far is the incoherent scatter radar which makes possible the continuous monitoring of electron density, temperatures, motions, electric fields, neutral density etc., up to altitudes well in excess of 1000 km.[1] The further studies of whistlers[2] and of radar echoes from the moon[3] in addition to observations of propagation effects on transmissions from space vehicles have greatly contributed to the present much improved state of knowledge about the ionosphere.

To the radio astronomer the ionosphere may be a source of degradation

---

[1] J. V. Evans, *Proc. IEEE* **57**, 496 (1969).
[2] D. L. Carpenter, *J. Geophys. Res.* **71**, 693 (1966).
[3] V. R. Eshleman, P. B. Gallagher, and R. C. Barthle, *J. Geophys. Res.* **65**, 3079 (1960).

---

*Chapter 2.1 is by **Tor Hagfors**.

in his observations. The ionosphere causes radio waves to be attenuated through absorption or through refraction effects. Phase and group delays may be introduced and in many cases serious changes both in angular position and in polarization may result. These effects are highly variable and unpredictable. It is difficult to take corrective measures without extensive ancillary observations of ionospheric properties, and even given all the methods available substantial residual effects may still occur.

It is the purpose of the present chapter to provide the radio astronomer with a brief review of the properties of the ionospheric layers, their origin, their effect on the propagation of electromagnetic waves, and the methods available for observing their properties so that corrective measures can be taken. In view of the highly variable and unpredictable nature of the ionosphere this seems to be a more sensible approach than to set down rigid rules for avoiding or removing ionospheric effects.

2.1.2. Propagation in the Regular Ionosphere†

The ionosphere, as we shall understand it here, is that part of the upper atmosphere where free electrons and ions exist in quantities such that they can substantially affect the propagation of radio waves. The part of the ionosphere below an altitude of 90 km is referred to as the D region. This region primarily affects radio waves of interest in radio astronomy through absorption effects, since the neutral density is so high that the electron-neutral collision frequency is substantial. The part from 90 to 120 km where ionization by X rays and ionization of O_2 by Lyman-β radiation predominate, is referred to as the E region. The E region usually does not contribute appreciably to either absorption or refraction effects under normal conditions. The part of the ionosphere where the highest production rate occurs, i.e., from 120 to 180 km, is commonly referred to as the $F1$ region. Above 180 km altitude diffusion and transport effects begin to have a profound effect on the electron density distribution. The part of the ionosphere between 180 and, say, 1000 km we shall refer to as the $F2$ region. The $F1$ and the $F2$ regions usually merge into one region and cannot be distinguished on the basis of the electron density profile. The F region may exert strong effects on the propagation of radio waves, both as regards refraction, phase and group delays, polarization and even absorption. Scintillation effects usually originate in the F region because of spatially irregular distribution of ionization.

The electron density in the D region during daytime is typically on the order of 10^8 to 10^9 m^{-3}, in the E region during the daytime electron densities of 10^{11} m^{-3} may often be found. F region electron densities are typically a

---

† See also Vol. 9B (*Plasma Physics*) of this treatise, Part 14.

few times 10^{12} m^{-3} during the daytime and typically a factor of three to four smaller during night-time conditions. The reader should remember that there are large variations in these numbers due to variations in solar activity and zenith angle, geographic variations in the neutral atmosphere, winds, and electric fields. An example of an electron density profile is shown in Fig. 1.[4]

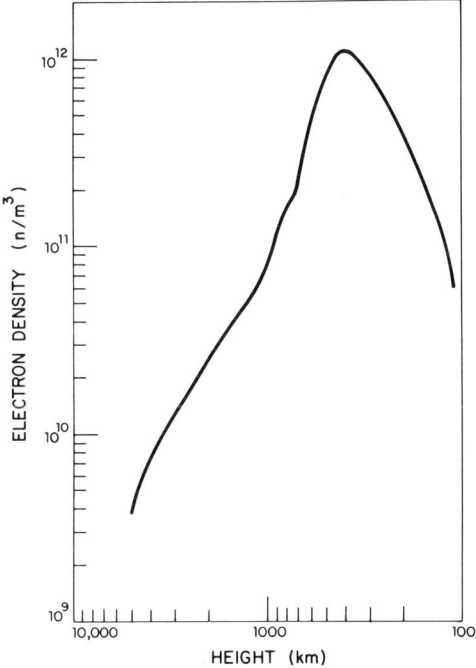

FIG. 1. An example of an electron density profile plotted as a function of height above the earth's surface.

Thrane and Piggott[5] have discussed the collision frequency v_M in the lower ionosphere. Figure 2 shows an example of the collision frequency versus height between 50 and 90 km altitude. This collision frequency varies somewhat geographically and with season. Further discussion of this variation is given by Thrane and Piggott. It is also possible to relate the collision frequency v_M to the neutral gas pressure through the relation

$$v_M = 6 \times 10^5 \, p/\text{sec}, \tag{2.1.1}$$

[4] K. L. Bowles, Nat. Bur. Std. Tech. Note 169 (1963).
[5] E. V. Thrane and W. R. Piggott, *J. Atmos. Terr. Phys.* **28**, 721 (1966).

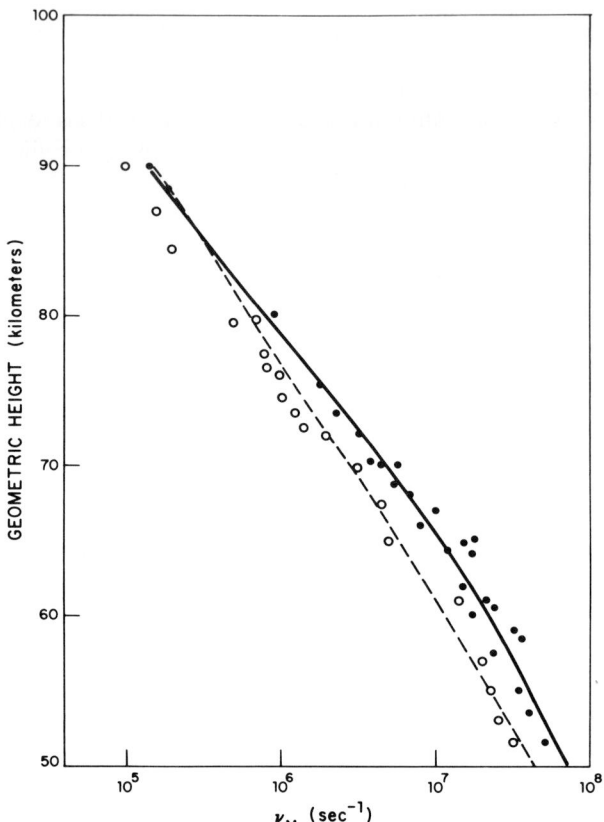

FIG. 2. An example of collision frequency versus height in the lower ionosphere. The curves represent winter (---) and summer (—) calculations based on winter (○) and summer (●) data points.

where the atmospheric pressure p must be expressed in newtons per meter squared. The atmospheric pressure as a function of height may be obtained from various model atmospheres such as the U.S. Standard Atmosphere Supplements, 1966,[6] or from Jacchia's model.[7]

The geomagnetic field is often approximated by that of a magnetic dipole of moment $a = 8.06 \times 10^{15}$ W m ($= 8.06 \times 10^{25}$ G cm^3). The magnetic dipole which gives the best fit to the geomagnetic field is tilted by 11° with respect to the geographic axis, and displaced from the geographic center of the earth

---

[6] "U. S. Standard Atmosphere Supplement 1966." U.S. Government Printing Office, Washington, D.C.

[7] L. G. Jacchia, Smithsonian Astrophysical Observatory, Spec. Rep. N332 (May 5, 1971).

by 436 km toward a point 6.5°N, 161.8°E. The north geomagnetic pole lies at 81°N latitude and 84.7°W longitude. The south geomagnetic pole is at 75°S latitude and 120.4°E longitude. Note that the North pole is actually a south magnetic pole so that the field lines run from the south to the north geomagnetic pole. The electron gyrofrequency $f_H = \omega_H/2\pi$ over the pole is approximately

$$f_H = 1.52 \times (6671/R)^3 \quad \text{MHz}, \tag{2.1.2}$$

where R is the distance from the dipole in kilometer units. In magnetic dipole coordinates the components of the magnetic flux density are

$$B_r = 2a \sin \Lambda/R^3, \quad B_\lambda = -a \cos \Lambda/R^3, \quad B_\phi = 0, \tag{2.1.3}$$

where Λ is the magnetic latitude. At times the magnetic field must be known with greater precision, and reference must be made to the International Geomagnetic Reference Field[8] or to more recent refinements.[9] Note that in magnetic field computations the x-axis points toward North, the y-axis toward West, and the z-axis toward local nadir. The declination D and inclination I are defined in Fig. 3.

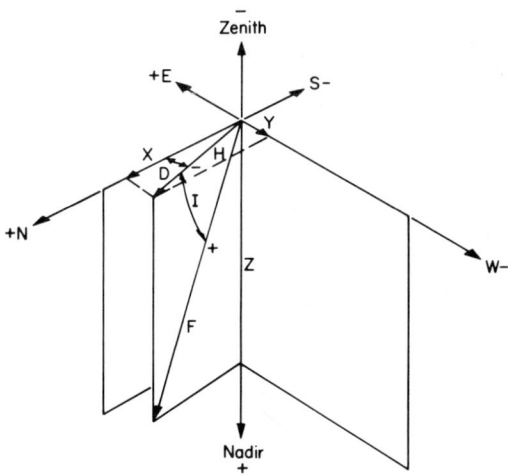

FIG. 3. Coordinate system for describing the geomagnetic field indicating the declination D, horizontal intensity H, vertical intensity Z, N–S component X, E–W component Y; total intensity F, and inclination I.

[8] J. C. Cain and S. J. Cain, GSFC Document X-612-68-501 (1968).
[9] J. C. Cain and R. E. Sweeney, *J. Geophys. Res.* **75**, 4360 (1970).

In order to describe the propagation of radio waves in the ionosphere we shall use the following standard notations[10]:

$$X = \frac{\omega_p^2}{\omega^2} = \frac{f_p^2}{f^2}, \qquad Z = \frac{\nu}{\omega}$$

$$Y = \frac{\omega_H}{\omega} = \frac{f_H}{f}, \qquad Y_L = Y \cos\theta, \qquad Y_T = Y \sin\theta \qquad (2.1.4)$$

Here $\omega_p^2 = ne^2/m\varepsilon_0 =$ (angular plasma frequency)^2, n the electron density (m^{-3}), ν the collision frequency, $\omega_H = |eB/m|$ the angular gyrofrequency, ω the angular frequency of wave, and θ the angle between wave normal and direction of magnetic field.

The various fundamental quantities entering into the above definitions are

$m = 9.11 \times 10^{-31}$ kg $\qquad e = 1.60 \times 10^{-19}$ C,
$\varepsilon_0 = 8.859 \times 10^{-12}$ C/(V m), $\qquad B =$ magnetic induction of geomagnetic field.

One should note the following useful relationship between the electron density (in units of reciprocal meters cubed) and the plasma frequency (in Hertz)

$$f_p = \omega_p/2\pi = 8.984\sqrt{n} \qquad (2.1.5)$$

With these definitions, assuming the plasma to be cold and the oscillatory motion of the ions to be negligible (not allowed in the whistler mode and in Alfven waves!) one obtains for the squared refractive index

$$\mu^2 = 1 - X\{1 - iZ - \tfrac{1}{2}Y_T^2/(1 - X - iZ) \pm [\tfrac{1}{4}Y_T^4/(1 - X - iZ)^2 + Y_L^2]^{1/2}\}^{-1}. \qquad (2.1.6)$$

The highest of the characteristic frequencies ω_p, ω_H and ν of the plasma in the ionosphere is normally the plasma frequency. In the ionosphere the plasma frequency $f_p = \omega_p/2\pi$ rarely exceeds 15 MHz. It follows that, for all frequencies of interest in radio astronomy applications the parameters X, Y and Z are small quantities and useful simplifications can be found. In the so-called quasi-longitudinal case one has

$$\mu^2 = 1 - \frac{X}{1 \pm Y_L - iZ}. \qquad (2.1.7)$$

This approximation is valid when the following inequality is satisfied[10]:

$$(\sin\theta \tan\theta)^2 = 4 \frac{|(1 - X - iZ)^2|}{Y^2}. \qquad (2.1.8)$$

[10] J. A. Ratcliffe, "Magnetoionic Theory and Its Application to the Ionosphere." Cambridge Univ. Press, London and New York, 1959.

2.1. THE IONOSPHERE

As an example, consider propagation at 100 MHz when the plasma frequency is 10 MHz. In this case the quasilongitudinal approximation applies well for $0° < \theta < 89°$.

Although rarely of interest in radio astronomy we also quote the two values of squared refractive indices for transverse propagation, i.e., for $\theta = 90°$. In this case we have

$$\mu^2 = 1 - [X/(1 - iZ)] \qquad (2.1.9a)$$

$$\mu^2 = 1 - X/\{1 - iZ - Y^2/(1 - X - iZ)\} \qquad (2.1.9b)$$

In (2.1.9a) the electric field of the wave is lined up with the external magnetic field and the wave is independent of that field. In (2.1.9b) the magnetic field of the wave is aligned with the external magnetic field.

2.1.3. Ionospheric Absorption of Electromagnetic Waves

Having stated the formulas for the refractive index and discussed the order of magnitude of the various quantities going into those expression we can specialize it so as to discuss the effect on the radio waves in the ionosphere. If the amplitude of the wave is attenuated with distance ℓ according to

$$E = E_0 \exp(-\ell\alpha) \qquad (2.1.10)$$

we find from (2.1.7) for α, assuming Y, X and Z small:

$$\alpha \approx \frac{\pi}{\lambda\mu} \frac{XZ}{Z^2 + (1 \pm Y_L)^2}, \qquad (2.1.11)$$

where λ is the free space wavelength and μ the real part of the refractive index. The other quantities were defined above. Note that the power absorption coefficient is twice this and that the attenuation per unit length is proportional to λ^2.

In the D and lower E region the collision frequency to be used to evaluate Z is determined from the collision frequency for momentum transfer v_M by the expression

$$v \simeq 2.5 v_M \qquad (D \text{ and } E \text{ region}). \qquad (2.1.12)$$

In the F region where the plasma density is relatively high the effective collision frequency is determined by collisions between electrons and ions:

$$v_{ei} = 10^{-6}[59 + 4.15 \log_{10}(T_e^2 T_i/n)]n T_e^{-3/2} \text{ sec}^{-1}, \qquad (2.1.13)$$

where the electron density is expressed in reciprocal meters cubed and where the temperatures are in degrees Kelvin.

The absorption of the waves in going through the ionosphere expressed in decibels is given by

$$L = 8.68 \int \alpha \, ds = 8.68 \int \alpha \frac{dh}{\cos i}$$

$$\approx \frac{1.16 \times 10^{-6}}{(f \pm f_L)^2} \int_0^\infty nv \frac{dh}{\cos i} \quad \text{dB}, \qquad (2.1.14)$$

where i is the angle between the ray and the local vertical and where we have assumed that $Z \ll Y_L$. In radio astronomy observations usually $Y_L \ll 1$ also so that f_L can be ignored in the denominator. Strictly speaking f_L should also have been retained under the integral sign since the angle between the ray and the magnetic field changes along the integration path.

At temperate latitudes Lawrence et al.[11] estimate that under daytime conditions the absorption at vertical incidence ($i = 0°$) at a frequency of 100 MHz amounts to 0.05 dB by day and 0.005 dB by night. Exceptional absorption may take place during polar cap events or during auroral events. This is not of great importance in the present discussion since one should avoid making radio astronomy observations in polar regions for other reasons as well. Enhanced absorption may take place following intense solar flares after which the vertical absorption may be as high as 1 dB at 100 MHz. Such enhancements are usually caused mostly by increases in D-region ionization. Under normal conditions at frequencies well above the F-region plasma frequency the D- and F-region absorptions are comparable. For further information the reader is referred to the literature.[5,12-14]

2.1.4. Faraday Rotation

In the study of polarized radio sources it is important to understand the effect of the ionosphere on the state of polarization. The magnetized ionospheric plasma is a doubly refractive medium and will support two orthogonal circularly polarized waves traveling with slightly different phase velocities. The refractive indices for these two modes, under the assumptions $X, Y, Z \ll 1$, are

$$\mu_\pm \approx 1 - \frac{1}{2} \frac{X}{1 \pm Y_L}. \qquad (2.1.15)$$

The difference in wave vectors for the two modes is

$$\Delta k = k_- - k_+ = \frac{2\pi}{\lambda} \frac{Y_L X}{1 - Y_L^2} \approx \frac{2\pi}{\lambda} X Y_L. \qquad (2.1.16)$$

---

[11] R. S. Lawrence, C. G. Little, and H. J. A. Chivers, *Proc. IEEE* **52**, 4 (1964).
[12] R. Heisler and G. L. Hower, *J. Atmos. Terr. Phys.* **32**, 1755 (1970).
[13] A. K. Saha, *J. Atmos. Terr. Phys.* **29**, 1261 (1967).
[14] E. V. Appleton and W. R. Piggott, *J. Atmos. Terr. Phys.* **5**, 141 (1954).

2.1. THE IONOSPHERE

The rate of rotation of the plane of polarization of a linearly polarized wave is

$$\frac{d\Omega}{ds} = \frac{1}{2}\Delta k = \frac{\pi}{\lambda} XY_L = \frac{\pi}{f^2 c} f_H f_p^2 \cos\theta, \qquad (2.1.17)$$

where $c = 2.998 \times 10^8$ m/sec is the speed of light in a vacuum. The total Faraday rotation through the ionosphere becomes

$$\Omega = \frac{\pi}{f^2 c} \int f_H f_p^2 \frac{\cos\theta}{\cos i} dh. \qquad (2.1.18)$$

Faraday rotation does not depolarize an arbitrary ray, it only rotates the axis of the polarization ellipse by the amount given in (2.1.18) but without changing the shape of the ellipse. Equation (2.1.18) shows that we must know the plasma frequency, the ray direction, and the direction and strength of the magnetic field along the ray path in order to predict the Faraday rotation angle.

In temperate latitudes during the winter a wave at a frequency of 100 MHz may be rotated some 15 times in traversing the ionosphere. At night this number may decrease to 1.5 to 3 times. This number may be scaled to other frequencies by observing the f^{-2} frequency dependence of the rotation. The amount of Faraday rotation may change with time either because the object under study changes position relative to the ionosphere or because of changes with time of the ionosphere, particularly at sunrise and sunset.[15-17]

2.1.5. Phase and Group Delays in the Ionosphere

Phase and group delays may be a cause for concern in some interferometric observations in radio astronomy. For sufficiently high frequencies we can ignore the magnetic field and collisions in these calculations and obtain for the excess phase path

$$\delta L_{\text{phase}} = \int (\mu - 1) \, ds \approx -\frac{1}{2} \int X \, ds = -\frac{1}{2} \int X \frac{dh}{\cos i}. \qquad (2.1.19)$$

The negative sign indicates that the phase is decreased by the presence of the plasma. The group delay, however, is increased by the same amount, i.e.,

$$\delta L_{\text{group}} = +\frac{1}{2} \int X \frac{dh}{\cos i} = \frac{40.28}{f^2} \int \frac{n \, dh}{\cos i}, \qquad (2.1.20)$$

---

[15] A. V. DaRosa and O. Almeida, Stanford Electron. Lab., Quart. Progr. Rep., Project 3304, April (1968).
[16] O. K. Garriott, F. L. Smith III, and P. C. Yuen, *Planet. Space Sci.* **13**, 829 (1965).
[17] D. H. Smith, *J. Geophys. Res.* **75**, 823 (1970).

where n must be in reciprocal meters cubed and dh is in meters. At 100 MHz, since the integrated electron content may vary from 5×10^{17} m^{-2} by day to 5×10^{16} by night[18-20] the excess group delay may amount to

$$\delta L_{\text{group}} \approx \begin{cases} (2000/\cos i) & \text{m} & \text{by day,} \\ (200/\cos i) & \text{m} & \text{by night.} \end{cases} \quad (2.1.21)$$

One should note that this full change, particularly during the morning, occurs in the course of 2 to 4 hr of local time change. This must, of course, be kept in mind in very long baseline interferometry (VLBI) interferometry.

A time rate of change of the phase path will cause a frequency shift which is proportional to the time rate of change of the excess phase path:

$$\Delta f = -\frac{f}{c}\frac{d}{dt}(\delta L_{\text{phase}}) = +\frac{40.28}{fc}\frac{d}{dt}\left(\int \frac{n\,dh}{\cos i}\right). \quad (2.1.22)$$

Probably the highest systematic frequency shifts take place during the passage of traveling ionospheric disturbances (TIDs) which may have their origin in the auroral zone. TIDs may cause the electron content to oscillate by several percent over a time period on the order of minutes. Hence, at 100 MHz one may encounter frequency shifts of

$$\Delta f \sim \begin{cases} (0.1/\cos i) & \text{Hz} & \text{by day,} \\ (0.01/\cos i) & \text{Hz} & \text{by night.} \end{cases} \quad (2.1.23)$$

Further details may be found in the "Handbook of Geophysics and Space Environment."[21]

2.1.6. Refraction Effects

In a horizontally stratified ionosphere one can compute the amount of refraction at infinity using geometrical optics. If the zenith angle for a particular object is denoted by i_0 at the observation point we have the following expression for the total bending expressed in radians:

$$\delta = \frac{\sin i_0 \, e^2}{2r_0 \, m\varepsilon_0 \, \omega^2} \int_0^\infty \frac{dh[1 + h/r_0]n(h)}{\{(1 + h/r_0)^2 - \sin^2 i_0\}^{3/2}}$$

$$= \frac{40.28 \sin i_0}{r_0 f^2} \int_0^\infty \frac{dh[1 + h/r_0]n(h)}{\{[1 + h/r_0]^2 - \sin^2 i_0\}^{3/2}}, \quad (2.1.24)$$

---

[18] O. K. Garriott, *J. Geophys. Res.* **65**, 1139 (1960).

[19] W. J. Ross, *J. Geophys. Res.* **65**, 2601 (1960).

[20] J. V. Evans and G. N. Taylor, *Proc. Roy. Soc. London* **263**, 189 (1961).

[21] S. L. Valley (ed.), "Handbook of Geophysics and Space Environment." McGraw-Hill, New York, 1965.

2.1. THE IONOSPHERE

where r_0 is the radius of the earth and where h is the height above ground level. Note that this refraction error is caused by the spherical shape of the earth since $\delta \to 0$ when we make $r_0 \to \infty$ and thus approach a flat earth. Curves of refraction effects to be expected have previously been computed on the basis of three rectangular slabs for E, F1 and F2 regions.[22] Other authors[23,24] assumed the electron density to be parabolic of the form:

$$n(h) = \begin{cases} n_{max}[1 - (h - h_m)^2/d^2] & \text{for } |h - h_m| < d, \\ 0 & \text{otherwise.} \end{cases} \quad (2.1.25)$$

Substitution of this, and approximately integrating over h in (2.1.24) leads to the result[23]:

$$\delta = \frac{2d \sin i_0}{3r_0} (f_p/f)^2 (1 + h_m/r_0)(\cos^2 i_0 + 2h_m/r_0)^{-3/2}, \quad (2.1.26)$$

where f_p is the plasma frequency corresponding to n_{max}. It would seem that, with present-day general availability of digital computers, it would be an easy task to use (2.1.24) directly to compute the refraction from the electron density profile. Figure 4 shows an example of the computation of the refraction on the basis of the electron density profile of Fig. 1 for a frequency of

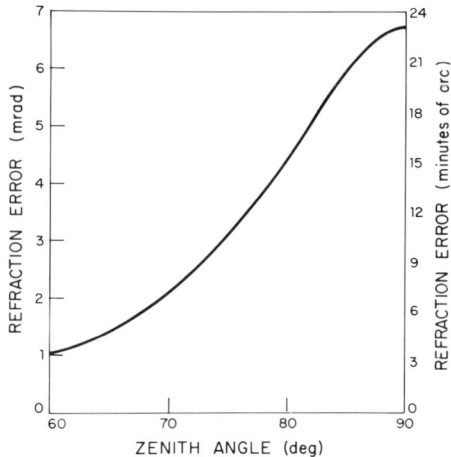

FIG. 4. An example of ionospheric refraction error (bending) as a function of zenith angle computed for the electron density profile in Fig. 1. The assumed wave frequency is 100 MHz.

[22] W. Pfister and T. J. Keneshea, AFCRL TN-56-203 (1956).
[23] D. K. Bailey, *J. Terr. Magn. Atmos. Elec.* **53**, 41 (1948).
[24] F. J. Kerr and C. A. Shain, *Proc. IRE* **39**, 230 (1951).

100 MHz. As can be seen the refraction increases very drastically below 20° altitude. By measurements of electron density profiles, however, a major fraction of this error can be removed.

In addition to the changes in elevation angle brought about by the regular horizontally stratified ionosphere, there are effects which may be associated with horizontal gradients in electron density. These horizontal gradients may have been caused by sunrise–sunset effects or by latitude-dependent electron density variation. This situation has been discussed in connection with satellite tracking for an assumed plane-earth geometry.[25] If cylindrical coordinates are introduced with the z-axis along the vertical at the observer and with R and ϕ for the other coordinates, one finds

$$\delta = + \frac{40.28}{f^2 \cos^2 i_0} \frac{dN}{dR}, \qquad \Delta\alpha = \frac{40.28}{f^2 \cos i_0} \frac{dN}{d\phi} \frac{1}{R_0}, \qquad (2.1.27)$$

where N is the vertically integrated electron content and R_0 the horizontal distance to the penetration point through the ionosphere. We see that the amount of elevation and azimuth deviation depends on the orientation of the projection of the path to the source with respect to the gradient of vertically integrated electron content. During the morning hours according to the data quoted in Section 2.1.3, which may be extreme, one could encounter ray bending corresponding to 1.5 mrad or several minutes of arc at a frequency of 100 MHz and at an elevation angle of 45°.

If greater accuracy is required in the calculations one may have to resort to ray tracing on a digital computer.[26,27] Usually the ionospheric information available for the analysis of a particular set of astronomical data is insufficient to warrant this complexity.

2.1.7. Propagation in the Irregular Ionosphere

Small-scale irregularities in the F and E regions can seriously affect radio wave propagation. The chief observed E-region effect is a strong enhancement of scattering from that region at frequencies which are normally not returned. The plasma density apparently is also enhanced in patches. The E-region irregularities are, however, normally not strong enough to affect seriously the propagation of waves traversing the region at frequencies far in excess of the F-region plasma frequency. We shall, therefore, only be concerned with the F-region phenomena.

There are several manifestations of the presence of these F-region irregularities. The ionosonde records (see Section 2.1.9) bear evidence of echoes

---

[25] T. Hagfors, J. Mann, and R. A. Power (in preparation).
[26] J. Haselgrove, *Proc. Phys. Soc. Lond.* **70**, 653 (1957).
[27] R. S. Lawrence and D. J. Pasakony, *in* "Space Research" (H. C. Van de Hulst, C de Jager and A. F. Moore, eds.), Vol. II, p. 258. North-Holland Publ., Amsterdam, 1961.

from within a range of angles. Signals from satellites and from radio stars show signs of amplitude and phase fluctuation as if a multipath mechanism is in operation. Recent observations of the electron density by satellite probes have provided direct evidence of the presence of an irregular structure which may explain the radio manifestations.[28]

The occurrence of spread F on ionosonde records shows that it is a night-time phenomenon centred on the magnetic equator and that there is another maximum of occurrence in the polar regions to the north of 60° geomagnetic latitude.[29] The occurrence of scintillation of signals from satellites has been extensively studied by Aarons and co-workers and shows a very similar dependence on time of day and geomagnetic coordinates.[30]

In addition to the small-scale irregularities which give rise to scintillation effects the so-called "traveling ionospheric disturbances" (TID) will often give rise to serious refraction of a traversing signal. These disturbances are thought to be caused by atmospheric gravity waves in the neutral atmosphere. The origin of these waves is probably related to either heating effects or hydromagnetic interactions in the auroral zone. Observations of electron density variations have shown peak changes of electron content of up to 8%.[31] The wavelenths of these variations vary from several tens to several hundreds of kilometers. The direction of travel of the waves apparently is preferentially toward the equator. For further information on the properties the reader is referred to a considerable literature on the subject.[32-35]

2.1.8. The Effect of Small-Scale Irregularities

The small-scale irregularities impose on an incoming wave a certain amount of phase perturbation which will vary from point to point across the wave front. This will cause the angular spectrum to appear widened. If the phase perturbations are shallow, i.e., only introduce phase perturbations of less than 1 rad, there will be a plane wave present in the transmitted radiation. The scattering introduced by the irregularities will only appear as a noise contribution on this wave. When the phase perturbations become larger, the "systematic" wave will disappear and the signal received on the ground will have the properties of a superposition of an angular spectrum of waves all of comparable strength. Depending on a combination of the distance

[28] P. L. Dyson, *J. Geophys. Res.* **74**, 6291 (1969).
[29] W. Calvert and C. W. Schmid, *J. Geophys. Res.* **69**, 1839 (1964).
[30] J. Aarons, H. E. Whitney, and R. S. Allen, *Proc. IEEE* **59**, 159 (1971).
[31] G. Vasseur and P. Waldteufel, *J. Atmos. Terr. Phys.* **31**, 885 (1969).
[32] K. L. Chan and O. G. Villard, *J. Geophys. Res.* **68**, 3197 (1963).
[33] M. J. Davis and A. V. DaRosa, *J. Geophys. Res.* **74**, 5741 (1969).
[34] N. N. Rao, G. F. Lyon, and J. A. Klobuchar, *J. Atmos. Terr. Phys.* **31**, 539 (1961).
[35] G. D. Thome, *J. Geophys. Res.* **73**, 6319 (1968).

from the irregular region, the size of the irregularities and the depth of the modulation, there will be an amplitude fluctuation associated with the phase fluctuations also.

Following the work of Booker[36] we write for the mean square phase fluctuation introduced along a distance D of the radio path containing irregularities of size $L \ll D$:

$$\langle \Delta\phi^2 \rangle = \frac{D}{L} \left(\frac{2\pi L}{\lambda}\right)^2 \left\langle \left(\frac{\Delta\mu}{\mu}\right)^2 \right\rangle \qquad (2.1.28)$$

where $\Delta\mu$ is the deviation in the refractive index μ and λ the radio wavelength. Introducing the simplified refractive index for no collisions or magnetic field we obtain

$$\langle \Delta\phi^2 \rangle = \frac{D}{L} \left(\frac{2\pi L}{\lambda}\right)^2 \frac{f_p^4}{4f^4} \left(\frac{\Delta n}{n}\right)^2, \qquad (2.1.29)$$

where Δn is the deviation in electron density from the mean value. The mean square angular fluctuation is

$$\langle \Delta\psi^2 \rangle = -\frac{1}{4} \left(\frac{\partial^2 \rho}{\partial x^2}\right)_0 DL \left(\frac{f_p}{f}\right)^4 \left\langle \left(\frac{\Delta n}{n}\right)^2 \right\rangle, \qquad (2.1.30)$$

where $\rho(x)$ is the correlation function of the radiation measured normal to the direction to the source and ψ is measured in the x-direction. Lawrence et al.[37] have observed rms angular fluctuations on the order of 0.5 mrad or 1.8 arc min at a frequency of 108 MHz. In a great many cases studies of scintillation effects have been carried out using the scintillation index S obtained from the signal power I through

$$S^2 = \frac{\langle I^2 \rangle - \langle I \rangle^2}{\langle I \rangle^2}, \qquad (2.1.31)$$

Briggs and Parkin[38] have shown how this quantity is related to the mean square phase fluctuation $\langle \Delta\phi^2 \rangle$ as

$$S^2 = 2\langle \Delta\phi^2 \rangle \bigg/ \left(1 + \frac{\pi^2 L^4}{4\lambda^2 Z^2}\right), \qquad (2.1.32)$$

where Z is the distance to the region containing the irregularities.

There seems to be little hope that one can ever correct for scintillation effects and recover the waves which existed above the ionosphere. The most

[36] H. G. Booker, Proc. IRE **46**, 298 (1958).
[37] R. S. Lawrence, J. L. Jespersen, and R. C. Lamb, J. Res. Nat. Bur. Std. **65D**, 333 (1961).
[38] B. H. Briggs and I. A. Parkin, J. Atmos. Terr. Phys. **25**, 339 (1963).

one can do is to avoid making low-frequency radio astronomy observations within 15° of the magnetic equator and in the arctic regions beyond about 60° geomagnetic latitude where these scintillation effects are most serious. In addition one should keep a watch by independent means of the presence of scintillating irregularities in order to assess their effect on the particular type of observation.

2.1.9. The Effect of Large-Scale Irregularities

We shall be particularly concerned with the effects which may be introduced as a result of TIDs. The actual variation in electron content brought about by a gravity or acoustic wave traveling through the ionosphere has been studied in some detail.[39] Here we shall only assume that a horizontally periodic variation in electron density has somehow been induced. The integrated electron content deviation may be represented as

$$n(x, y) = n_0 \sin(K_x x + K_y y). \qquad (2.1.33)$$

The wavelenth of the density wave is

$$\Lambda = 2\pi/(K_x^2 + K_y^2)^{1/2}. \qquad (2.1.34)$$

If we assume that the earth is plane and introduce a unit vector $\mathbf{e}_h$ along the projection of the radius vector to the source on this plane, we obtain for the elevation and azimuth deviation brought about by the density wave

$$\Delta \gamma = + \frac{40.28 n_0}{\cos^2 i_0 f^2} (\mathbf{K} \cdot \mathbf{e}_h) \cos(\mathbf{K} \cdot \mathbf{r}_0), \qquad (2.1.35)$$

where

$$\mathbf{e}_t = \mathbf{e}_h \times \mathbf{e}_z,$$

where $\mathbf{e}_z$ is the vertical unit vector at the observer and $\mathbf{r}_0$ a radius vector toward the source but of length corresponding to the distance to the irregular region. As a numerical example consider the case where $i_0 = 45°$, $f = 100$ MHz, and where n_0 is 4% of the high total content of 5×10^{17} m^{-2}. We also assume $\mathbf{K}$ to be directed along $\mathbf{e}_h$ so that we have only elevation fluctuations. For a density wavelength of 50 km one then obtains

$$\Delta \gamma_{max} = 20 \quad \text{mrad} = 1.14°.$$

Tracking of satellites through what appears to be TIDs has indeed often been observed to produce peak deviations at 400 MHz of 0.05° which would scale to 0.8° at 100 MHz (A. Freed, private communication). TIDs hence

---
[39] T. M. Georges and W. H. Hooke, *J. Geophys. Res.* **75**, 6295 (1970).

could seriously affect accurate position determinations of radio sources. Variations in phase path may also be associated with the TID, but the amount will depend strongly on the relative orientation of the density wave and the ray to the source.

2.1.10. Ionospheric Measurements

In this section we shall very briefly summarize a number of observations which can be made to provide information on ionospheric parameters. In some cases this information can be used to make corrections to the astronomical data, and in other cases they may only be suitable as a warning that the astronomical data may be of limited validity. Most of the equipment used in monitoring the ionosphere represents a large investment and most radio-astronomy observatories probably would prefer to cooperate with a nearby ionospheric observatory rather than acquire their own equipment.

The oldest and most widely used method for studying the ionosphere up to the peak of the electron density in the F layer is the vertical incidence sounder. This is a vertical looking HF radar where the frequency is swept from about 1 up to 20 MHz. The time delays are recorded as a function of frequency as a virtual height $h'(f)$. The virtual height corresponds to the time delay up to a point in the ionosphere where the refractive index μ goes to zero. From this information, it is possible to derive the true electron densities by inverting an integral equation.[40]

The so-called "topside sounder" is similar, but mounted in a satellite. The Canadian "Alouette" satellite series provides regular ionograms which may be analyzed in the same way as the bottom-side ionograms to connect with these and provide a complete spectrum density profile. Difficulties do arise in the HF sounder analysis because of changes in sign of the slope of the density versus height curve. In spite of this, it is probably possible to use this kind of data to remove the major effects of the quiet ionosphere.

A better, but vastly more costly and complicated, method for measuring the electron density is the incoherent scatter technique.[1] The technique relies on weak scattering from the ionosphere at frequencies which are far in excess of the plasma frequency. The scattering then appears to come from nearly free electrons and the density can be derived from the strength of the signal or from certain plasma oscillation effects. Electron densities have been derived by this method from less than 100-km altitude to several-thousand-km altitude, an example of which is shown in Fig. 1.

There are several satellites available for monitoring ionospheric effects. Measurements of the Faraday rotation of signals from geostationary satellites can be used to monitor changes in total electron content in a particular

[40] J. O. Thomas, *Proc. IRE* **47**, 162 (1959).

direction with respect to the observing station. The absolute value of the electron content can be obtained if Faraday rotation measurements are made at several different frequencies in order to remove ambiguities.[18,41,42] The total integrated electron content can also be measured by comparing the phase paths of signals transmitted coherently at different frequencies from a satellite.[43,44] Profiles of electron density as a function of height can also be made using a combination of rockets and the Faraday or phase path measurements. Several methods are available for the direct observation of electron densities by means of probes mounted on rockets and satellites.[45]

The lower ionosphere, primarily of concern because of the absorption which may be produced, can also be explored by a number of methods. Electron densities may be studied by wave interaction methods (Luxemburg effect),[46] by partial reflections,[47] by studies of the phase of long wavelength reflections,[48] and by riometer[49] (relative ionospheric opacity meter) techniques. The latter method measures the absorption of cosmic noise caused primarily by D-region effects. By monitoring this absorption it is possible to scale the absorption to other frequencies and zenith angles and make appropriate correction.

[41] K. C. Yeh and G. W. Swenson, *J. Geophys. Res.* **66**, 1061 (1961).
[42] R. V. Bhonsle, A. V. DaRosa, and O. K. Garriott, *Radio Sci.* **69D**, 929 (1965).
[43] B. Burgess, *in* "Electron Density Profiles in the Ionosphere and Exosphere" (B. Maehlum, ed.), p. 224, Pergamon Press, Oxford, 1962.
[44] J. L. Alpert, *Space Sci. Rev.* **4**, 5 (1965).
[45] E. R. Schmerling, *Rev. Geophys.* **4**, 329 (1966).
[46] R. E. Barrington and E. V. Thrane, *J. Atmos. Terr. Phys.* **24**, 31 (1962).
[47] F. F. Gardner and J. L. Pawsey, *J. Atmos. Terr. Phys.* **3**, 321 (1953).
[48] D. G. Deeks, *Proc. R. Soc. Lond.* **A291**, 413 (1:66).9
[49] G. G. Little and H. Leinbach, *Proc. IRE* **46**, 334 (1958).

2.2. Structure of the Neutral Atmosphere*

2.2.1. Introduction

The neutral atmosphere affects the amplitude, phase, and direction of propagation of microwaves. Propagation effects are caused by atmospheric gases and by condensed water. These effects will be discussed in the chapters that follow. Each of the effects may be related to distributions of meteorological parameters—temperature, pressure, relative humidity and the liquid water content of clouds and hydrometers. In this chapter, the distributions of the meteorological parameters will be discussed. For more information, texts on climatology should be consulted.[1,2]

The atmosphere is a mixture of gases. The principal constituents are nitrogen, oxygen, argon, and carbon dioxide constituting 78.09, 20.95, 0.93, and 0.03% by volume of the mixture, respectively, for altitudes below 80 km. The minor constituents excluding water vapor are listed in Table IV of Chapter 2.3. Several of the gases change in relative percentage of the total mixture. Of particular importance are the variable constituents—water vapor, carbon dioxide, and ozone. Below 80 km the relative concentrations of the fixed constituents are maintained by turbulent mixing. Above this height, the effects of diffusion begin to change the relative concentrations of the gases. The properties of the atmosphere at heights above 80 km are of importance to propagation at frequencies below 1 GHz, but are usually not of importance to microwave propagation. See Chapter 2.1 for a discussion of propagation in the ionosphere.)

2.2.2. Atmospheric Temperature

The changes in temperature of the atmosphere between the surface and 80 km are characterized by a general lapse or decrease of temperature with height to the tropopause, a general increase in temperature to the stratopause, and a general decrease again to the mesopause. The region between

---

[1] R. G. Barry and R. J. Chorly, "Atmosphere, Weather, and Climate," Holt, New York, 1970.
[2] H. L. Crutcher, in "Climate of the Free Atmosphere" (D. F. Rex, ed.). Elsevier, New York, 1969.

---

* Chapter 2.2 is by R. K. Crane.

2.2. STRUCTURE OF THE NEUTRAL ATMOSPHERE

the surface and the tropopause is commonly called the *troposphere*,[3] the region between the tropopause and the stratopause is called the *stratosphere*, and the region between the stratopause and the mesopause is called the *mesosphere*. The several atmospheric regions and a temperature profile that represents the average properties of the atmosphere for July at 45° N latitude[4] are depicted in Fig. 1.

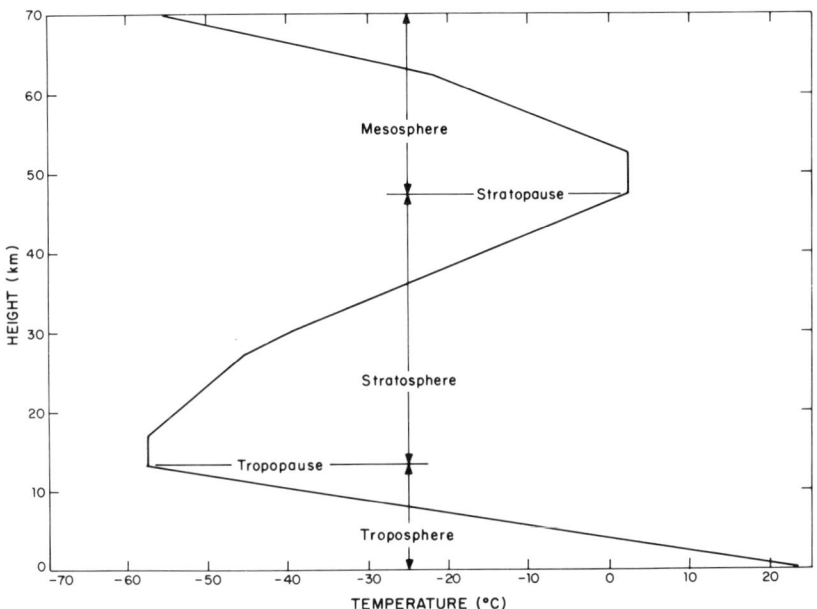

FIG. 1. Temperature profile for 45° N latitude, July, U.S. Standard Atmosphere [U.S. Standard Atmosphere Supplements, 1966. Environmental Sci. Serv. Administration, Dept. of Commerce, Washington, D.C. (1966)].

The temperature profile represents the average properties of the atmosphere at a 45° N latitude for July. The average properties change with season, with the day, and with the time of day. In addition to changes in the average properties of the temperature profile, changes at a height may occur over time scales ranging from minutes to hours. Soundings taken at Albany, New York on August 12, 1966 at 0000 UT and February 1, 1966 at 0000 UT are depicted on Fig. 2 together with the U.S. Standard atmosphere profiles for 45° N latitude for January and July.[4] These profiles were obtained from radiosonde

[3] R. A. Craig, "The Upper Atmosphere: Meteorology and Physics." Academic Press, New York, 1965.
[4] U.S. Standard Atmosphere Supplements, 1966. Environmental Sci. Serv. Administration, Dept. of Commerce, Washington, D.C. (1966).

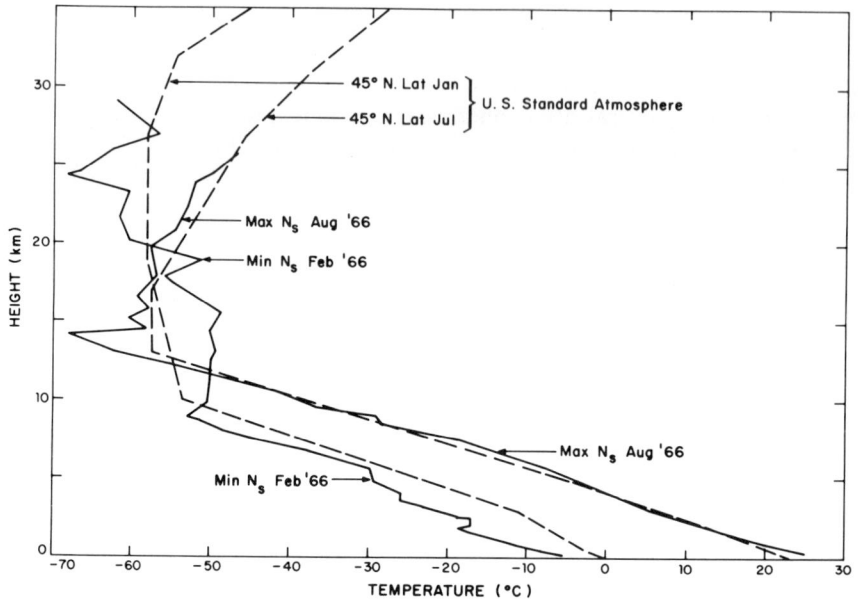

FIG. 2. Temperature profiles at Albany, New York and U.S. Standard Atmosphere [U.S. Standard Atmosphere Supplements, 1966. Environmental Sci. Serv. Administration, Dept. of Commerce, Washington, D.C. (1966)]. The quantity N_s refers to the index of refraction at the surface of the earth.

measurements made by the National Weather Service (NWS) and represent the profile for Albany, New York from February 1966 with the lowest radio refractivity value at the surface (see Section 2.5.1) and the soundings from August 1966 with the highest surface radio refractivity value. The soundings show that large departures may occur from the monthly average for that latitude. The short time period changes in the profile will generally be significantly smaller than the differences between the individual soundings and the model profiles shown in Fig. 2.

2.2.3. Atmospheric Pressure

Atmospheric pressure decreases in a roughly exponential manner with height. The temperature and pressure profiles generally vary little from hour to hour and day to day for soundings made within the same air mass.[1,5] The changes from air mass to air mass that occur across fronts are significant. These variations occur on a large or synoptic scale spanning hundreds of

[5] S. Petterssen, "Weather Analysis and Forecasting," Vol. II, 2nd ed. McGraw-Hill, New York, 1956.

kilometers. Synoptic scale changes mainly affect the troposphere. Smaller scale variation may be of importance near the surface. Diurnal changes in temperature occur due to surface heating and cooling. Relatively small changes in temperature may also occur in the free atmosphere (a kilometer or more above the surface) due to internal atmospheric waves, turbulence, convection, and changes in the state of water in the atmosphere.

2.2.4. Atmospheric Water Vapor

The water vapor concentration in the lower troposphere is highly variable especially within 1 to 2 km of the surface. The average midlatitude moisture profile is given in Figs. 3 and 4.[6,7] The mixing ratio profile corresponding to a

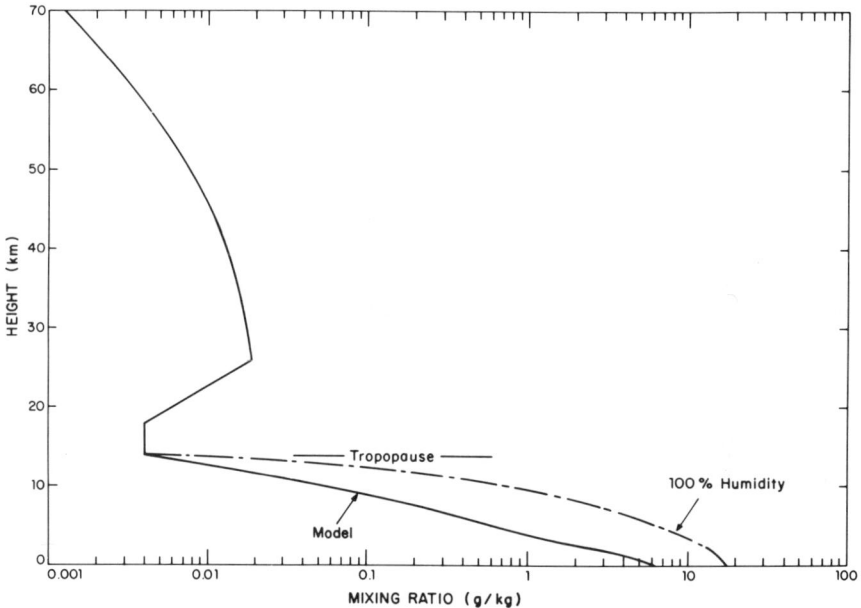

FIG. 3. Model mixing-ratio profiles.

saturated troposphere for the 45° N. latitude July U.S. Standard Atmosphere is also shown in Fig. 3. Data for the two Albany radiosonde profiles depicted in Fig. 2 are shown in Fig. 4. Again the departure from the model profile and the variability with height are evident. Significant variation in water vapor

[6] N. Sissenwine, D. D. Grantham, and H. A. Salmela, AFCRL-68-0550, Air Force Cambridge Res. Lab., Bedford, Massachusetts (October, 1968).
[7] M. Gutnick, AFCRL-62-681, Air Force Cambridge Res. Lab., Bedford, Massachusetts (July 1962).

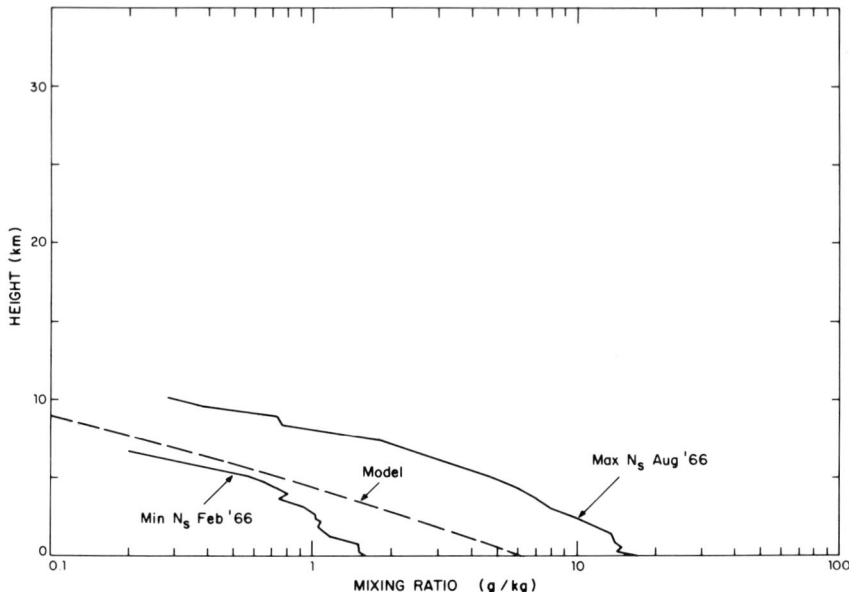

FIG. 4. Mixing-ratio profiles for Albany, New York. The extreme cases corresponding to maximum and minimum values of the surface index of refraction N_s are shown. The model profile from Fig. 3 appears here for comparison.

content occur both on the synoptic scale and on the smaller scale sizes associated with clouds and groups of clouds (mesoscale). The air within a cloud is generally saturated and outside the clouds lower humidity values are usually observed.

2.2.5. Propagation Effects

For use in estimating the effects of the atmosphere on microwave propagation we require profiles of temperature, pressure, and water vapor content. Climatological averages are useful in estimating typical propagation effects but for a specific set of measurements, the detailed profiles for that time and place are required. The only data readily available for the generation of profiles for particular times and places are obtained from radiosonde observations. Throughout the United States and many other parts of the world radiosonde observations are made everyday at 0000 UT and 1200 UT. The radiosonde observations are made from stations located several hundreds of kilometers from each other. The data from an isolated observation is generally not useful for the estimation of propagation effects unless it is made within several hours and about 50 km of the time and location of the required

2.2. STRUCTURE OF THE NEUTRAL ATMOSPHERE

propagation effects prediction. Several weather services routinely prepare maps that interpolate the radiosonde data and extrapolate the measurements in time using numerical methods and the equations of hydrodynamics.

Radiosonde data are generally reported in terms of temperature and dew point depression as a function of pressure. To relate these data to temperature and dew point as a function of height, the height must be computed from the hydrostatic equation

$$h_N = h_0 + \sum_{i=1}^{N} \Delta h_i = 14.645 \sum_{i=1}^{N} (T_i + T_{i-1}) \log_e \left(\frac{P_{i-1}}{P_i}\right) + h_0,$$

$$T_i = (t_i + 273.16)\left(1 + 0.388 \frac{e_i}{P_i}\right)$$

(2.2.1)

where h_N is height (in meters) of the top of layer N, h_0 the start height in meters, t_i the temperature (°C) at the top of layer i, P_i the pressure (in millibars) at the top of layer i, and e_i is partial pressure (in millibars) of water vapor at the top of layer i. The layers used in the computations are between the reported temperature, pressure, and moisture-content values. The properties of the atmosphere change most rapidly with height and the layers are usually assumed to be spherically symmetric (horizontally stratified) about the center of the earth.

Condensed water particles either in clouds or rain are also of importance to microwave propagation. For precise prediction of propagation effects, the amount of condensed water at each point in the atmosphere would have to be determined. Weather radar data are available in some locations for use in estimating the rain effects. Rain, however, varies rapidly in time and space, and estimation of propagation effects for a particular time and location are impossible unless simultaneous radar observations are made. The statistical properties of rain are considered in Chapter 2.4.

2.3. Absorption and Emission by Atmospheric Gases*

2.3.1. Introduction†

At centimeter and shorter wavelengths, absorption and emission by atmospheric gases can significantly affect the propagation of electromagnetic radiation through the atmosphere. At frequencies from 1 to 300 GHz, which are considered here, and which shall be referred to as *microwave frequencies*, absorption by atmospheric gases is dominated by water vapor lines at 22 and 183 GHz, oxygen lines near 60 GHz and at 118 GHz, and relatively narrow and weaker ozone lines above 100 GHz. Nonresonant absorption by water vapor and oxygen also has significant effects in the window regions away from the dominant lines. Other atmospheric molecules have spectral lines in the microwave frequency region, but the expected strength of these lines is too small to affect propagation significantly.

In the following sections, we shall first formulate the radiative transfer expressions appropriate for calculating absorption and emission by atmospheric gases. Then the general expressions for the spectral-line absorption coefficient will be given, followed by specific expressions for calculating absorption by water vapor and oxygen. Problems associated with the calculation of absorption by these two molecules will also be discussed. The absorption by microwave lines of ozone and other minor constituents will then be considered briefly. Finally, the results of calculations of atmospheric absorption and emission will be given, as well as measured values.

2.3.2. Radiative Transfer at Microwave Frequencies

The equation of radiative transfer gives the mathematical description of atmospheric absorption and emission, and the classic work on radiative transfer is that of Chandrasekhar.[1] For the nonscattering, nonrefractive atmosphere in thermal equilibrium, which is the case for gaseous absorption at microwave frequencies, the radiative transfer equation takes a particularly simple form, which, as given by Chandrasekhar, is

$$I_\nu(s) = I_\nu(0) \exp[-\tau_\nu(0, s)] + \int_0^s B_\nu(T) \exp[-\tau_\nu(s', s)] k_\nu(s')\, ds', \quad (2.3.1)$$

---

[1] S. Chandrasekhar, "Radiative Transfer." Dover, New York, 1960.

† See also Vol. 3A (*Molecular Physics*), Chapter 2.1.

* Chapter 2.3 is by J. W. Waters.

2.3. ABSORPTION AND EMISSION BY ATMOSPHERIC GASES

where $I_\nu(s)$ is the intensity of radiation at frequency ν and position s, $B_\nu(T)$ the Planck function giving the intensity of radiation at frequency ν from a blackbody at temperature T, $k_\nu(s')$ the volume absorption coefficient at frequency ν and position s', and

$$\tau_\nu(s', s) = \int_{s'}^{s} k_\nu(s'') \, ds'' \tag{2.3.2}$$

is the optical depth, or opacity, between points s and s'. The integrals are taken along the path of observation, and the geometry is sketched in Fig. 1.

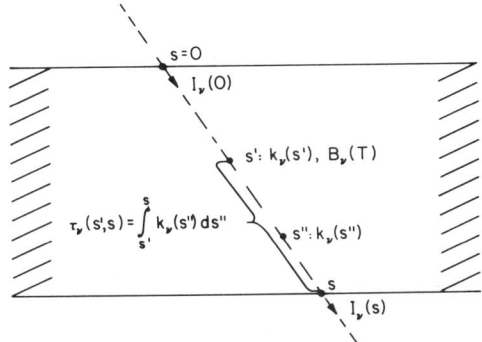

FIG. 1. Geometry for the radiative transfer equation.

The absorption coefficient k_ν in the preceding two equations is the volume absorption coefficient, which differs from the mass absorption coefficient in Chandrasekhar's expression by a factor of ρ^{-1}, where ρ is the density of the absorbing substance. The absorption coefficient describes the interaction of radiation with the absorbing matter, and is, in general, a function of the density of the absorbing substance, and the temperature and pressure of the atmosphere, all of which can vary along the path of observation. The volume absorption coefficient has dimensions of reciprocal centimeters, which are related to values of absorption in decibels per kilometer by 1 cm^{-1} = 10^6 · $\log_{10} e$ dB/km = 4.34×10^5 dB/km. Optical depths are dimensionless, and can be expressed in the dimensionless "units" of nepers (logarithms to base e) or decibels (logarithms to base 10); the two are related by 1 Np = 10 $\log_{10} e$ dB = 4.34 dB.

The two terms on the right side of (2.3.1) have simple physical interpretations. The first describes the attenuation of the radiation $I_\nu(0)$ by the factor $\exp[-\tau_\nu(0, s)]$ as it passes from $s = 0$ to s. The second term gives the emission $B_\nu(T)k_\nu(s') \, ds'$ from path element ds', attenuated by $\exp[-\tau_\nu(s', s)]$, as it passes from s' to s.

In the frequency region considered here, the Planck function $B_\nu(T)$ reduces, for the range of neutral atmospheric temperatures, to

$$B_\nu(T) = 2kT\nu^2/c^2, \tag{2.3.3}$$

where k is Boltzmann's constant, and c the speed of light. It is then convenient to rewrite (2.3.1) in terms of the brightness temperature T_B

$$T_B(s) = T_B(0)\exp[-\tau_\nu(0,s)] + \int_0^s T(s)\exp[-\tau_\nu(s',s)]k_\nu(s')\,ds', \tag{2.3.4}$$

where T_B is the equivalent blackbody temperature of the radiation, and is defined by the equation

$$T_B(s) = (c^2/2k\nu^2)I_\nu(s). \tag{2.3.5}$$

It is instructive to consider the simple situation where the temperature of the absorber is independent of position s. Since $d\tau_\nu(s',s) = -k_\nu(s')\,ds'$, the integral in (2.3.4) can then be performed simply, and that equation becomes

$$T_B(s) = T_B(0)\exp[-\tau_\nu(0,s)] + T(1 - \exp[-\tau_\nu(0,s)]), \tag{2.3.6}$$

where the fact that $\tau(s,s) = 0$ has been used. We now consider three situations. First, for a transparent medium, $\tau(0,s) = 0$, and (2.3.6) reduces to $T_B(s) = T_B(0)$; the observed brightness temperature at s is the incident brightness temperature, as it must be for a transparent medium. Second, for an opaque medium, $\tau_\nu(0,s) \gg 1$, and we obtain $T_B(s) = T$; the brightness temperature at s is the temperature of the medium. Third, for a medium in which the temperature is equal to the incident brightness temperature, i.e., $T = T_B(0)$, (2.3.6) gives $T_B(s) = T = T_B(0)$; the medium emits as much radiation as it absorbs, and the intensity of the radiation at s is the same as at $s = 0$.

The opacity of the atmosphere is usually expressed in terms of its value τ_ν^0 for a zenith path through the atmosphere. For stratified conditions, the opacity for nonzenith paths is then given by

$$\tau_\nu = m\tau_\nu^0 \tag{2.3.7}$$

where m is the air mass, or amount of air in the path of observation, relative to a zenith path. For zenith angles less than 70° (elevation angles above 20°) the air mass is simply the secant of the zenith angle (cosecant of the elevation angle) to an accuracy better than 1%. The atmospheric opacity can be determined by measuring the absorption of a signal originating outside the atmosphere, or, at frequencies where the atmosphere is not opaque, by measuring the thermal emission from the atmosphere and relating it to the opacity by using (2.3.6). In the latter case, it is necessary to assume an equivalent emission temperature for the atmosphere. The equivalent emission

2.3. ABSORPTION AND EMISSION BY ATMOSPHERIC GASES

temperature depends slightly on frequency, since the atmospheric emission from different levels (with different temperatures) is weighted differently depending on the opacity, which is a function of frequency.

2.3.3. Microwave Spectral Line Absorption

Spectral line absorption and emission occur when a quantized system, such as a molecule, interacts with an electromagnetic radiation field and makes a transition between two quantum states of the system. In the absence of line-broadening mechanisms, interactions can only occur at frequencies $v_{\ell m}$ given by the Bohr formula

$$v_{\ell m} = (E_m - E_\ell)/h$$

where h is Planck's constant, and E_m and E_ℓ are the internal energies of the upper and lower molecular states, respectively, which are involved in the transition. The theoretical problem of obtaining the resonance frequencies $v_{\ell m}$ of a molecule amounts to finding the quantized energy levels for that molecule, and determining which transitions between the levels can interact with an electromagnetic field. Several references,[2-6] for example, are available which discuss this problem and molecular spectroscopy in general, and the interested reader is referred to them.

The general expression for the absorption coefficient $(k_v)_{\ell m}$ between states ℓ and m may be written as

$$(k_v)_{\ell m} = \frac{8\pi^3 N v \mu^2}{3hcQ} \left\{ e^{-E_\ell/kT} - e^{-E_m/kT} \right\} g_\ell |\phi_{\ell m}|^2 f(v, v_{\ell m}). \quad (2.3.8)$$

Here N is the number of absorbing molecules per unit volume, μ the total dipole moment, g_ℓ the statistical weight of the lower state, $\phi_{\ell m}$ the transition matrix element, $f(v, v_{\ell m})$ a function describing the line shape, and Q the partition function.

For temperatures encountered in the atmosphere, the partition function is given accurately by the following approximate expressions:

---

[2] G. Herzberg, "Spectra of Diatomic Molecules." Van Nostrand-Reinhold, Princeton, New Jersey, 1950.
[3] G. Herzberg, "Infrared and Raman Spectra of Polyatomic Molecules." Van Nostrand-Reinhold, Princeton, New Jersey, 1945.
[4] C. H. Townes and A. L. Schawlow, "Microwave Spectroscopy." McGraw-Hill, New York, 1955.
[5] W. Gordy, W. V. Smith, and R. F. Trambarulo, "Microwave Spectroscopy." Dover, New York, 1966.
[6] W. Gordy and R. L. Cook, "Microwave Molecular Spectra." Wiley (Interscience), New York, 1970.

(a) for linear molecules

$$Q = \frac{kTG}{\sigma hB},\qquad(2.3.9a)$$

(b) for nonlinear molecules

$$Q = \frac{G}{\sigma}\left[\frac{\pi}{ABC}\left(\frac{kT}{h}\right)^3\right]^{1/2},\qquad(2.3.9b)$$

where σ is the symmetry number, depending on the point group of which the molecule is a member, and A, B, and C the rotational constants of the molecule. The common factor G is given by

$$G = \prod_i (2I_i + 1) \Big/ \prod_j [1 - \exp(-\omega_j hc/kT)]^{d_j},$$

where I_i is the nuclear spin of the ith atom in the molecule and ω_j and d_j, respectively, the frequencies and degeneracies of the jth normal mode of vibration. Here the products over i and j are taken, respectively, over all atoms in the molecule and all normal modes of vibration. A general discussion of these partition functions is given by Herzberg,[3] pp. 505–510.

We now consider the line-shape function $f(v, v_{\ell m})$. The mechanism that dominates spectral line broadening in the lower atmosphere is collisional interaction between molecules. Doppler, or thermal, broadening of microwave lines becomes important only at altitudes above approximately 70 km where the molecular collisional frequency is comparable to, or smaller than, the Doppler linewidth. Natural line broadening resulting from the finite lifetime of molecular states because of spontaneous transitions is of the order of 10^{-9} to 10^{-4} Hz for microwave electric-dipole transitions (and approximately 10^4 smaller for magnetic dipole transitions) and is completely negligible.

From a classical point of view, the effect of a collision on a radiating or absorbing molecule may be considered a disturbance of the interaction between the molecule and the radiation field. This disturbance tends to broaden the spectral distribution of the radiated or absorbed energy. For example, consider an extremely hypothetical situation in which a molecule radiates perfectly monochromatically, except that collisions alternately turn the radiation on and off. The collision frequency is assumed much smaller than the molecular radiation frequency. Examination of the spectrum of radiation in this hypothetical case shows that the collisions have smeared the "monochromatic" molecular radiation over a frequency range approximately equal to the collision frequency.

The mathematical description of collisional broadening of spectral lines can be shown to be formally similar to the description of the effect of collisions on a classical oscillator. The classical oscillator model also provides physical insights into the broadening process. A collision may affect an oscillator by changing its amplitude, phase, orientation, or momentum. Taking into account all these effects, Ben-Reuven[7,8] derived the line shape†

$$f(v, v_{\ell m}) = \frac{1}{\pi}\left(\frac{v}{v_{\ell m}}\right)\left\{\frac{2(\gamma - \zeta)v^2 + 2(\gamma\zeta)[(v + v_{\ell m} + \delta)^2 + \gamma^2 - \zeta^2]}{[(v_{\ell m} + \delta)^2 - v^2 + \gamma^2 - \zeta^2]^2 + 4v^2\gamma^2}\right\}, \quad (2.3.10)$$

where γ is a parameter describing the collisional damping of the oscillator, ζ a parameter describing the mean rate of momentum-reverting collisions, and δ a parameter that vanishes either if there is no average change in oscillator amplitude by collisions, or if all changes in phase or orientation are equally probable. In some cases these parameters may be calculated by integrating the classical equations of motion of the colliding molecules treated as classical particles. This calculation requires knowledge of the intermolecular potential.

If there are no momentum-reverting collisions, and if all phase changes induced by collisions are equally probable, then $\zeta = 0$ and $\delta = 0$, and, after some algebraic manipulation, (2.3.10) can be rewritten as

$$f(v, v_{\ell m}) = \frac{1}{\pi}\left(\frac{v}{v_{\ell m}}\right)\left\{\frac{\Delta v}{(v - v_{\ell m})^2 + \Delta v^2} + \frac{\Delta v}{(v + v_{\ell m})^2 + \Delta v^2}\right\}, \quad (2.3.11)$$

where $\Delta v = \gamma$. Equation (2.3.11) is the line shape derived first by Van Vleck and Weisskopf.[9] If there are only momentum-reverting collisions, then $\gamma = \zeta$, $\delta = 0$, and the Ben-Reuven shape becomes

$$f(v, v_{\ell m}) = \frac{1}{\pi}\frac{4vv_{\ell m}\Delta v}{(v_{\ell m}^2 - v^2)^2 + 4v^2\Delta v^2}, \quad (2.3.12)$$

where $\Delta v = \zeta$. Equation (2.3.12) is sometimes referred to as the "kinetic line shape," and was independently derived by Gross,[10] and by Zhevakin and

---

[7] A. Ben-Reuven, *Phys. Rev. Lett.* **14**, 349 (1965).
[8] A. Ben-Reuven, *Advan. Atom. Mol. Phys.* **5**, 201 (1969).
[9] J. H. Van Vleck and V. F. Weisskopf, *Rev. Mod. Phys.* **17**, 227 (1945).
[10] E. P. Gross, *Phys. Rev.* **97**, 395 (1955).

---

† There is a typographical error in Ben-Reuven's[8] Eq. (63); the second factor of $(\gamma - \zeta)$ in the numerator of that equation should be $(\gamma + \zeta)$. Equation (2.3.10) differs from Ben-Reuven's expression by the factor $v/v_{\ell m}$, due to the way in which the line shape is defined. This difference in definition may be seen by comparing (2.3.8) for the absorption coefficient with Ben Reuven's Eq. (17) when $hv/kT \ll 1$. In that case, the overall frequency dependence of Ben-Reuven's absorption coefficient expression is $k_v \propto v^2 F(v)$, where $F(v)$ is his line shape. Equation (2.3.9), however, has the frequency dependence $k_v \propto vv_{\ell m}f(v, v_{\ell m})$. The two expressions for the absorption coefficient must agree, which requires that $f(v, v_{\ell m}) = (v/v_{\ell m})F(v)$.

Naumov.[11] Both the Van Vleck–Weisskopf and the kinetic line shapes have been used for calculations of the atmospheric microwave spectrum, but the more general Ben-Reuven shape has not yet been applied to atmospheric problems. It should be noted that the line shapes given above make the impact approximation for collisions, which neglects any effects that occur within the duration of the time of collision.

The collisional linewidth parameter Δv which appears in (2.3.11) and (2.3.12) is a function of both temperature and pressure. Assuming that Δv is proportional to the frequency of collisions, and that the molecules are "hard" so that the collisional cross section is independent of velocity (and, therefore temperature) leads to $\Delta v \propto p T^{-1/2}$, where p is pressure. Collision cross sections for the broadening of microwave lines are, however, sometimes several times larger than the "hard" cross sections calculated from kinetic theory. This is seen to be feasible by considering the energies involved in the kinetic theory and in the microwave-line-broadening collisions.[4] A kinetic-theory collision supposes that the molecules come close enough for interactions involving most of the kT kinetic energy. Microwave lines are disturbed by much smaller energies $hv \ll kT$, and, therefore, affected by molecular separations much larger than affect kinetic-theory collisions. Because of this "long-range" effect, the collision cross section (and, consequently, the linewidth parameter) is not expected to be independent of the molecular velocity. A slow molecule spends more time near the radiating molecule, and causes more disturbance than one which passes more quickly. The temperature dependence of the linewidth will depend on the type of intermolecular force involved, and, for a force which varies with separation r as r^{-n}, the temperature dependence of the linewidth can be shown to be $\Delta v \propto T^{-(n+1)/2(n-1)}$. Typical intermolecular forces have $n = 3$ to $n = 6$, and the linewidth parameter can be written as

$$\Delta v = \Delta v^0 (p/p_0)(T/T_0)^{-x}, \qquad (2.3.13)$$

where x is usually between 0.7 and 1.0 and Δv^0 is the value of the linewidth parameter at $p = p_0$ and $T = T_0$. Values of Δv^0 and of x may be different for collisions involving different types of molecules, because different types of intermolecular forces may be involved. The preceding equation for the linewidth parameter is sometimes written more conveniently in terms of the number density N of molecules, rather than the pressure:

$$\Delta v = \Delta v^0 (N/N_0)(T/T_0)^{1-x}, \qquad (2.3.14)$$

where, again, Δv^0 is the value of Δv at $T = T_0$ and $N = N_0$.

[11] S. A. Zhevakin and A. P. Naumov, *Izv. Vyssh. Uchebn. Zaved.* [*Sov. Radiophys.*] **6**, 674 (1963).

2.3. ABSORPTION AND EMISSION BY ATMOSPHERIC GASES

A useful expression describes the absorption coefficient at the line center of a collisionally broadened spectral line. In this case the absorption coefficient can be written as

$$(k_\nu)_{\ell m} = \alpha(\nu_{\ell m}, T) N/N_{\text{air}}, \qquad (2.3.15)$$

where we use the linewidth given by (2.3.14) with $x = 1$, and we assume broadening dominated by collisions with "air" molecules so that the N in (2.3.14) refers to the molecular number density of air N_{air}. An "air molecule" is a fictitious molecule having line-broadening properties of the mixture of molecules in air. The factor $\alpha(\nu_{\ell m}, T)$, sometimes called the "maximum absorption coefficient," depends only on the spectral-line frequency and temperature, and is given by

$$\alpha(\nu_{\ell m}, T) = \frac{8\pi^2 g_\ell |\mu_{\ell m}|^2 \nu_{\ell m}^2 e^{-E_\ell/kT}}{3ckTQ} \left(\frac{N_0}{\Delta\nu^0}\right). \qquad (2.3.16)$$

It is worthwhile to note that at line center of a collisionally broadened spectral line the absorption coefficient is proportional to the mixing ratio N/N_{air} of the absorbing molecule and not to its absolute concentration.

At sufficiently low pressures where collisional broadening is small, Doppler broadening of the atmospheric lines must be considered. Because molecular velocities have a Gaussian distribution, the Doppler line shape is also Gaussian

$$f(\nu, \nu_{\ell m}) = \frac{1}{\Delta\nu_D} \left(\frac{\ln 2}{\pi}\right)^{1/2} \exp\left[-\ln 2\left(\frac{\nu - \nu_{\ell m}}{\Delta\nu_D}\right)^2\right], \qquad (2.3.17)$$

where the Doppler linewidth is given by

$$\Delta\nu_D = \frac{\nu_{\ell m}}{c}\left(\frac{2kT}{m}\ln 2\right)^{1/2} = 3.58 \times 10^{-7} \nu_{\ell m} \left(\frac{T}{M}\right)^{1/2}, \qquad (2.3.18)$$

and where m is the molecular mass in grams, M the molecular weight in grams per mole, and $\Delta\nu_D$ and $\nu_{\ell m}$ have the same units. At $T = 273°K$, Doppler widths are 0.03 MHz for the 22-GHz water-vapor line, 0.06 MHz for a 60-GHz oxygen line, and 0.09 MHz for a 100-GHz ozone line. These widths are of the order of collisionally broadened widths at pressures corresponding to altitudes of approximately 80 km in the terrestrial atmosphere. To account for simultaneous Doppler and collisional broadening, the resonance frequency in the collision-shape expression should be convolved with the Doppler shift probability distribution over all possible Doppler shifts.[12]

---

[12] B. H. Armstrong and R. W. Nicholls, "Emission, Absorption and Transfer of Radiation in Heated Atmospheres," pp. 218–229. Pergamon, Oxford. 1972.

So far, only the absorption coefficient due to a single transition has been considered, but in reality a quantum system usually has many allowed transitions. If there is no interaction between the various transitions, and the radiation from different transitions is incoherent, then the total absorption coefficient at any given frequency is simply the summation of the absorption coefficients for the individual transitions

$$k_v = \sum_{\substack{\text{all} \\ \text{transitions}}} (k_v)_{\ell m}. \qquad (2.3.19)$$

This is the expression which has, until now, been used almost exclusively for atmospheric microwave calculations.

When spectral lines significantly overlap, however, there may be coherent effects, and collisions may transfer intensity from one spectral line to another without broadening the lines. In this case, one is not justified in simply summing the absorption calculated for individual transitions. In order to calculate the total absorption coefficient one needs a collisional transfer matrix[13] whose ijth element gives the amplitude in line j transferred from unit amplitude in line i before the collision. Calculation of such a transfer matrix is a very tedious undertaking, at best, but in certain cases can be done for a given intermolecular potential by following classical collision trajectories, and averaging over the initial conditions of the collisions. When the collisional transfer between lines is sufficiently rapid, the lines blend together into a band whose overall width is less than that expected by a superposition of individual lines having linewidth parameters determined at lower pressures where the individual lines are resolved. This "relative narrowing" of absorption bands with increasing pressure has been observed in the 60-GHz oxygen-absorption band. The problem of overlapping lines is an active research area, and the interested reader is referred to journal articles[8,13] for more information.

We shall now consider in detail the spectral line absorption by water vapor, oxygen, and ozone, and briefly discuss microwave lines of other atmospheric molecules.

2.3.3.1. Water Vapor. The microwave spectrum of water vapor H_2O is caused by electric dipole transitions between rotational states of the molecule. H_2O is a nonlinear molecule and an asymmetric rotor which means that no two of its three principal moments of inertia are equal, and its rotational spectrum has been extensively studied.[14–45] Rotational spectral lines of water

[13] R. G. Gordon, *J. Chem. Phys.* **45**, 1649 (1966).
[14] D. M. Dennison, *Rev. Mod. Phys.* **12**, 175 (1940).
[15] B. T. Darling and D. M. Dennison, *Phys. Rev.* **57**, 128 (1940).
[16] G. E. Becker and S. H. Autler, *Phys. Rev.* **70**, 300 (1946).
[17] R. L. Kyle, R. H. Dicke, and R. Beringer, *Phys. Rev.* **69**, 694 (1946).
[18] C. H. Townes and F. R. Merritt, *Phys. Rev.* **70**, 558 (1946).

2.3. ABSORPTION AND EMISSION BY ATMOSPHERIC GASES 151

vapor are distributed somewhat randomly from microwave through far-infrared frequencies; the ten lowest frequency lines of $H_2{}^{16}O$, the dominant isotopic form, are given in Table I. The quantum numbers for the transitions producing the lines are also given in the J_{K_{-1},K_1} notation, where J is the quantum number of total rotational angular momentum and K_{-1}, K_1 are quantum numbers associated, respectively, with corresponding states of limiting prolate and limiting oblate symmetric rotors.[4]

The rotational constants[32] for $H_2{}^{16}O$ are $A = 835.7$ GHz, $B = 434.9$ GHz, and $C = 278.4$ GHz, and the normal vibrational frequencies[15] are $\omega_1 = 3693.8$ cm^{-1}, $\omega_2 = 1614.5$ cm^{-1}, and $\omega_3 = 3801.7$ cm^{-1}. $H_2{}^{16}O$ is a member of the C_{2v} point group,[3] which has symmetry number $\sigma = 2$. The two hydrogen atoms each have nuclear spin $I = \tfrac{1}{2}$, and the oxygen atom has $I = 0$. At temperatures encountered in the atmosphere the vibrational part of

---

[19] J. H. Van Vleck, *Phys. Rev.* **71**, 425 (1947).
[20] G. W. King, R. M. Hainer, and P. C. Cross, *Phys. Rev.* **71**, 433 (1947).
[21] W. S. Benedict, H. H. Claassen, and J. H. Shaw, *J. Res. Nat. Bur. Std.* **49**, 91 (1952).
[22] W. S. Benedict and L. D. Kaplan, *J. Chem. Phys.* **30**, 388 (1959).
[23] S. A. Zhevakin and A. P. Naumov, *Izv. Vyssh. Uchebn. Zaved. Radiofiz.* [*Sov. Radiophys.*] **6**, 674 (1963).
[24] W. S. Benedict and L. D. Kaplan, *J. Quant. Spectrosc. Radiat. Transfer* **4**, 453 (1964).
[25] J. R. Rusk, *J. Chem. Phys.* **42**, 493 (1965).
[26] V. Ya. Ryadov and N. I. Furashov, *Izv. Vyssh. Uchebn. Zaved. Radiofiz.* [*Sov. Radiophys.*] **9**, 1073 (1966); *Opt. Spektrosk.* **24**, 186 (1968).
[27] M. Lichtenstein, V. E. Derr, and J. J. Gallagher, *J. Mol. Spectrosc.* **20**, 391 (1966).
[28] L. Frenkel and D. Woods, *Proc. IEEE* **54**, 498 (1966).
[29] H. Bluyssen, A. Dymanus, and J. Verhoeven, *Phys. Lett.* **24A**, 482 (1967).
[30] S. A. Zhevakin and A. P. Naumov, *Radio Eng. Electron. Phys.* **12**, 1067 (1967).
[31] J. T. Hall, *Appl. Opt.* **6**, 1391 (1967).
[32] R. T. Hall and J. M. Dowling, *J. Chem. Phys.* **47**, 2454 (1967).
[33] L. Frenkel, *J. Mol. Spectrosc.* **26**, 227 (1968).
[34] N. E. Gaut, MIT RLE Tech. Rep. 467 (1968).
[35] D. E. Burch, *J. Opt. Soc. Amer.* **58**, 1383 (1968).
[36] H. J. Liebe, M. C. Thompson, and T. A. Dillon, *J. Quant. Spectrosc. Radiat. Transfer* **9**, 31 (1969).
[37] W. J. Burroughs, R. G. Jones, and H. A. Gebbie, *J. Quant. Spectrosc. Radiat. Transfer* **9**, 809 (1969).
[38] T. W. Whaley and B. M. Fannin, *IEEE Trans. Antennas Propagat.* **AP-17**, 682 (1969).
[39] C. O. Hemmi and A. W. Straiton, *Radio Sci.* **4**, 9 (1969).
[40] H. J. Liebe and T. A. Dillon, *J. Chem. Phys.* **50**, 727 (1969).
[41] P. E. Fraley and K. N. Rao, *J. Mol. Spectrosc.* **29**, 312 (1969).
[42] T. A. Dillon and H. J. Liebe, *J. Quant. Spectrosc. Radiat. Transfer* **11**, 1803 (1971).
[43] N. E. Gaut and E. C. Reifenstein III, Environmental Res. and Tech. Rep. No. 13, Lexington, Massachusetts (1971).
[44] Yu. A. Dryagin, A. G. Kislyakov, L. M. Kukin, A. I. Naumov, and L. I. Fedoseev, *Izv. Vyssh. Uchebn. Zaved. Radiofiz.* [*Sov. Radiophys.*] **9**, 1078 (1966).
[45] R. Emery, *Infrared Phys.* **12**, 65 (1972).

the partition function differs from unity by less than 4×10^{-4} and can be neglected. Gaut's[34] direct evaluation of the partition function for this case gives

$$Q = 172.4(T/293)^{3/2},$$

which agrees to within 2% with the general expression in Eq. (2.3.9b). The water-vapor molecule has an electric dipole moment[27] of 1.88×10^{-18} esu cm. Using the theory outlined previously, we now consider calculation of the water-vapor attenuation coefficient k_v by summing expression (2.3.8) over the appropriate set of transitions. For the frequency range up to 300 GHz it is sufficiently accurate to make the summation over the ten lowest frequency transitions listed in Table I. These transitions are well separated in frequency so that no coherence effects are expected. With the specific information in the previous paragraphs we find the following expression for k_v in units of reciprocal centimeters:

$$k_v = 1.44 \rho v T^{-3/2} \times \sum_{\substack{\text{all} \\ \text{transitions}}} \{e^{-E_\ell/kT} - e^{-E_m/kT}\} g_\ell |\phi_{\ell m}|^2 f(v, v_{\ell m}). \tag{2.3.20}$$

Here the water-vapor concentration is expressed in terms of density (grams per meter cubed) and the other units are v (gigahertz), f (reciprocal gigahertz), T (degrees Kelvin), and the transition parameters are as given in Table I. Several workers[11,31,44] have found better agreement between calculated and measured water-vapor absorption by using the kinetic line shape rather than the Van Vleck–Weisskopf shape. There is still, however, a discrepancy in the window regions of the spectrum, in that the calculated values are generally less than those measured. This behavior was evident when Van Vleck[19] compared his initial calculations of water-vapor absorption with the early measurements of Becker and Autler.[16] Using the kinetic line shape with linewidth parameters for transition ℓm given by

$$\Delta v_{\ell m} = \Delta v_{\ell m}^0 \left(\frac{p}{1013}\right)\left(\frac{T}{300}\right)^{-x} \left[1 + 4.6 \times 10^{-3} \frac{\rho T}{p}\left(\frac{\Delta v_{\ell m}(H_2O)}{\Delta v_{\ell m}^0} - 1\right)\right], \tag{2.3.21}$$

where ρ is the water-vapor density in grams per meter cubed, p the total pressure in millibars and $\Delta v_{\ell m}^0$, $\Delta v_{\ell m}(H_2O)$ and x for the lower transitions are given in Table I, Gaut and Reifenstein[43] found that the discrepancy between calculated and measured absorption for frequencies below 1000 GHz could be fitted to within 10% accuracy by an additive correction term proportional to the square of the frequency. The measured discrepancy and fitted correction term are shown in Fig. 2. Frenkel[33] has found similar behavior, and has also

TABLE I. Lower-Frequency Spectral Lines of H_2O^{16}

Frequencies[a] $\nu_{\ell m}$ (GHz)	Transitions[b] Upper state			Lower state			g_ℓ	$|\phi_{\ell m}|^2$	Energy levels[c] E_m (cm^{-1})	E_ℓ (cm^{-1})	Linewidth parameters[d] $\Delta\nu^0_{\ell m}$ (GHz)	$\Delta\nu_{\ell m}(H_2O)$ (GHz)	x
	J	K_{-1}	K_1	J	K_{-1}	K_1							
22.23515[e]	6	1	6	5	2	3	3	.0549	447.30	446.56	2.85[f]	13.68[f]	.626
183.31012[g]	3	1	3	2	2	0	1	.1015	142.27	136.16	2.68[g]	14.49[g]	.649
(323.)	10	2	9	9	3	6	3	.0870	1293.80	1283.02	2.30	12.04	.420
325.1538[h]	5	1	5	4	2	2	1	.0891	326.62	315.78	3.03[i]	15.21	.619
380.1968[h]	4	1	4	3	2	1	3	.1224	224.84	212.16	3.19[i]	15.84	.630
(390.)	10	3	7	11	2	10	1	(.0680)	1538.31	1525.31	2.11	11.42	.330
(436.)	7	5	3	6	6	0	1	(.0820)	1059.63	1045.03	1.50	7.94	.290
(438.)	6	4	3	5	5	0	3	.0987	756.76	742.11	1.94	(10.44)	(.360)
(442.)	7	5	2	6	6	1	3	(.0820)	1059.90	1045.11	1.51	8.13	.332
448.0008[h]	4	2	3	3	3	0	3	.1316	300.37	285.42	2.47	14.24	.510

[a] Frequency values in parentheses are calculated by S. N. Ghosh and H. D. Edwards [U.S. Air Force Survey in Geophysics, No. 82, AFCRL, Bedford, Massachusetts (1956)], and should be accurate to ~4 GHz, as determined by comparison of their calculated values with the measured frequencies.

[b] Matrix elements $|\phi_{\ell m}|^2$ are from G. W. King et al. [*Phys. Rev.* **71**, 433 (1947)], except for values in parentheses which were estimated by N. E. Gaut [M.I.T. Res. Lab. Electron. Tech. Rep. 467 (1968)] from computations given in *Microwave Spectral Tables* [U.S. Dept. Commerce, NBS Monograph 70].

[c] From R. T. Hall and J. M. Dowling [*J. Chem. Phys.* **47**, 2454 (1967)], and D. M. Dennison [*Rev. Mod. Phys.* **12**, 175 (1940)]. The energy levels in Table I are in wavenumber units of reciprocal centimeters and should be multiplied by Planck's constant for use in (2.3.20).

[d] Except where noted linewidths $\Delta\nu_{\ell m}$ and temperature exponents x are from W. S. Benedict and L. D. Kaplan [*J. Chem. Phys.* **30**, 388 (1959); *J. Quant. Spectrosc. Radiat. Transfer* **4**, 453 (1964)], with values of $\Delta\nu^0_{\ell m}$ computed for nitrogen broadening only. Values in parentheses were estimated by N. E. Gaut [M.I.T. Res. Lab. of Electron. Tech. Rep. 467 (1968)] based on averages of values for similarly arising lines.

[e] H. J. Liebe et al., *J. Quant. Spectrosc. Radiat. Transfer* **9**, 31 (1969). Hyperfine components of this line have been measured by H. Bluyssen et al., *Phys. Lett.* **24A**, 482 (1967).

[f] H. J. Liebe and T. A. Dillon, *J. Chem. Phys.* **50**, 727 (1969).

[g] J. R. Rusk, *J. Chem. Phys.* **42**, 493 (1965).

[h] M. Lichtenstein, V. E. Derr, and J. J. Gallagher, *J. Mol. Spectrosc.* **20**, 391 (1966).

[i] V. Ya. Ryadov, N. I. Furashov, *Izv Vyssh. Uchebn. Zaved. Radiofiz.* [*Sov. Radiophys.*] **9**, 1073 (1966); *Opt. Spektrosk.* **24**, 186 (1968).

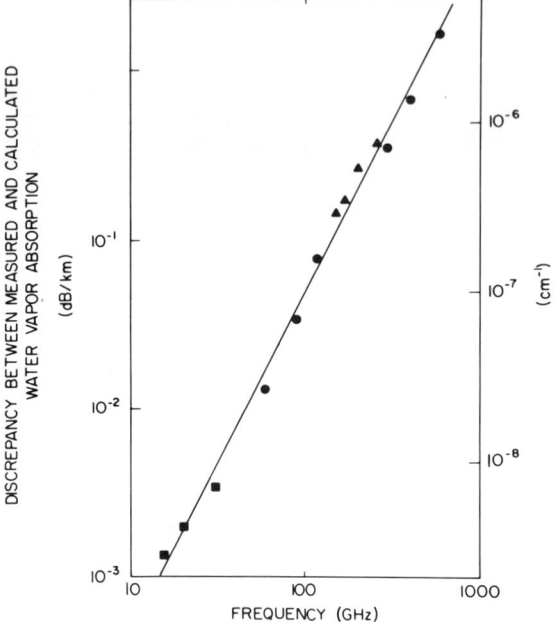

FIG. 2. Discrepancy between measured and calculated water-vapor absorption, and the empirical correction term (solid line) given by Eq. (2.3.22) [from N. E. Gaut and E. C. Reifenstein III, Environmental Res. and Tech. Rep. No. 13, Lexington, Massachusetts (1971)]. The plotted points are (■) from G. E. Becker and S. H. Autler [*Phys. Rev.* **70**, 300 (1946)], (▲) from L. Frenkel and D. Woods [*Proc. IEEE* **54**, 498 (1966)], and (●) from D. E. Burch [*J. Opt. Soc. Amer.* **58**, 1383 (1968)], for $T = 300°K$, $p = 1000$ mbar, and $\rho = 1$ gm/m³.

examined the discrepancy at higher frequencies. Sufficient experimental data are not available to determine completely how the discrepancy varies with water-vapor density, temperature, and pressure. As a working assumption for the empirical correction, Gaut and Reifenstein use

$$\Delta k_v = 1.08 \times 10^{-11} \rho \left(\frac{300}{T}\right)^{2.1} \left(\frac{p}{1000}\right) v^2 \quad \text{cm}^{-1}, \qquad (2.3.22)$$

where the quantities are the same as in (2.3.21). Figure 3 shows the absorption coefficient calculated for water vapor from the preceding expressions, both with and without the empirical correction. Also shown are measured values. As can be seen, the calculation with the empirical correction agrees fairly well with the measurements.

For frequencies below approximately 100 GHz, the water-vapor absorption coefficient expression can be simplified by grouping together in a single term

2.3. ABSORPTION AND EMISSION BY ATMOSPHERIC GASES

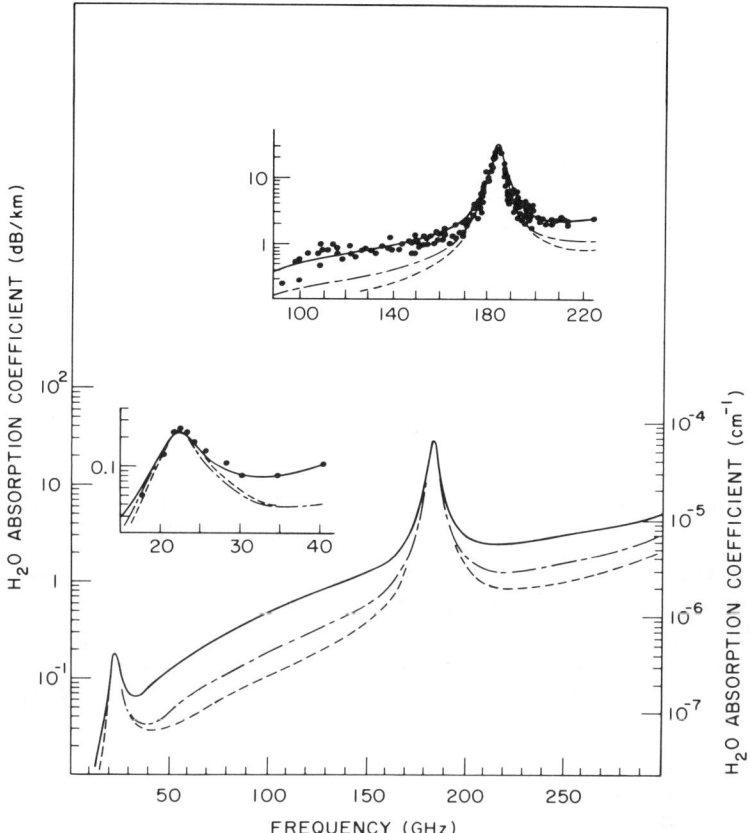

FIG. 3. Measured and calculated water-vapor absorption. Calculations are shown for the Van Vleck–Weisskopf line shape (- - -), the kinetic line shape (— - —) and the kinetic line shape with the added empirical correction discussed in the text (—), and where $T = 300°K$, $p = 1013$ mbar, and $\rho = 7.5$ gm/m³. Points in the 20–40 GHz inset are measurements of G. E. Becker and S. H. Autler [*Phys. Rev.* **70**, 300 (1946)], where $T = 318°K$, $p = 1013$ mbar, and $\rho = 10$ gm/m³. Points in the 100–220 GHz inset are measurements quoted by Yu. A. Dryagin, A. G. Kislyakov, L. M. Kukin, A. I. Naumov, and L. I. Fedoseev [*Izv. Vyssh. Uchebn. Zaved. Radiofiz.* (*Sov. Radiophys.*) **9**, 1078, (1966)], where $T = 300°K$, $p = 1013$ mbar, and $\rho = 7.5$ gm/m³.

the contributions from lines other than the one at 22 GHz, and can be written

$$k_\nu = \rho \nu^2 \, \Delta \nu_1 \, T^{-3/2} \left\{ \frac{7.18 e^{-644/T}}{T} \frac{1}{(494.40190 - \nu^2)^2 + 4\nu^2 \, \Delta \nu_1^2} \right.$$

$$\left. + 2.77 \times 10^{-8} \right\} \quad \text{cm}^{-1}. \qquad (2.3.23\text{a})$$

where

$$\Delta v_1 = 2.96 \left(\frac{p}{1013}\right)\left(\frac{300}{T}\right)^{0.626} \left(1 + 0.018 \frac{\rho T}{p}\right) \quad \text{GHz}, \quad (2.3.23\text{b})$$

and ρ is in grams per meter cubed, T in degrees Kelvin, p in mbars, and v in gigahertz. The empirical correction term has been included in the additive constant within the braces in Eq. (2.3.23a). Equations (2.3.23) give values which agree to better than 5% for $v \leq 100$ GHz, $270 \leq T \leq 320°$K, and $300 \leq p \leq 1050$ mbar, with computations of Gaut and Reifenstein[43] which included the empirical correction term.

The fact that the excess measured absorption varies as the square of the frequency suggests that it is due at least in part to the impact assumption used in the derivation of the collision line shapes. The absorption coefficient can be shown to be proportional to the product of the square of the frequency and the Fourier transform of the autocorrelation function of the dipole moment of the molecule.[8] Figure 2 can then be interpreted as indicating that the excessive absorption arises from a dipole moment whose Fourier transform is constant over a frequency range greater than 10^{12} sec^{-1}. This implies that an excessive dipole moment exists for a time shorter than 10^{-12} sec, which is typical of the duration time of a collision. The impact approximation, however, assumes instantaneous collisions. That this approximation leads to the excess measured absorption was suggested by Van Vleck[19] in his initial calculations, and has been discussed in more detail by Frenkel.[33]

There are other strong indications that some of the excess measured water-vapor absorption is, at least partially, due to water-vapor dimers $(H_2O)_2$. Dimers have been observed in water vapor by mass spectroscopy,[46] and, in 1967, Viktorova and Zhevakin[47] proposed absorption by dimers as the explanation of the excessive water-vapor absorption. They predicted a dimer spectral line at approximately 210 GHz (and other lines at higher frequencies), and calculated that the dimer absorption exceeded that of the monomer at microwave frequencies away from the monomer lines. More recent calculations by the same authors,[48,49] which include internal rotation of the dimer and a reduced binding energy, give many more dimer lines blended together in the microwave range and an overall absorption that is lower by a factor of approximately four than that given by their earlier calculations.

Observations of absorption at frequencies near that of the dimer line predicted by Viktorova and Zhevakin have since been reported in solar

[46] R. E. Leckenby and E. J. Robbins, *Proc. Roy. Soc. Lond.* **A291**, 389 (1966).
[47] A. A. Viktorova and S. A. Zhevakin, *Sov. Phys.-Dokl.* **11**, 1065 (1967).
[48] A. A. Viktorova and S. A. Zhevakin, *Sov. Phys.-Dokl.* **15**, 836 (1971).
[49] A. A. Viktorova and S. A. Zhevakin, *Sov. Phys.-Dokl.* **15**, 852 (1971).

2.3. ABSORPTION AND EMISSION BY ATMOSPHERIC GASES 157

absorption measurements and in the laboratory.[50-58] The laboratory measurements[53,54] show that the strength of this feature varies as the square of the water vapor concentration, as expected for dimers since their concentration is expected to be proportional to the number of monomer collisions and, therefore, to the square of the monomer concentration. The laboratory measurements[54] also indicate that the dimer absorption accounts for about half the observed "excess" absorption at a frequency of approximately 900 GHz. Recent solar absorption measurements[58] show that the feature attributed to the dimer actually is two features at approximately 230 and 260 GHz, respectively, which are very broad and merged together. This atmospheric measurement is not inconsistent with laboratory measurements[45] in the 220–225-GHz range, which gave a negative result for the dimer feature. The excess absorption by H_2O cannot, however, be said to be satisfactorily understood, and more experimental and theoretical work is needed.

2.3.3.2. Oxygen. The oxygen molecule has no electric dipole moment, and, consequently, no spectral lines caused by electric dipole transitions. It does, however, have a magnetic dipole moment arising from the combined spins of two unpaired electrons in the $^3\Sigma_g^-$ electronic ground state of the molecule, and magnetic dipole transitions give rise to spectral lines in the microwave frequency range which have been studied by many workers.[59-82] Changes in

[50] H. A. Gebbie and W. J. Burroughs, *Nature (London)* **217**, 1241 (1968).
[51] H. A. Gebbie, J. Chamberlain, and W. J. Burroughs, *Nature (London)* **220**, 893 (1968).
[52] W. J. Burroughs, R. G. Jones, and H. A. Gebbie, *J. Quant. Spectrosc. Radiat. Transfer* **9**, 809 (1969).
[53] J. E. Harries, W. J. Burroughs, and H. A. Gebbie, *J. Quant. Spectrosc. Radiat. Transfer* **9**, 799 (1969).
[54] H. A. Gebbie, W. J. Burroughs, J. Chamberlain, J. E. Harries, and R. G. Jones, *Nature (London)* **221**, 143 (1969).
[55] J. E. Harries and W. J. Burroughs, *Infrared Phys.* **10**, 165 (1970).
[56] R. A. Bohlander, H. A. Gebbie, and G. W. F. Pardoe, *Nature (London)* **228**, 156 (1970).
[57] I. G. Nolt, T. Z. Martin, C. W. Wood, and W. M. Sinton, *J. Atmos. Sci.* **28**, 238 (1971).
[58] J. E. Harries and P. A. R. Ade, *Infrared Phys.* **12**, 81 (1972).
[59] J. H. Van Vleck, *Phys. Rev.* **71**, 413 (1947).
[60] B. V. Gokhale and M. W. P. Strandberg, *Phys. Rev.* **84**, 844 (1951).
[61] R. Beringer and J. G. Castle, Jr., *Phys. Rev.* **81**, 82 (1951).
[62] R. S. Anderson, W. V. Smith, and W. Gordy, *Phys. Rev.* **87**, 561 (1952).
[63] J. O. Artman and J. P. Gordon, *Phys. Rev.* **96**, 1237 (1954).
[64] R. M. Hill and W. Gordy, *Phys. Rev.* **93**, 1019 (1954).
[65] M. Mizushima and R. M. Hill, *Phys. Rev.* **93**, 745 (1954).
[66] M. Tinkham and M. W. P. Strandberg, *Phys. Rev.* **97**, 937 (1955).
[67] M. Tinkham and M. W. P. Strandberg, *Phys. Rev.* **97**, 951 (1955).
[68] M. Tinkham and M. W. P. Strandberg, *Phys. Rev.* **99**, 537 (1955).
[69] R. W. Zimmerer and M. Mizushima, *Phys. Rev.* **121**, 152 (1961).
[70] A. E. Schulze and C. W. Tolbert, *Nature (London)* **200**, 747 (1963).

the orientation of the electronic spin, relative to the orientation of the molecular rotation, produce a band of spin–rotation spectral lines near 60 GHz and a single line at 118 GHz. Absorption by these transitions was first calculated by Van Vleck.[59] Transitions between different rotational states of O_2 produce spectral lines at frequencies of 367 GHz and higher.[74,75] Even though magnetic dipole transitions are typically 10^4 less intense than electric dipole transitions, the O_2 transitions produce quite strong atmospheric absorption, because of the large abundance of oxygen in the atmosphere.

The dominant isotopic species of molecular oxygen $^{16}O_2$ has total nuclear spin $I = 0$, which requires the complete wave function for the molecule to be symmetric in nuclear coordinates. The electronic wave function for the $^3\Sigma_g^-$ ground state of the molecule is antisymmetric in nuclear coordinates, and, therefore, the remaining wave function must also be antisymmetric in order to satisfy the overall symmetry requirements. Rotational wave functions that satisfy the overall symmetry requirements are those for which the rotational quantum number N is odd.[2] Consequently, only rotational states corresponding to $N = 1, 3, 5, \ldots$ are allowed for $^{16}O_2$. The electronic spin quantum number for the ground electronic state is $S = 1$, and the spin can have three allowed orientations with respect to the rotation. The total angular momentum quantum number J can then have the values $J = N - 1$, N, $N + 1$, and the selection rules on J allow the transition $\Delta J = \pm 1, 0$.

The rotational constant[76,77] for $^{16}O_2$ is $B = 43.1$ GHz, and its normal vibrational frequency[4] is $\omega_1 = 1580.36$ cm^{-1}. It is a member of the $D_{\infty h}$ point group[2] for which the symmetry number is $\sigma = 2$, and the zero spin of the nuclei give $\prod_i (2I_i + 1) = 1$. Under conditions in the atmosphere the vibration factor in the partition function reduces to a single factor whose value at $T = 300°$K differs from unity by 5×10^{-4}, and can therefore be replaced by unity. There is an additional factor of three required in the parti-

---

[71] L. F. Stafford and C. W. Tolbert, *J. Geophys. Res.* **68**, 3431 (1963).
[72] C. A. Zhevakin and A. P. Naumov, *Radio Eng. Electron. Phys.* **10**, 844 (1965).
[73] B. G. West and M. Mizushima, *Phys. Rev.* **143**, 31 (1966).
[74] J. S. McKnight and W. Gordy, *Phys. Rev. Lett.* **21**, 1787 (1968).
[75] H. A. Gebbie, W. J. Burroughs, and G. R. Bird, *Proc. Roy. Soc. London* **A310**, 579 (1969).
[76] T. T. Wilheit, Jr. and A. H. Barrett, *Phys. Rev.* **A1**, 213 (1970).
[77] W. M. Welch and M. Mizushima, *Phys. Rev.* **A5**, 2692 (1972).
[78] M. Mizushima, J. S. Wells, K. M. Evenson, and W. M. Welch, *Phys. Rev. Lett.* **29**, 831 (1972).
[79] R. G. Gordon, *J. Chem. Phys.* **46**, 448 (1967).
[80] U. Mingelgrin, U. S. Dept. Commerce, Telecommun. Res. Eng. Rep. 32 (1972).
[81] T. A. Dillon and J. T. Godfrey, *Phys. Rev.* **A5**, 599 (1972).
[82] H. J. Liebe and W. M. Welch, U.S. Dept. of Commerce, Office Telecommun. Rep. 73-10, 1973.

2.3. ABSORPTION AND EMISSION BY ATMOSPHERIC GASES

tion function because of the three possible spin orientations for each rotational state. The partition function Q for O_2 from Eq. (2.3.9a) may be written as

$$Q = \frac{3}{2}\frac{kT}{hB} = 212.5\left(\frac{T}{293}\right), \tag{2.3.24}$$

where the numerical factor has been evaluated from the constants given previously. With sufficient accuracy for calculating the Boltzmann factor, the energy for rotational state N is $E_N = hBN(N+1)$, which leads to

$$\frac{E_N}{kT} = 2.07\frac{N(N+1)}{T}. \tag{2.3.25}$$

The magnetic dipole moment for $^{16}O_2$ is two Bohr magnetons,[59] and has the value $\mu = 1.85 \times 10^{-20}$ erg/G.

The transitions $\Delta J = \pm 1$ produce spectral lines at microwave frequencies. In these transitions, the state $J = N$ is the state of higher energy; transitions to the state N are *absorption* transitions, whereas transitions from the state N are *emission* transitions. The respective frequencies are usually labeled N_+ or N_-, where the subscript indicates a transition $(J = N) \leftrightarrow (J = N \pm 1)$, and are given in Table II. The angular momenta associated with the electronic spin and molecular rotation of $^{16}O_2$ are coupled according to Hund's case (b),[4] and the transition matrix elements can be evaluated from the general expressions for this type of coupling.[59] The matrix elements, including the degeneracy factor $g_J = 2J + 1$, are

$$\Delta J = -1: \quad g_{N_-}|\phi_{N_-}|^2 = \frac{(N+1)(2N-1)}{N}, \tag{2.3.26a}$$

$$\Delta J = +1: \quad g_{N_+}|\phi_{N_+}|^2 = \frac{N(2N+3)}{N+1}. \tag{2.3.26b}$$

In addition to the microwave spectral line absorption caused by the transitions $\Delta J = \pm 1$, there is "nonresonant" absorption by the oxygen molecule because of nonvanishing matrix elements for the $\Delta J = 0$ transition. These matrix elements[59] are

$$\Delta J = 0: \quad g_{N_0}|\phi_{N_0}|^2 = \frac{2(N^2+N+1)(2N+1)}{N(N+1)}. \tag{2.3.26c}$$

The expression derived by Van Vleck[59] for the nonresonant absorption is

$$(k_\nu)_{N_0} = \frac{4\pi^3 N\mu^2 \nu}{3ckTQ} g_{N_0}|\phi_{N_0}|^2 \exp(-E_N/kT)F(\nu), \tag{2.3.27}$$

where

$$F(\nu) = \lim_{\nu_{\ell m} \to 0} [\nu_{\ell m} f_{\text{vvw}}(\nu, \nu_{\ell m})] = \frac{2\nu\,\Delta\nu}{\pi(\nu^2 + \Delta\nu^2)} \tag{2.3.28}$$

TABLE II. Microwave Transition Lines of $^{16}O_2$

N	ν_{N-} (GHz)^a	ν_{N+} (GHz)^a
1	118.750343^b	56.264766^d
3	62.486255^c	58.446580^c
5	60.306044^c	59.590978^c
7	59.164215^c	60.434776^c
9	58.323885^c	61.150570^c
11	(57.612488)^d	61.800169^d
13	56.968180^d	62.411223^d
15	56.363393^d	(62.997991)^d
17	55.783819^d	63.568520^c
19	55.221372^d	64.127777^d
21	54.671145^d	(64.678914)^d
23	54.1302^e	65.22412^c
25	53.5959^e	65.764744^d
27	53.0669^e	(66.30206)^f
29	52.5424^e	(66.83677)^f
31	52.0214^e	(67.36951)^f
33	(51.50302)^f	(67.90073)^f
35	(50.9873)^f	(68.4308)^f
37	(50.4736)^f	(68.9601)^f
39	(49.9618)^f	(69.4887)^f

^{a} Frequencies in parenthesis are calculated rather than measured.
^{b} J. S. McKnight and W. Gordy, *Phys. Rev. Lett.* **21**, 1787 (1968).
^{c} R. W. Zimmerer and M. Mizushima, *Phys. Rev.* **121**, 152 (1961).
^{d} B. G. West and M. Mizushima, *Phys. Rev.* **143**, 31 (1966).
^{e} J. W. Waters, *Nature* (London), **242**, 506 (1973).
^{f} H. J. Liebe and W. M. Welch, U.S. Dept. of Commerce, Office Telecommun. Rep. 73-10 (1973).

and $f_{vvw}(\nu, \nu_{\ell m})$ is the Van Vleck–Weisskopf line shape. This description of the nonresonant absorption has the same form as that developed by Debye, and, for a reasonable choice of $\Delta \nu$, gives values in agreement with measurement. Its derivation, however, has been criticized.[72]

At pressures higher than approximately 100 mbar, corresponding to altitudes in the terrestrial atmosphere below approximately 16 km, individual spin–rotation lines of O_2 have significant overlapping. Gordon[79] has theoretically studied the problem of overlapping spin–rotation lines in pure O_2,

2.3. ABSORPTION AND EMISSION BY ATMOSPHERIC GASES

and his calculations of absorption at several atmospheres pressure give good agreement with measured values, whereas calculations which simply sum the absorption by individual lines do not give agreement. The calculations which include overlapping effects are quite complicated, and have not yet been applied to atmospheric examples where the situation is complicated by collisions with several types of molecules. Gordon's theory has been used[80] to calculate O_2 absorption for atmospheric pressures and collisions with argon molecules, and the results predict the relative narrowing of the 60-GHz band which has been observed in the atmospheric spectrum. By "relative narrowing," we mean that the overall width of the band is less than that expected by summing the absorption from individual lines that have linewidth parameters extrapolated from values measured at lower pressures where the individual lines are resolved.

The relative narrowing of the 60-GHz O_2-band was observed before overlapping line interactions were considered as an explanation of the effect.[63,83,84] It has been common practice to fit the measured absorption to the absorption calculated by summing that of the individual transitions with the linewidth parameter made an empirical function of pressure, or of altitude, in the atmosphere.[85-92] Although this procedure is unsatisfying from a theoretical point of view, it leads to values that are in fair agreement with measurement, and it greatly simplifies the calculation.†

Agreement with atmospheric opacity measurements made at various altitudes below 12 km was obtained by Reber[92] to within approximately 8% rms discrepancy in the 48–72-GHz frequency range by summing the individual transitions and using the kinetic line shape with the empirical linewidth parameter given by

$$\Delta v = \begin{cases} \Delta v_1 = 1.41 \times 10^{-3} p(300/T) & \text{for} \quad \Delta v \leq \Delta v_B \quad (2.3.29\text{a}) \\ \tfrac{2}{3} \Delta v_B + \tfrac{1}{3} \Delta v_1 & \text{for} \quad \Delta v \geq \Delta v_B, \quad (2.3.29\text{b}) \end{cases}$$

[83] A. B. Crawford and D. C. Hogg, *Bell Syst. Tech. J.* **35**, 907 (1956).
[84] C. W. Tolbert and A. W. Straiton, *IRE Trans. Antennas Propagat.* **AP-5**, 239 (1957).
[85] M. L. Meeks, *J. Geophys. Res.* **66**, 3749 (1961).
[86] M. L. Meeks and A. E. Lilley, *J. Geophys. Res.* **68**, 1683 (1963).
[87] A. H. Barrett, J. W. Kuiper, and W. B. Lenoir, *J. Geophys. Res.* **71**, 4723 (1966).
[88] C. J. Carter, R. L. Mitchell, and E. E. Reber, *J. Geophys. Res.* **73**, 3113 (1968).
[89] E. E. Reber, R. L. Mitchell, and C. J. Carter, *IEEE Trans. Antennas Propagat.* **AP-18**, 472 (1970).
[90] D. L. Croom, *Planet. Space Sci.* **19**, 777 (1971).
[91] V. J. Falcone, Jr., K. N. Wulfsberg, and S. Gitelson, *Radio Sci.* **6**, 347 (1971).
[92] E. E. Reber, *J. Geophys. Res.* **77**, 3831 (1972).
[92a] P. W. Rosenkranz, *IEEE Trans. Antennas Propaga* **AP-23**, 498 (1975).

† P. W. Rosenkranz has recently developed an expression for the oxygen absorption coefficient that includes first-order coherence effects in overlapping lines.[92a]

where Δv is in gigahertz, $\Delta v_B = 0.0527$ GHz, p the total atmospheric pressure in millibars, and T in degrees Kelvin. Variation in the linewidth parameter for different transitions[61-64] is neglected. A special case must be made for the 1_- line at 118 GHz, however. This line is isolated from the band of lines near 60 GHz, overlapping effects are not expected, and (2.3.29a), which gives the measured linewidth, should be used for all values of Δv.

By using this procedure to account empirically for line-overlapping effects, we can write the oxygen absorption coefficient (units of reciprocal centimeters) in the atmosphere as

$$k_v = 1.44 \times 10^{-5} p T^{-3} v \sum_{N \text{ odd}} \{g_{N_+} |\phi_{N_+}|^2 v_{N_+} f(v, v_{N_+})$$
$$+ g_{N_-} |\phi_{N_-}|^2 v_{N_-} f(v, v_{N_-}) + \tfrac{1}{2} g_{N_0} |\phi_{N_0}|^2 F(v)\} e^{-E_N/kT}, \qquad (2.3.30)$$

where (2.3.8) has been used and summed over all transitions, and the number density of oxygen molecules has been converted to total atmospheric pressure p (which adds a factor T^{-1}) using the atmospheric O_2 volume mixing ratio of 0.21. In (2.3.30), p is in millibars, T in degrees Kelvin, v in gigahertz, and the numerical factor has been evaluated from the constants given previously. The kinetic line shape factor $f(v, v_{N_\pm})$ is given by (2.3.12), and the nonresonant shape $F(v)$ is given by (2.3.28); the linewidth is given by (2.3.29) with the 1_- line treated as discussed in the text following that equation. For lack of a better expression, (2.3.29b) can be used for the linewidth in the nonresonant shape $F(v)$, as it gives values consistent with measurement. At $p = 1013$ mbar, (2.3.29b) gives $\Delta v = 0.51$ GHz, which compares with the value of 0.60 GHz used for the nonresonant linewidth in calculations by Zhevakin and Naumov.[72] The calculated O_2 absorption coefficient is given in Fig. 4 where it is compared with measured values.

The terrestrial magnetic field removes the $2J + 1$ degeneracy of the rotational states of O_2, and spreads the energy levels over a frequency range of ~ 1 MHz. In general, an $N_\pm$ line is split into a total of $3(2N \pm 1)$ Zeeman components in the presence of a magnetic field where the factor of three is due to the three allowed changes in the magnetic quantum number $M: \Delta M = \pm 1, 0$. Only at pressures below approximately 3 mbar, corresponding to altitudes above approximately 40 km in the terrestrial atmosphere, where the collisional broadening is of the order of, or smaller than, ~ 1 MHz, is the Zeeman splitting noticeable. The absorption and emission by the Zeeman components is both polarized and isotropic, and the scalar radiative transfer equation given by (2.3.4) is not an adequate description of the absorption and emission process. Lenoir[93] developed a matrix equation of radiative transfer for an anisotropic and polarized medium and evaluated the transition matrix

---

[93] W. B. Lenoir, *J. Geophys. Res.* **73**, 361 (1968).

2.3. ABSORPTION AND EMISSION BY ATMOSPHERIC GASES

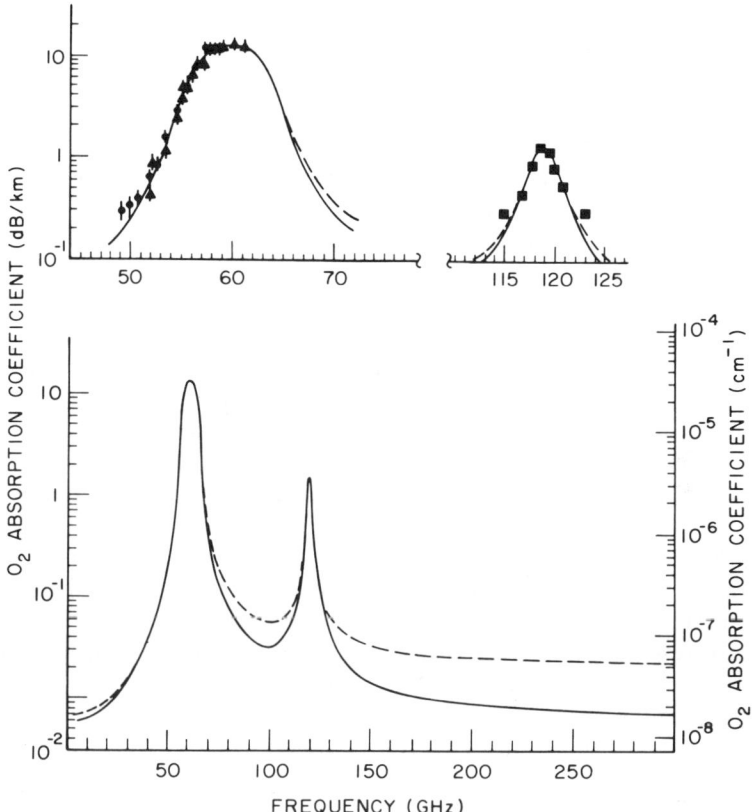

FIG. 4. Measured and calculated oxygen absorption. Calculations are shown for the kinetic line shape (—) with the empirical linewidth of E. E. Reber [*J. Geophys. Res.* **77**, 3831 (1972)] as discussed in the text, and for the Van Vleck–Weisskopf line shape (– – –) with the empirical linewidth expression given by A. H. Barrett, J. W. Kuiper, and W. B. Lenoir [*J. Geophys. Res.* **71**, 4723 (1966)], and where $T = 300°K$, $p = 1013$ mbar, and the O_2 volume mixing ratio is 0.21. Points in the 50–70 GHz inset are (▲), measurements of J. O. Artman and J. P. Gordon [*Phys. Rev.* **96**, 1237 (1954)], and (●), measurements of A. B. Crawford and D. C. Hogg [*Bell Syst. Tech. J.* **35**, 907 (1956)], while those in the 115–125 GHz inset are (■), measurements of A. E. Schulze and C. W. Tolbert [*Nature (London)* **200**, 747 (1963)].

elements and frequency separations of the Zeeman components. He also calculated mesospheric emission in the 60-GHz O_2-band, assuming no interaction between the overlapping Zeeman components. Using the scalar radiative transfer equation, Croom[90] also calculated mesospheric absorption and emission by the Zeeman components of the 118-GHz line.

2.3.3.3. Ozone. Before discussing the microwave spectrum of ozone O_3, we shall briefly summarize the features of ozone in the atmosphere.

Ozone is formed in the atmosphere by a three-body molecular collision in which one of the colliding molecules is atomic oxygen. Because atomic oxygen increases with altitude, whereas the number of molecules available for a collision decreases with altitude, the vertical distribution of ozone exhibits a maximum. The altitude of maximum ozone concentration is usually found between 20 and 30 km. Craig[94] gives a summary of the photochemical processes affecting ozone, and describes the techniques of measurements. Synoptic data are available for ozone up to altitudes of approximately 30 km, and the total amount is maximum in spring and minimum in autumn, with the largest variations at high latitudes. At 45° latitude the seasonal variation in total ozone is approximately $\pm 20\%$ of the mean value. Significant daily variations also occur, and values of total ozone in autumn may exceed values in spring for individual daily values. Variation also occurs in the vertical distribution, and profiles may show a single well-defined maximum, a broad flat maximum, or occasionally one or two secondary maxima. Above approximately 55 km, time constants for photochemical reactions are sufficiently small that a significant diurnal variation in ozone may exist, and there may be a nocturnal increase in ozone concentration above this altitude.[95] This is still a subject of research, however, as the reaction rates are not well known, and measurements at these altitudes are few. Profiles of ozone distribution with altitude are shown in Fig. 5.

The microwave spectrum of $^{16}O_3$, the dominant isotopic form of ozone, like that of H_2O, is due to electric dipole transitions between rotational states of the molecule. Unlike H_2O, however, O_3 has many transitions which give rise to frequencies in the microwave, as well as the submillimeter, frequency range. Also unlike H_2O, O_3 is only slightly asymmetric, being a nearly prolate symmetric rotor, and amenable to certain simplifying approximations in computing the rotational energies of the states. A thorough study of the rotational spectrum of ozone was made by Gora[97] who calculated values of frequencies and absorption coefficients for the rotational transitions. Many of the microwave lines of ozone have been measured recently.[98]

The stronger ozone lines below 300 GHz are listed in Table III. The values of the maximum absorption coefficients for the lines are from Gora, and were calculated for a temperature of 220°K, the approximate temperature of the

[94] R. A. Craig, "The Upper Atmosphere: Meteorology and Physics." Academic Press, New York, 1965.

[95] J. London, *Space Res.* **7**, 172 (1967).

[96] "Handbook of Geophysics and Space Environments." McGraw-Hill, New York, 1965.

[97] E. K. Gora, *J. Mol. Spectrosc.* **3**, 78 (1959).

[98] M. Lichtenstein, J. J. Gallagher, and S. A. Clough, *J. Mol. Spectrosc.* **40**, 10 (1971).

2.3. ABSORPTION AND EMISSION BY ATMOSPHERIC GASES

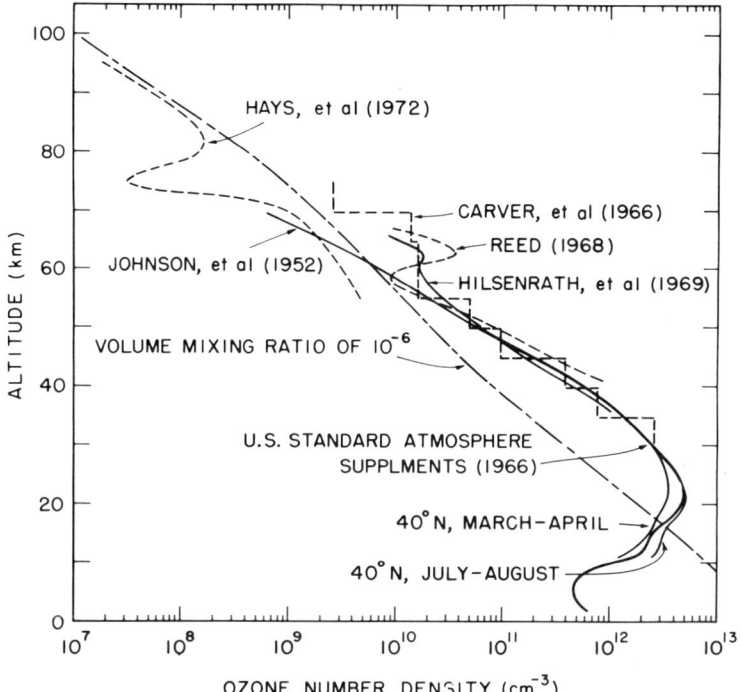

FIG. 5. Atmospheric ozone profiles. The 40°N distributions are from the "Handbook of Geophysics and Space Environments" [McGraw-Hill, New York, 1965], and other distributions are from "U.S. Standard Atmosphere Supplements, 1966" [U.S. Printing Office, Washington, D.C., 1966], F. S. Johnson, J. D. Purcell, R. Tousey, and K. Watanabe [*J. Geophys. Res.* **57**, 157 (1952)], J. H. Carver, B. H. Horton, and F. G. Burger [*ibid.* **71**, 4189 (1966)], E. I. Reed, *ibid.* **73**, 2951 (1968)], E. Hilsenrath, L. Seiden, and P. Goodman [*ibid.* **74**, 6873 (1969)], and P. B. Hays, R. G. Roble, and A. N. Shah [*Science* **176**, 793 (1972)].

ozone layer in the atmosphere. The value of the linewidth parameter used by Gora for the calculations is 2.31 MHz/mbar, which was inferred from measurements[99] in the 9.6-μ ozone band for an ozone–air environment at 293°K. Measurements[98] of resolved rotational transitions of ozone give a linewidth parameter of 3.48 MHz/mbar for pure ozone at 300°K. These two values of linewidth parameters are not necessarily inconsistent, as one expects ozone–"air" collisions to be less effective in broadening a line than ozone–ozone collisions.

The last two columns of Table III give approximate values of the total atmospheric zenith opacity on the line centers, and values of expected zenith

[99] C. D. Walshaw, *Proc. Phys. Soc.* **A68**, 530 (1955).

Table III. Stronger Microwave Lines of $^{16}O_3$ below 300 GHz[a]

Frequencies	Transition						$\alpha(\nu_{\ell m}, T)$	$\sim(\Delta\tau)_{O_3}$	$\sim$Zenith$(\Delta T_B)_{O_3}$ (°K)	
	Upper state			Lower state				at zenith		
$\nu_{\ell m}$ (GHz)	J	K_{-1}	K_1	J	K_{-1}	K_1	$(10^{-5}$ cm$^{-1})$	(Np)	No H$_2$O	2 gm/cm^2 H$_2$O
67.35624	6	0	6	5	1	5	50	0.010	0.5	0.5
96.22834	2	1	1	2	0	2	98	0.020	4.1	3.2
101.73687	4	1	3	4	0	4	185	0.037	7.7	5.8
103.87839	14	3	11	15	2	14	53	0.011	2.3	1.7
109.55933	20	2	18	19	3	17	58	0.012	2.5	1.8
110.83604	6	1	5	6	0	6	284	0.057	12.	7.8
118.36434	1	1	1	0	0	0	61	0.012	0.0	0.0
124.08746	8	1	7	8	0	8	402	0.080	16.	10.
125.38958	8	0	8	7	1	7	236	0.047	9.5	6.0
125.41319	20	4	16	21	3	19	55	0.011	2.3	1.4
128.31385	6	2	4	7	1	7	56	0.011	2.3	1.4
136.86024	19	4	16	20	3	17	69	0.014	3.0	1.7
142.17512	10	1	9	10	0	10	537	0.11	22.	12.
144.91944	14	1	13	13	2	12	186	0.037	7.8	4.0
148.74485	12	3	9	13	2	12	106	0.021	4.5	2.2
154.04643	11	3	9	12	2	10	113	0.023	4.9	2.2
164.95182	3	1	3	2	0	2	232	0.046	9.7	3.0
165.78445	12	1	11	12	0	12	686	0.14	28.	7.8
167.57271	4	2	2	5	1	5	62	0.012	2.6	0.7
175.18635	25	5	21	26	4	22	59	0.012	2.6	0.1

175.44565	18	4	14	19	3	17	117	0.023	4.9	0.2
184.37831	10	0	10	9	1	9	636	0.13	26.5	0.0
184.74884	22	2	20	21	3	19	153	0.031	6.6	0.0
193.35130	17	4	14	18	3	15	148	0.030	6.4	0.5
195.43051	14	1	13	14	0	14	866	0.17	34.	3.4
195.72119	10	3	7	11	2	10	168	0.036	7.7	0.8
(199.38477)	30	3	27	29	4	26	61	0.012	2.6	0.4
(206.13195)	24	5	19	25	4	22	89	0.018	3.9	0.9
208.64244	5	1	5	4	0	4	545	0.11	23.	5.7
210.80380	9	3	7	10	2	8	180	0.036	7.7	1.9
(214.95548)	16	1	15	15	2	14	440	0.088	18.	4.6
(226.05412)	16	4	12	17	3	15	205	0.041	8.7	1.8
(229.57488)	23	5	19	24	4	20	120	0.024	5.2	1.1
(231.28125)	16	1	15	16	0	16	990	0.20	39.	8.1
(235.70964)	16	2	14	16	1	15	1430	0.28	53.	9.8
(237.14600)	14	2	12	14	1	13	1430	0.28	53.	9.3
(238.43195)	30	6	24	31	5	27	50	0.010	2.2	0.4
(239.09303)	18	2	16	18	1	17	1370	0.27	52.	9.0
(242.31860)	12	2	10	12	1	11	1390	0.28	53.	9.3
243.45370	12	0	12	11	1	11	1290	0.26	50.	8.3
244.15804	8	3	5	9	2	8	210	0.042	9.0	1.5
247.76122	15	4	12	16	3	13	250	0.050	11.	1.8
248.18332	20	2	18	20	1	19	1310	0.26	50.	8.3
(249.78846)	7	1	7	6	0	6	1010	0.20	39.	6.6
(249.96190)	10	2	8	10	1	9	1310	0.26	50.	8.3

[a] Frequencies are from M. Lichtenstein et al. [J. Mol. Spectrosc. 40, 10 (1971)], where parentheses indicate values calculated rather than measured. The maximum absorption coefficients $\alpha(\nu_{l_m}, T)$ were calculated by E. K. Gora [J. Mol. Spectrosc. 3, 78 (1959)] for $T = 220°K$, and this table includes lines with $\alpha > 50 \times 10^{-5}$ cm^{-1}. Values for $\Delta\tau$ and ΔT_B due to ozone were calculated as described in the text.

TABLE III (continued)

Frequencies	Transition							$\alpha(\nu_{\ell m}, T)$ (10^{-5} cm^{-1})	$\sim (\Delta\tau)_{O_3}$ at zenith (Np)	$\sim$Zenith$(\Delta T_B)_{O_3}$ (°K)	
	Upper state			Lower state							
$\nu_{\ell m}$ (GHz)	J	K_{-1}	K_1	J	K_{-1}	K_1				No H$_2$O	2 gm/cm^2 H$_2$O
(258.20206)	22	5	17	23	4	20		160	0.032	6.9	0.9
(258.71610)	8	2	6	8	1	7		1150	0.23	45.	6.1
(262.85807)	24	2	22	23	3	21		280	0.056	12.	1.6
(263.69236)	22	2	20	22	1	21		1240	0.25	48.	6.0
(263.88606)	29	6	24	30	5	25		70	0.014	3.0	0.4
(264.92605)	7	3	5	8	2	6		210	0.042	9.0	1.1
(267.26654)	6	2	4	6	1	5		930	0.19	38.	4.7
(273.05063)	18	1	17	18	0	18		1110	0.22	43.	4.8
(274.47842)	4	2	2	4	1	3		650	0.13	27.	3.0
276.92378	14	4	10	15	3	13		310	0.062	13.	1.3
279.48590	2	2	0	2	1	1		270	0.054	11.	1.0
279.89348	32	3	29	31	4	28		90	0.018	3.9	0.4
282.83766	21	5	17	22	4	18		210	0.042	9.0	0.7
286.08720	18	1	17	17	2	16		800	0.16	32.	2.7
286.15650	24	2	22	24	1	23		1150	0.23	45.	3.7
286.29420	3	2	2	3	1	3		520	0.10	21.	1.7
288.95895	9	1	9	8	0	8		1620	0.32	60.	4.5
(290.97495)	28	6	22	29	5	25		90	0.018	3.9	0.3
293.17125	5	2	4	5	1	5		830	0.17	34.	2.3
(293.54842)	6	3	3	7	2	6		210	0.042	9.0	0.6

2.3. ABSORPTION AND EMISSION BY ATMOSPHERIC GASES

emission brightness temperatures of the lines. The values of opacity were obtained from the expression

$$(\Delta\tau)_{O_3} = \int (k_v)_{O_3} \, d\ell \approx \alpha(v, T) \int (N_{O_3}/N_{\text{air}}) \, d\ell, \tag{2.3.31}$$

where $\alpha(v, T)$ is the maximum absorption coefficient for O_3 given in Table III, and a typical value of 20 cm was used for the integral $\int (N_{O_3}/N_{\text{air}}) \, d\ell$. The brightness temperatures $(\Delta T_B)_{O_3}$ of the thermal emission of the ozone lines was calculated from

$$(\Delta T_B)_{O_3} = T_{O_3}[1 - \exp(-\Delta\tau_{O_3})] \exp(-\tau_L), \tag{2.3.32}$$

which assumes a two-layer model of the atmosphere, where the ozone is located in the upper layer with temperature T_{O_3} and the lower layer has opacity τ_L. A value of 220°K was used for T_{O_3} and values of τ_L which correspond to no water vapor and to 2 gm/cm² water vapor, respectively, in the atmosphere. The values of τ_L were taken from Fig. 6, which appears in section 2.3.4.

2.3.3.4. Other Minor Constituents. Table IV gives the surface concentration of atmospheric trace constituents, and indicates which of these have spectral lines in the microwave frequency range. We shall not attempt here to review the present knowledge concerning the distribution of these molecules in the atmosphere, and shall only briefly discuss their spectra. The discussion should by no means be considered complete. For N_2O and CO, approximate values for the maximum absorption coefficient will be calculated. Ozone O_3 was discussed in the previous section.

2.3.3.4.1. NITROUS OXIDE N_2O AND CARBON MONOXIDE CO. Both N_2O and CO are linear molecules and belong to the point group $C_{\infty v}$ for which the symmetry number is $\sigma = 1$. The partition function for such molecules is then kT/hB, and, neglecting the vibrational portion of the partition function, the maximum absorption coefficient from (2.3.16) can be written as

$$\alpha(v_J, T) = \frac{8\pi^2 \mu^2 v_J^2 hB}{3c(kT)^2} \left(\frac{N_0}{\Delta v^0}\right) g_J |\phi_J|^2 \exp(-E_J/kT), \tag{2.3.33}$$

where Δv^0 is the linewidth for a number density N_0 of molecules, and it has been assumed that the temperature exponent in the linewidth expression (2.3.14) is $x = 1$. The microwave spectral lines of N_2O and CO are produced by the rotational transitions $J \to J + 1$, and have frequencies given by $v_J = 2B(J + 1)$, where J is the angular momentum quantum number that can have the values 0, 1, 2, ..., and B is the molecular rotational constant. The matrix elements for these transitions are $|\phi_J|^2 = (J + 1)/(2J + 1)$, and the lower state degeneracies are $g_J = 2J + 1$. The energy for the lower state of the

TABLE IV. Atmospheric Gases Other Than Nitrogen Oxygen, and Water Vapor

Name (chemical formula)		Relative concentration by volume at surface[a]
Argon	(Ar)	9.3×10^{-3}
Carbon dioxide	(CO_2)	$(2-4) \times 10^{-4}$
Neon	(Ne)	1.8×10^{-5}
Helium	(He)	5.2×10^{-6}
Methane	(CH_4)	$(1.2-1.5) \times 10^{-6}$
Krypton	(Kr)	1.1×10^{-6}
Hydrogen	(H_2)	$(0.4-1.0) \times 10^{-6}$
Nitrous oxide[b]	(N_2O)	$(2.5-6.0) \times 10^{-7}$
Carbon monoxide[b]	(CO)	$(0.1-2.0) \times 10^{-7}$
Xenon	(Xe)	8.6×10^{-8}
Ozone[b]	(O_3)	$(0-5) \times 10^{-8}$
Ammonia[b]	(NH_3)	$(0-2) \times 10^{-8}$
Sulphur dioxide[b]	(SO_2)	$(0-2) \times 10^{-8}$
Hydrogen sulfide[b]	(H_2S)	$(0.2-2) \times 10^{-8}$
Formaldehyde[b]	(CH_2O)	$(0-1) \times 10^{-8}$
Nitrogen dioxide[b]	(NO_2)	$(0-3) \times 10^{-9}$
Chlorine	(Cl_2)	$(0.3-1.5) \times 10^{-9}$
Iodine	(I_2)	$(0.4-4) \times 10^{-11}$

[a] C. E. Junge, "Air Chemistry and Radioactivity." Academic Press, New York, 1963.
[b] Molecular specie has a microwave spectrum.

transition is $E_J = BhJ(J + 1)$. By substituting these expressions in (2.3.33) and numerically evaluating the constants, we obtain for the maximum absorption coefficient in units of reciprocal centimeters,

$$\alpha(v_J, T) = 4.41 \times 10^{-7} \frac{\mu^2 B^3 (J+1)^3}{\Delta v^0 (T/273)^2}$$

$$\times \exp[-1.76 \times 10^{-4} BJ(J+1)/(T/273)], \quad (2.3.34)$$

where μ is in Debye units (1 D = 10^{-18} esu cm), B in gigahertz, and Δv^0 is the linewidth in gigahertz for standard atmosphere pressure (1013 mbar) and $T = 273°K$, corresponding to Loschmidt's number of molecules $N_0 = 2.69 \times 10^{19}$ cm^{-3}.

The rotational constant and dipole moment for $^{14}N_2{}^{16}O$, the dominant isotopic form of nitrous oxide, are $B = 12.6$ GHz and $\mu = 0.166$ D; for $^{12}C^{16}O$, the dominant isotope of carbon monoxide, these constants are $B = 57.9$ GHz, and $\mu = 0.10$ D.[4] Spectral line frequencies for these two

2.3. ABSORPTION AND EMISSION BY ATMOSPHERIC GASES

TABLE V. Microwave Lines of $^{14}N_2\,^{16}O$

Transition $J \to J'$	Frequency[a] (GHz)	Calculated maximum absorption coefficient at $T = 273°K$ (cm^{-1})	Calculated opacity for 10-km path with 4×10^{-7} mixing ratio (Np)
$0 \to 1$	25.12325	8.1×10^{-6}	3.2×10^{-6}
$1 \to 2$	50.24603*	6.4×10^{-5}	2.6×10^{-5}
$2 \to 3$	75.3696	2.2×10^{-4}	8.8×10^{-5}
$3 \to 4$	100.49174*	5.0×10^{-4}	2.0×10^{-4}
$4 \to 5$	125.61373*	9.7×10^{-4}	3.9×10^{-4}
$5 \to 6$	150.73513*	1.6×10^{-3}	6.4×10^{-4}
$6 \to 7$	175.85572	2.5×10^{-3}	1.0×10^{-3}
$7 \to 8$	200.97526	3.7×10^{-3}	1.5×10^{-3}
$8 \to 9$	226.09381	5.0×10^{-3}	2.0×10^{-3}
$9 \to 10$	251.21133	6.6×10^{-3}	2.7×10^{-3}
$10 \to 11$	276.32750	8.4×10^{-3}	3.4×10^{-3}
$11 \to 12$	301.44238	1.0×10^{-2}	4.0×10^{-3}

[a] Frequencies are from *Microwave Spectral Tables* [U.S. Dept. Commerce, NBS Monograph 70], with the exception of that for the $2 \to 3$ transition which was calculated from the rotational constant. An asterisk (*) indicates that fine structure or excited vibrational transitions have been measured for these lines.

molecules and the maximum absorption coefficient calculated from (2.3.34) are given in Tables V and VI. The calculations assumed a value of 3 GHz for Δv^0. Also given in these tables are the calculated absorption by the transitions over a 10-km path for expected surface concentrations in the atmosphere.

2.3.3.4.2. AMMONIA NH$_3$. The microwave spectrum of ammonia results from the tunneling of the nitrogen atom through the plane of the hydrogen atoms, and is discussed in detail by Townes and Schawlow.[4] Approximately

TABLE VI. Microwave Lines of $^{12}C\,^{16}O$

Transition $J \to J'$	Measured frequency[a] (GHz)	Calculated maximum absorption coefficient at $T = 273°K$ (cm^{-1})	Calculated opacity for 10-km path with 10^{-7} mixing ratio (Np)
$0 \to 1$	115.27120	2.9×10^{-4}	2.9×10^{-5}
$1 \to 2$	230.53797	2.2×10^{-3}	2.2×10^{-4}

[a] Frequencies are from *Microwave Spectral Tables* [U.S. Dept. Commerce, Monograph 70].

250 lines have been measured between 2 and 40 GHz, and Townes and Schawlow give calculated maximum absorption coefficients for 66 of these lines. The 10 strongest lines occur between 21 to 28 GHz with maximum absorption coefficients between 2×10^{-4} and 8×10^{-4} cm^{-1}. Using these absorption coefficients, and the concentrations of ammonia given in Table IV, we calculate that absorption by atmospheric ammonia over a 10-km path is expected to amount, at most, to 10^{-5} Np.

2.3.3.4.3. SULFUR DIOXIDE SO_2. Sulfur dioxide, like ozone, is a slightly (nearly prolate) asymmetric rotor. Thirty-one measured lines below 300 GHz of the dominant isotopic species are listed in the microwave spectral tables,[100] and an additional fourteen given by Barrett.[101] Ghosh and Edwards[102] calculate 77 lines with frequency below 300 GHz.

2.3.3.4.4. HYDROGEN SULFIDE H_2S. Two lines of H_2S below 300 GHz have been measured at frequencies of 168.76251 and 216.71042 GHz, respectively.[100]

2.3.3.4.5. FORMALDEHYDE CH_2O. Thirty-nine lines of the dominant isotope with frequencies below 226 GHz have been measured.[100]

2.3.3.4.6. NITROGEN DIOXIDE NO_2. Several lines of NO_2 have been measured near 15, 26, and 40 GHz.[100]

2.3.4. The Microwave Spectrum of the Terrestrial Atmosphere

The total absorption through the atmosphere and the thermal emision by water vapor and oxygen can be calculated by using the radiative transfer equation and the absorption coefficients for these gases given in the preceding sections. The absorption and emission, calculated by using the 1962 U.S. Standard Atmosphere model of temperature and pressure, are given in Figs. 6 and 7, respectively. (See Chapter 2.2 for a discussion of this model.) Some measured values of atmospheric absorption are also indicated in Fig. 6. Calculations are shown both for no water vapor and for 2 gm/cm^2 total water vapor in the atmosphere. The curves for 2 gm/cm^2 water vapor assume a water-vapor density of 10 gm/m^3 at the surface, exponentially decreasing with altitude with a 2-km scale height. Above 15 km a water-vapor distribution with a constant volume mixing ratio of 2 ppm was assumed for the calculation.

Calculations of Figs. 6 and 7 used (2.3.2) and (2.3.4) for the opacity and emission, respectively, and the integrals in these equations were numerically evaluated between 0 to 60-km altitude. In computing the water-vapor absorption coefficient we assumed the kinetic line shape in (2.3.12), used

---

[100] Microwave Spectral Tables. U.S. Dept. Commerce, NBS Monograph 70.

[101] A. H. Barrett, *Mem. Soc. Roy. Sci. Liege* **8**, 197 (1963).

[102] S. N. Ghosh and H. D. Edwards, U.S.A.F. Survey in Geophysics, No. 82, AFCRL, Bedford, Massachusetts (1956).

2.3. ABSORPTION AND EMISSION BY ATMOSPHERIC GASES

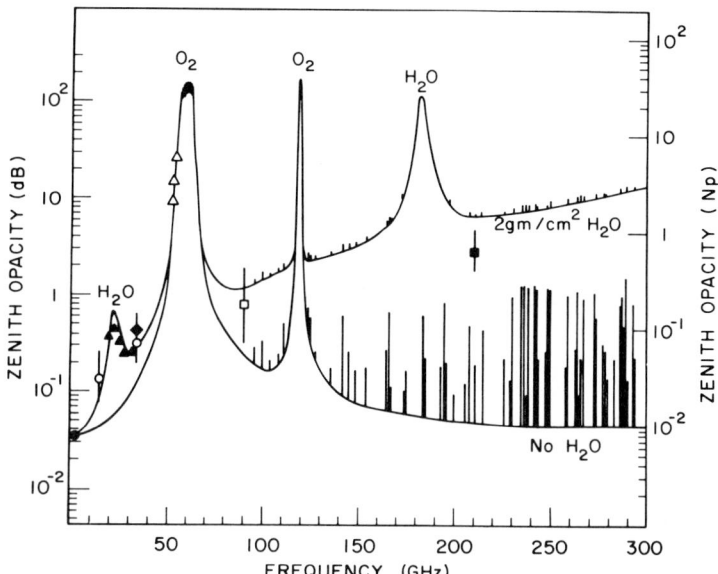

FIG. 6. Atmospheric zenith opacity. The two curves give the zenith opacity calculated for water vapor, oxygen, and ozone as discussed in the text. Also shown are measured values reported by P. J. Encrenaz, A. A. Penzias and R. W. Wilson (●) [*Astron. Astrophys.* **9**, 51 (1970)], E. E. Altshuler, V. J. Falcone, Jr., and K. N. Wulfberg (○)[*IEEE Spectrum* **5**, 83 (1968)], C. J. Carter, R. L. Mitchell, and E. E. Reber, (△) [*J. Geophys. Res.* **73**, 3113 (1968)], D. H. Staelin (▲) [*ibid.*] **71**, 2875 (1966)], F. Picherit, (◆), [*C. R. Acad. Sci. Paris* **266**, 784 (1968)], F. I. Shimabukuro and E. E. Epstein (□)[*IEEE Trans. Antennas Propagat.* **AP-18**, 485 (1970)], and W. A. Johnson, T. T. Mori, and F. I. Shimabukuro (■)[*ibid.* **AP-18**, 512 (1970)].

the linewidth parameter from (2.3.21), and included the empirical correction described by (2.3.22). The oxygen absorption coefficient given by (2.3.30) was used in the calculations with the kinetic line shape and linewidth parameters of (2.3.29) for the transitions around 60 GHz. For the 118-GHz oxygen line we used (2.3.29a) alone for all values of Δv, as discussed in the text following those equations.

Also shown in Figs. 6 and 7 are the absorption and emission by the ozone lines, with values taken from Table III. Water-vapor dimer absorption was not included in the calculations, except as it may have contributed to the empirical correction term in the water-vapor absorption coefficient expression. The reported dimer lines in the region of ∼230 to 260 GHz would lead to more absorption and emission than indicated in Figs. 6 and 7.

Variations in the clear-sky atmospheric absorption and emission result primarily from variations in the amount of water vapor in the atmosphere.

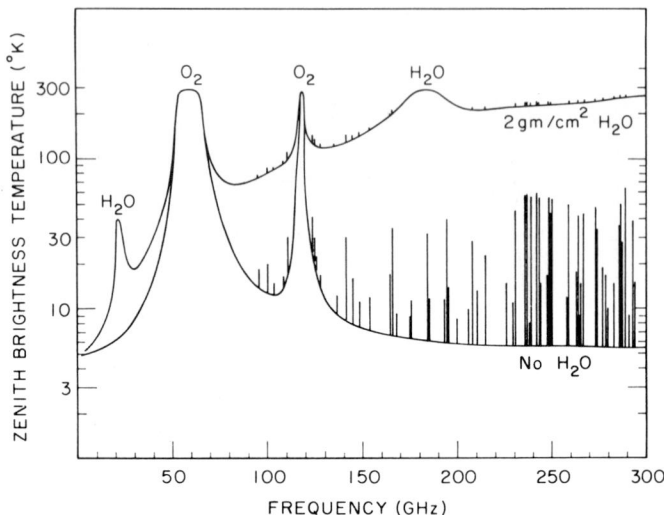

FIG. 7. Calculated atmospheric microwave emission. Calculations shown here are for observations looking at zenith from sea level, as discussed in the text. Cosmic radiation of 3°K incident on the atmosphere from the top was included in the calculations.

Variations in the amount of oxygen absorption, except in the most opaque regions of the atmosphere, are relatively small. Calculated values of absorption in the 60-GHz oxygen band for various atmospheric temperature models are given by Reber et al.[103]

The amount of water-vapor absorption for given atmospheric conditions can be estimated, to some extent, if the surface water-vapor density, which can be easily measured, is known. Figure 8 shows values of zenith opacity at frequencies of 15, 22.2, 35, and 90 GHz plotted against surface water-vapor density. The scatter of the points in this figure is due to variations in the water-vapor altitude distribution, and, to a lesser extent, to variations in the atmospheric temperature profile. The amount of scatter indicates the accuracy with which atmospheric opacity can be estimated from knowledge of the surface water-vapor density. The linear equations which best fit the points for the respective frequencies are given in the figure caption.

Reported measurements of relatively large amounts of water vapor in the stratosphere led to the prediction of a large, narrow spectral feature at 22.235 GHz resulting from the stratospheric water vapor superimposed on the broad feature due to water vapor in the troposphere.[104] Searches for this

[103] E. E. Reber, R. L. Mitchell, and C. J. Carter, *IEEE Trans. Antennas Propagat.* **AP-18**, 472 (1970).

[104] A. H. Barrett and V. K. Chung, *J. Geophys. Res.* **67**, 4259 (1962).

2.3. ABSORPTION AND EMISSION BY ATMOSPHERIC GASES

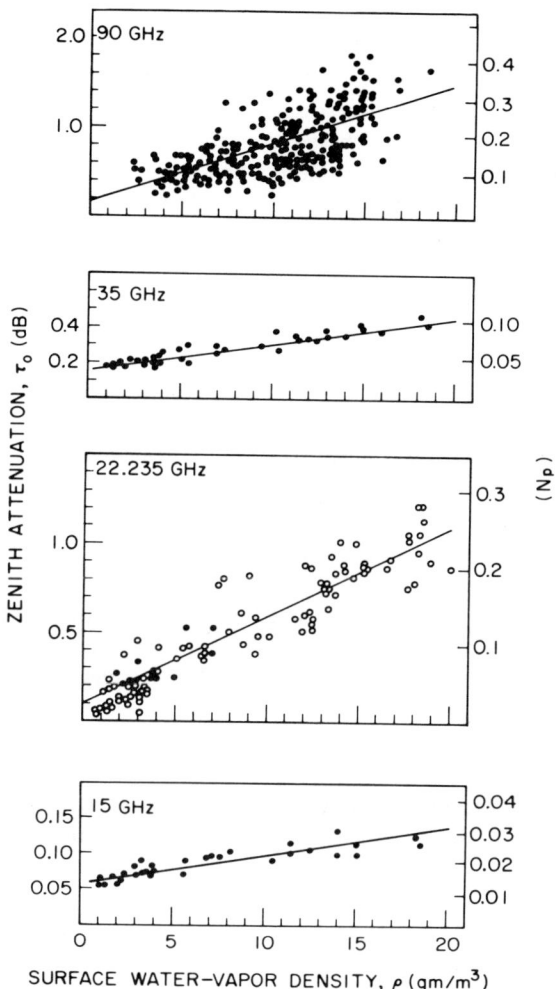

FIG. 8. Atmospheric zenith opacity versus surface water-vapor density. Values at 90 GHz are from F. I. Shimabukuro and E. E. Epstein [*IEEE Trans. Antennas Propagat.* **AP-18**, 485 (1970)], those at 35 and 15 GHz from E. E. Altshuler, V. J. Falcone, Jr., and K. N. Wulfsberg [*IEEE Spectrum* **5**, 83 (1968)]. At 22.235 GHz the open points are calculated and the closed points measured [J. W. Waters, Ph.D. Thesis, Massachusetts Inst. of Technol., Cambridge, Massachusetts (1970)]. The equations for the straight lines that best fit the points are: at 90 GHz, τ_0 (dB) = 0.17 + 0.06ρ (gm/m^3); at 35 GHz, τ_0 (dB) = 0.17 + 0.013ρ (gm/m^3); at 22.235 GHz, τ_0 (dB) = 0.11 + 0.048ρ (gm/m^3); and at 15 GHz, τ_0 (dB) = 0.055 + 0.004ρ (gm/m^3).

feature[34,105,106] were unsuccessful, however, and were consistent with other measurements of a relatively "dry" stratosphere containing a water-vapor mixing ratio of a few parts per million.[107–109]

On the edges of the 60-GHz O_2-band, relatively narrow spectral features appear in the atmospheric spectrum. These features are caused by high-rotational transitions of O_2 on the edge of the band, and are narrow because they originate in the upper stratosphere where collisional broadening is much less than in the lower atmosphere. Atmospheric emission from these lines on the low-frequency band edge has been measured.[110]

Microwave ozone lines that have been measured in the terrestrial atmosphere are those at 101 and 110 GHz. The 101-GHz measurement[111] gave 0.35-dB absorption at 64° zenith angle, which implies 0.035-Np opacity at zenith. This compares with 0.037 Np estimated for this line in Table III. The 110-GHz measurement[112] gave .03 ± .01-Np zenith opacity, which compares with 0.057 Np estimated in Table III. The discrepancy between the measured values and Table III values for the atmospheric opacity of these lines could be because the amount of atmospheric ozone is different for the measurements than that assumed in the calculations that led to Table III. The discrepancy could also result from uncertainties in the linewidth parameter for ozone.

Infrared techniques have recently been extended to millimeter wavelengths, and atmospheric absorption measurements[58] have, in addition to showing absorption attibuted to the water-vapor dimer, suggested the presence of unidentified weak absorption lines at ~135 and ~205 GHz. Aircraft measurements of submillimeter stratospheric emission[109,113,114] have also shown spectral features attributed to HNO_3, N_2O, NO_2, and SO_2. The identity of some of these features is not yet certain, however, and more refined measurements, both by microwave and infrared techniques should prove interesting.

[105] J. W. Waters, Ph.D. Thesis, Massachusetts Inst. of Technol., Cambridge, Massachussetts (1970).
[106] L. A. Bonvini, D. L. Croom, and A. C. Gordon-Smith, *J. Atmos. Terr. Phys.* **28**, 891 (1966).
[107] H. J. Mastenbrook, *J. Atmos. Sci.* **25**, 299 (1968).
[108] D. McKinnon and H. W. Morewood, *J. Atmos. Sci*, **27**, 483 (1970).
[109] J. E. Harries, *Nature (London)* **241**, 515 (1973).
[110] J. W. Waters, *Nature (London)* **242**, 506 (1973).
[111] W. M. Caton, G. G. Manella, P. M. Kalaghan, A. E. Barrington, and H. I. Ewen, *Astrophys. J.* **151**, L153 (1968).
[112] F. I. Shimabukuro and W. J. Wilson, *J. Geophys. Res.* **78**, 6136 (1973).
[113] J. E. Harries, N. R. W. Swann, J. E. Beckman, and P. A. R. Ade, *Nature (London)* **236**, 159 (1972).
[114] I. G. Nolt, T. Z. Martin, C. W. Wood, and W. M. Sinton, *J. Atmos. Sci.* **28**, 238 (1971).

2.4. Extinction by Condensed Water*

2.4.1. Introduction

Individual hydrometeors such as rain, hail, and snow particles both absorb and scatter incident radiation. The interaction between a particle and the incident field may be computed using electromagnetic scattering theory. Exact solutions are available only for simple shapes and distributions of dielectric properties within the scatterer. Rain may be reasonably modeled with spheres having the dielectric properties of water. More exact models of rain that take into account the nonspherical nature of the raindrops have been tried[1] but are not generally used because they require relatively large amounts of computer time and the differences between their results and results based on the exact solution for spherical particles are smaller than the statistical uncertainty of the results due to variations in the drop-size distributions. Approximate calculations of hail and snow scattering have also been made, but the problem of rain scattering has received most attention. From the point of view of both probability of occurrence and severity of effect, rain is most important.

2.4.2. Solution to the Scattering Problem

The theory of scattering by a single lossy dielectric sphere of size comparable to a wavelength is based on a solution to the scattering problem.[2] The solution which is attributed to Mie has been verified experimentally many times (see, for example, Gerhardt et al.).[3] The use of the scattering properties of a single raindrop in the description of the effects of rain also requires a description of the drop-size distribution and an assumption about the statistics of drop location within a volume. The use of measured drop-size distributions and the assumption that the drops are distributed throughout the volume in accordance with a Poisson process allows one to compute the per unit volume extinction (attenuation) and scattering cross sections of

---

[1] T. Oguchi, *Radio Sci.* **8**, 31 (1973).
[2] H. C. Van de Hulst, "Light Scattering by Small Particles." Wiley, New York, 1957.
[3] J. R. Gerhardt, C. W. Tolbert, S. A. Brunstein, and W. W. Bahn, *J. Meteorol.* **18**, 340 (1961).

---

* Chapters 2.4 and 2.5 are by R. K. Crane.

rain.[4] Computations of the extinction cross section per unit volume or the attentuation coefficient (specific attenuation) made using the average measured dropsize distribution reported by Laws and Parsons,[5] are given in Fig. 1. For comparison, the attenuation coefficients for liquid water clouds

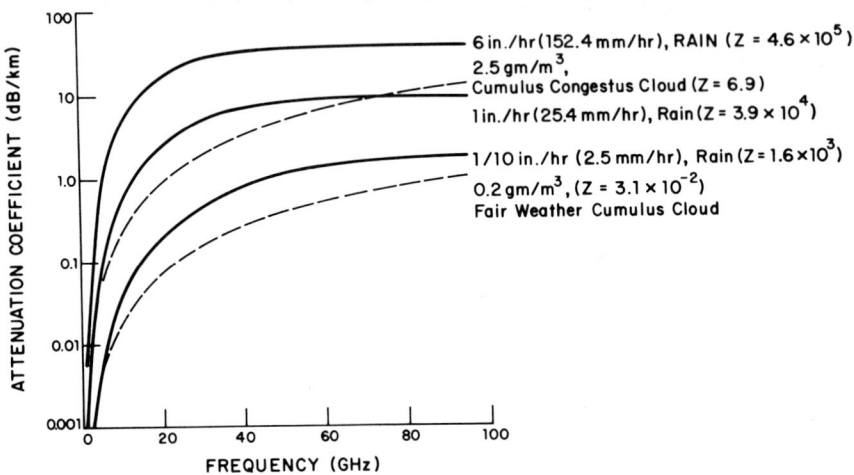

Fig. 1. Attenuation coefficient as a function of frequency for liquid scatterers, rain and clouds. The coefficients were computed using the Laws and Parsons drop-size distribution with a drop temperature of 0°C. The quantity Z is a measure of the radar cross section per unit volume.

are also presented. The cloud computations were made in an identical manner to those for rain with the exception that the cloud particle-size distributions of Weickmann and aufm Kampe[6] were used.

2.4.3. Effects of Drop-Size Distributions

Measurements of raindrop-size distributions show large variations for the same location, rain type, and rain rate. These variations imply that the attenuation and scattering properties of rain will also vary for a given rain rate. Figure 2 shows for a frequency of 16 GHz a scatter diagram which presents the results of computations of the attenuation coefficient and rain rate for 4741 drop-size distributions taken from December 1960 to March

[4] R. K. Crane, M.I.T. Lincoln Lab., Lexington, Massachusetts, Tech. Rep. 426, ASTIA Doc. AD-647798 (October 1966).
[5] J. O. Laws and D. A. Parsons, *Trans. Amer. Geophys. Un.* **24**, 452 (1943).
[6] H. K. Weickmann, and H. J. aufm Kampe, *J. Meteorol.* **10**, 204 (1943).

2.4. EXTINCTION BY CONDENSED WATER

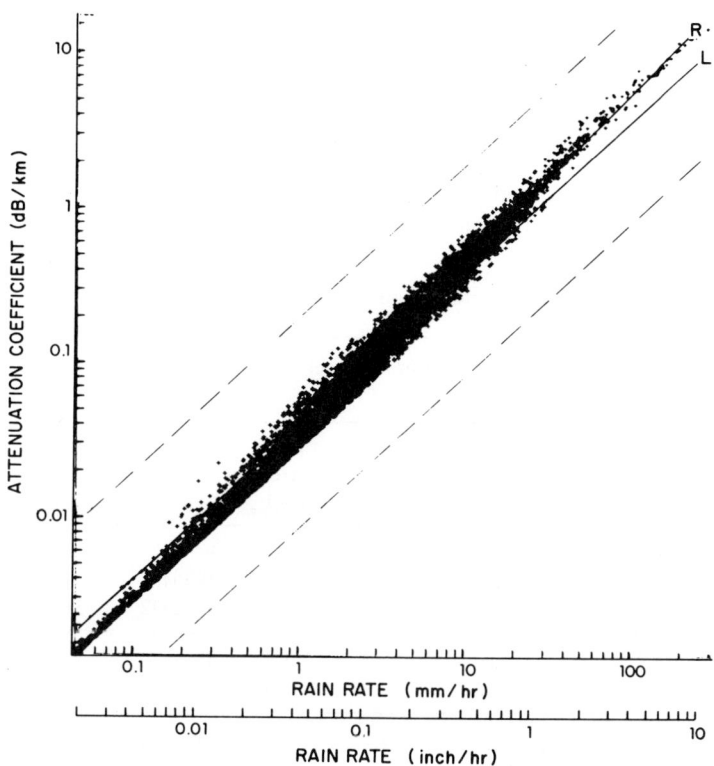

Fig. 2. Scatter diagram of attenuation versus rain rate for a frequency of 16.03 GHz. and drop temperature of 10°C. Regression curves shown are linear (L) and logarithmic (R),

1962 at Franklin, North Carolina.[7] The dashed lines indicate the maximum and minimum possible attenuation coefficients for a given rate if we assume the worst and best possible monodisperse drop-size distributions with the limitation that the drop size lie within the range of sizes attributable to rain. It is evident that, in natural rain, the fluctuations in attenuation due to variations in the drop-size distributions are significantly less than those theoretically possible. The root mean square (rms) variation in the data points about the linear least square fit or linear regression curve at a fixed rain rate is 28%, and the rms variation about the logarithmic regression line is 22%. Figure 2 shows that the linear and logarithmic lines are better fits at opposite ends of the rain-rate scale.

[7] E. A. Mueller and A. L. Sims, Illinois State Water Survey, Urbana, Illinois, Tech. Rep. TR-ECOM-02071-RR3 (September 1967).

2.4.4. Models for Attenuation Computations

The statistical analysis of many drop-size distributions provides models for use in calculating attenuation effects. The models are easier to use than the standard table of computations made from averaged drop-size distributions and have the advantage of also providing an estimate of the possible variation in the estimated attenuation. Linear and logarithmic regression models for several frequencies are listed in Table I. The data for North

TABLE I. Attenuation versus Rain-Rate Models

Frequency (GHz)	Logarithmic model[a] $A = \alpha R^\beta$			Linear model[a] $A = \alpha R$	
	α	β	rms error (%)	α	rms error (%)
2.8[b]	0.000459	0.954			
7.5	0.00459	1.06	28	0.00481	31
9.4	0.00870	1.10	30	0.00932	36
16.0	0.0374	1.10	22	0.0403	28
34.9	0.225	1.05	10	0.234	12
69.7[b]	0.729	0.893			

[a] Models computed from 4741 data points with Franklin, North Carolina, drop-size measurements, R in millimeters per hour and A in decibels per kilometer. [See E. A. Mueller and A. L. Sims, Illinois State Water Survey, Urbana, Illinois, Tech. Rep. TR-ECOM-02071-RR3 (September 1967)].
[b] Data based on 4590 drop-size measurements analyzed by E. A. Mueller and A. L. Sims [Illinois State Water Survey, Urbana, Illinois, Tech. Rep. TR-ECOM-0271-F (May 1967)].

Carolina are reasonably representative. The results of the analysis of the North Carolina data do not differ significantly from computations made with drop-size data from other areas.[8,9] Using the models given in Table I we computed the attentuation coefficients for a 100 mm/hr (4 in./hr) rain rate given in Fig. 3 together with the values computed using the Laws and Parsons and Marshall and Palmer drop-size distributions.[10] The scatter

[8] E. A. Mueller and A. L. Sims, Illinois State Water Survey, Urbana, Illinois, Tech. Rep. TR-ECOM-02071-F (May 1967).
[9] R. K. Crane, M.I.T. Lincoln Lab., Lexington, Massachusetts, Tech. Note 1968-33. ASTIA Doc. AD-678079 (September 1968).
[10] J. S. Marshall and W. McK. Palmer, *J. Meteorol.* **5**, 165 (1948).

2.4. EXTINCTION BY CONDENSED WATER

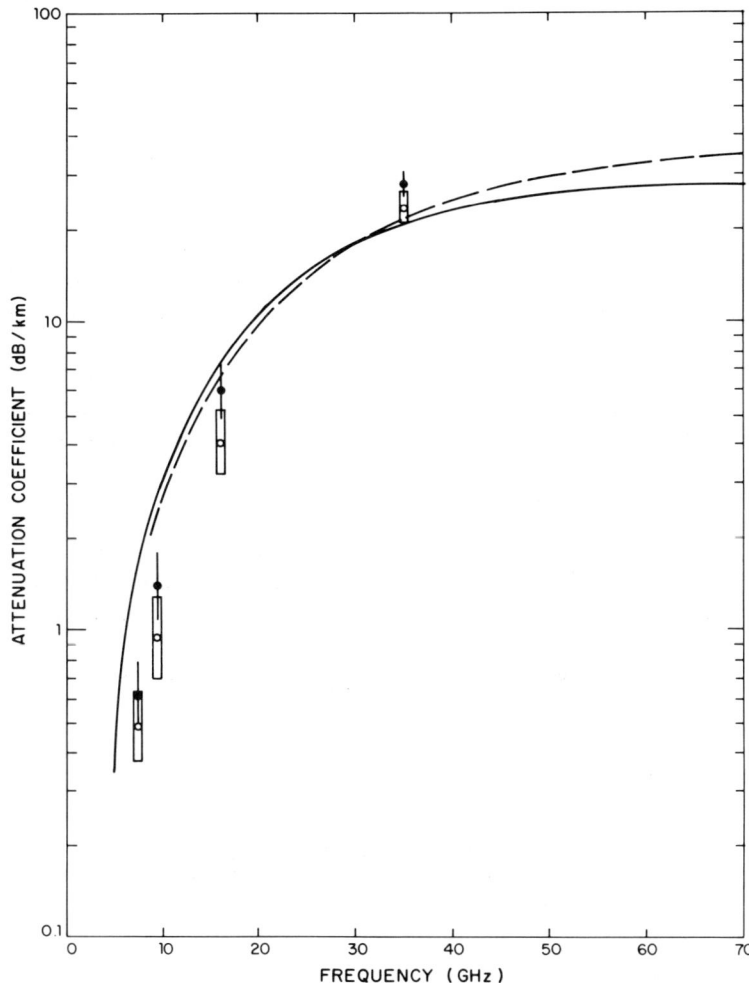

Fig. 3. Comparison of attenuation versus rain-rate models for a drop temperature of 10°C and a rate of 101.6 mm/hr (4 in./hr). Coefficients were computed using the Marshal and Palmer drop-size distribution (the dashed curve) and the Laws and Parsons drop-size distribution (the solid curve). Logarithmic and linear regression of attenuation on rain rate are indicated by (●) and (○), respectively, where wings on the points represent rms fluctuations of data about the model curve.

diagrams for each of the other frequencies are similar to the one presented in Fig. 2. For frequencies below 35 GHz and at the 100 mm/hr (4-in./hr) rain rate, the logarithmic regression model provides a better fit to the data than either of the averaged drop-size distributions. At 35 GHz, the linear model provided a fit that was as good as either of the averaged distributions.

Both the model computations and the computations based on averaged drop-size distributions provide an estimate of the attenuation coefficient. In general, however, it is not sufficient to compute the attenuation for a line-of-sight path through rain. Rain both scatters and absorbs the incident radiation. The scattered radiation may be singly or multiply scattered and eventually reach the receiving antenna. For incoherent measurement systems (radiometers) the radiative transfer equation governs propagation through the absorbing and scattering medium. The detailed mathematical analysis of this radiation-transfer problem has been summarized by Crane.[11] We shall consider here only the single-scattering approximation. For this purpose let us define the single-scattering albedo as the ratio of energy scattered to total energy lost from the incident wave including both absorption and scattering.

2.4.5. Single-Scattering Albedo

The single-scattering albedo is readily computed for a given drop-size distribution using Mie theory.[4] The results obtained using the Laws and Parsons drop-size distribution are given in Fig. 4. For rain, the single-

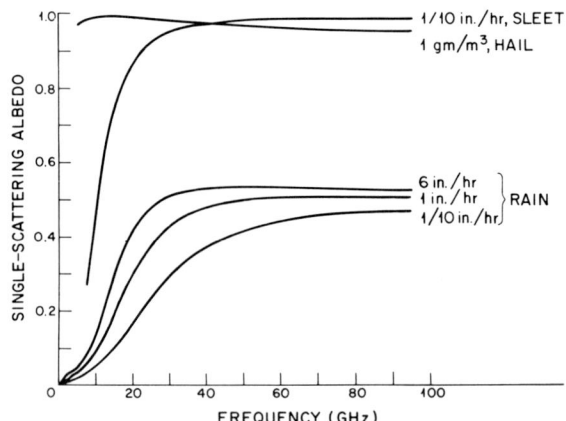

FIG. 4. Single-scattering albedo versus frequency. Albedo values were computed using the Laws and Parsons drop-size distribution with a drop temperature of 0°C. [Note that 1/10 in./hr = 2.5 mm/hr, 1 in./hr = 25 mm/hr, and 6 in./hr = 150 mm/hr.]

[11] R. K. Crane, *Proc. IEEE* **59**, 173 (1971).

scattering albedo is less the 0.6, and at frequencies below 10 GHz it is less than 0.1. At the lower frequencies, where the total attenuation is low, it is expected that the nonscattering-transmission equation will be sufficient. At higher frequencies, multiple scattering may be important. Using the criterion that the product of the single-scattering albedo and the optical depth should be less than unity, we can calculate the distance at which multiple scattering becomes important.

The results of computations using the Laws and Parsons distribution are given in Fig. 5. The computations show that in rain and at frequencies above 60 GHz, multiple-scattering effects may be important.

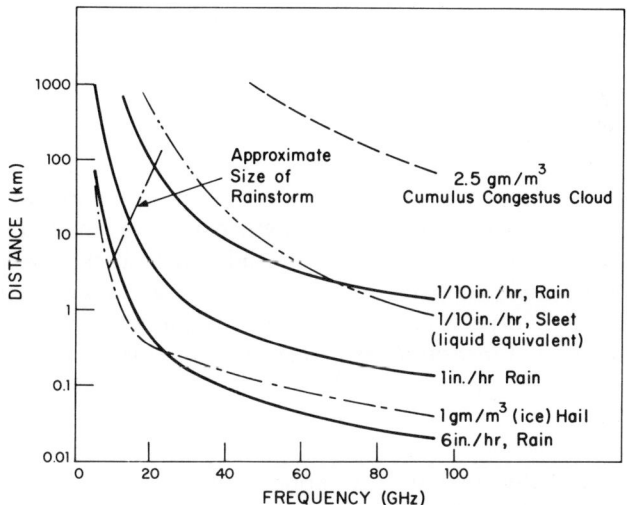

FIG. 5. Mean distance to multiple scattering as a function of frequency. Distance values were computed using the Laws and Parsons drop-size distribution with a drop temperature of 0°C. [Note that 1/10 in./hr = 2.5 mm/hr, 1 in./hr = 25 mm/hr, and 6 in./hr = 150 mm/hr.]

2.4.6. Multiple Scattering

Multiple scattering does not affect coherent transmission systems (interferometers) in the same way as incoherent systems. Except for scattering by drops spaced closer than a wavelength, the multiple-scattered signals have random phases when compared with the attenuated direct signal. Since the drops are separated by distances larger than a wavelength the effect of multiple scattering on the attenuation of the coherent signal is negligible.[4] The transmission equation for a single-scattering medium, therefore, applies to coherent transmission systems for all combinations of optical depth and single-scatter albedo. The multiply scattered signal is, however, present in the form of excess noise. The effect of multiple scattering on an incoherent transmission system is to increase the received signal when compared to that predicted on the basis of single scatter. The attenuation experienced by an incoherent transmission system, therefore, should be less than that for coherent systems. At present, neither adequate measurements nor an adequate theoretical treatment of multiple scattering effects have been made. The computations of attenuation using the calculated attenuation coefficients as reported in the literature are all made using the transmission equation for a single-scattering medium and strictly apply only to coherent transmission systems. No significant departures from estimates based on single-scattering theory have been reported.

2.4.7. Measured Attenuation

The first comprehensive tabulation of experimentally determined values of the attenuation coefficient versus rain rate for a large number of frequencies in the centimeter and millimeter wavelength bands was made by Ryde and Ryde.[12] Recently, Medhurst[13] corrected and extended the earlier work of Ryde and Ryde and compared the updated curves of attenuation versus rain-rate with the experimental data then available in the literature. A number of additional measurements were available for comparison with the calculated results. Comparison did not, however, validate the method. Medhurst observed a "marked tendency for the observed attenuations to fall *well above* levels which, according to the theory, cannot be exceeded." From this he concluded that "the applicability of the Mie theory to the practical rainfall situation cannot be said to be demonstrated." Since the publication of Medhurst's paper, several additional experiments have been performed and reported.[14,15] Except for the work of Semplak and Turrin,[16] each of the papers reported the same tendencies as Medhurst noted.

The measurements just cited and those reviewed by Medhurst tend to be crude in the handling of meteorological data. Generally long paths were used with relatively few rain gauges. The only data that show good agreement between estimated and measured attenuation were published by Usikov

---

[12] J. W. Ryde and D. Ryde, General Electric Co., Res. Lab., Wembley, England, Rep. 8516 (August 1944).

[13] R. G. Medhurst, *IEEE Trans. Antennas Propagat.* **AP-13**, 550 (1965).

[14] J. Bell, *Proc. Inst. Elec. Eng.* **114**, 545 (1967).

[15] J. F. Roche, H. Lake, D. T. Worthington, C. K. H. Tsao, and J. T. deBettencourt, *IEEE Trans. Antennas Propagat.* **AP-18**, 452 (1970).

[16] R. A. Semplak and R. H. Turrin, *Bell Syst. Tech. J.* **48**, 1767 (1969).

et al.[17] In this work an extremely short, 50 m path was used, and data were reported only when identical rain rates were observed at two gauges 30 m apart along the path. The tendency toward agreement in experiments with short paths and dense rain gauge networks together with differing amounts of disagreement in others suggests that the discrepancies are not due to the inapplicability of Mie theory as suggested by Medhurst but due to the inadequacy of meteorological data. More intensive radar observations[9] show that rain is composed of many small, relatively heavy showers imbedded in a larger area of light rain. A network of rain gauges spaced far apart could easily miss these localized showers.

Radar measurements may be used to observe the spatial changes in rain intensity that correspond to the temporal changes in rain rate at the surface. A radar measurement of the structure of a New England summer shower obtained on July 28, 1967 is shown in Fig. 6. The radar used was the

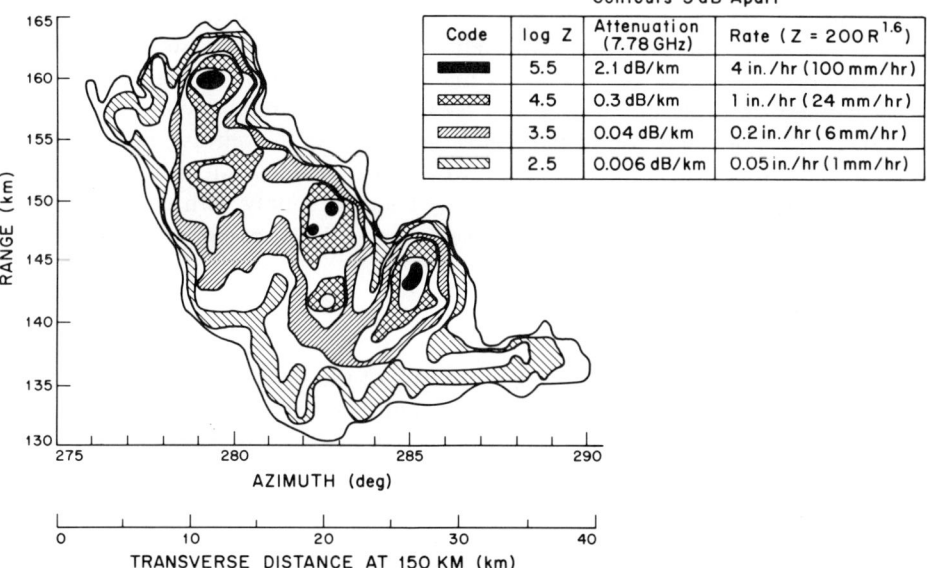

FIG. 6. Weather-radar observations of showery rain using the Millstone Hill L-band radar at 1.4° elevation and 275°–290° azimuth. Data was compiled on July 24, 1967.

Millstone Hill L-band radar which has approximately a 2-km³ resolution volume.[9] The radar map of rain intensity shows several small cells with widths the order of 5 km across and peak to minimum rain intensity distances the order of 3 km.

[17] O. Ya. Usikov, V. L. German, and I. Kh. Vakser, *Ukr. Fiz. Zh.* (*Ukr. Phys. J.*) **6**, 618 (1961).

2.5. Refraction Effects in the Neutral Atmosphere

2.5.1. Introduction

The atmosphere complicates measurements of astronomical radio sources in two ways. The pointing of a radio telescope is perturbed by atmospheric bending of the incidents rays, and the effective path length of waves passing through the atmosphere is altered by both refractive index and bending. In this chapter we consider in separate sections the index of refraction at radio frequencies (Section 2.5.2), the bending of rays (Section 2.5.3), and pathlength effects important in interferometry (Section 2.5.4). The approach followed in the latter two sections to a large extent involves the results of machine computation of refraction effects for representative sets of atmospheric profiles. These computations made use of the ray-tracing techniques of geometrical optics, described by Freehafer[1] and Born and Wolf.[2]

The index of refraction in the neutral atmosphere in contrast to the situation in the ionosphere (see Section 2.1.6) is nearly independent of frequency over the radio spectrum. This chapter is concerned only with effects in the neutral portion of the atmosphere. Refraction effects in the ionosphere are discussed in Chapter 2.1.

2.5.2. Radio Refractivity

Refraction effects depend on the real part of the index of refraction (complex) of the gases in the atmosphere. Computation of the imaginary part of the index of refraction (absorption coefficient) is discussed in Chapter 2.3. Except for the anomolous dispersion of the water-vapor lines and the oxygen lines, the real part of the index of refraction is essentially independent of frequency. The frequency-independent real part of the index of refraction is given by[3]

$$N = \text{Re}[n - 1] \times 10^6 = K_1 \frac{P_d}{T} + K_2 \frac{e}{T} + K_3 \frac{e}{T^2} + K_4 \frac{P_c}{T}, \quad (2.5.1)$$

where n is the index of refraction (complex), N the radio refractivity, P_d the partial pressure of the dry gases in the atmosphere, e the partial pressure of

---

[1] J. E. Freehafer, in "Propagation of Short Radio Waves" (D. E. Kerr, ed.), p. 41. McGraw-Hill, New York, 1951.
[2] M. Born and E. Wolf, "Principles of Optics," Sect. 3.1, Pergamon, Oxford, 1964.
[3] E. K. Smith and S. Weintraub, *Proc. IRE* **41**, 1035 (1953).

2.5. REFRACTION EFFECTS IN THE NEUTRAL ATMOSPHERE

water vapor, P_c the partial pressure of carbon dioxide, and T the absolute temperature. Equation (2.5.1) holds for frequencies less than 100 GHz with an error of less than 0.5%, due to anomolous dispersion.

The coefficients K_1–K_4 were determined empirically using laboratory measurements. Bean and Dutton[4] summarized the measurements giving the following best estimates for the coefficients K_1–K_4:

$$K_1 = 77.6 \pm 0.1 \quad (°\text{K/mbar}),$$
$$K_2 = 72 \pm 9 \quad (°\text{K/mbar}),$$
$$K_3 = (3.75 \pm 0.03) \times 10^5 \quad [(°\text{K})^2/\text{mbar}],$$
$$K_4 = (5/3)K_1.$$

Since the partial pressure due to carbon dioxide is approximately 0.03% of the total pressure and K_2 is approximately equal to K_1, (2.5.1) may be simplified as

$$N = 77.6 \frac{P}{T} + 3.73 \times 10^5 \frac{e}{T^2} \qquad (2.5.2)$$

where $P = P_d + P_c + e$ and P, e are expressed in millibars, T in degrees Kelvin. The radio refractivity N is commonly referred to as being expressed in N units to denote the relationship between N and n given by (2.5.1). Equation (2.5.2) is estimated to be within 0.5% of the true value for the ranges of atmospheric parameters normally encountered and frequencies below 30 GHz. The largest contribution to the uncertainty in (2.5.2) is measurement error in determining the coefficients. For frequencies between 30 and 100 GHz, (2.5.2) is within 1% of the true value, the added error being caused by anomolous dispersion.

Equation (2.5.2) relates the radio refractivity to the readily measured atmospheric parameters, pressure and temperature, and to the partial pressure of water vapor. Water vapor is typically measured by hygrometers (relative humidity), psychrometers (wet-bulb temperature), or by dew-point devices (dew-point temperature). With any of these measurements, the saturation vapor pressure is required to convert the measured value to vapor pressure. Tables of saturation vapor pressure are available.[5] For ease of calculation, the following approximation is within 0.4% of the tabulated values for temperatures between $-30°C$ to $+30°C$:

$$e_s(T) = 6.105 \exp\left[25.22 \left(\frac{T-273}{T}\right) - 5.31 \log_e\left(\frac{T}{273}\right)\right], \qquad (2.5.3)$$

---

[4] B. R. Bean and E. J. Dutton, Nat. Bur. Std. Monograph 92, U.S. Dept. of Commerce (1966).

[5] R. J. List, Smithsonian Meteorological Tables. Smithsonian Inst., Washington, D.C. (1958).

where $e_s(T)$ is saturation vapor pressure in millibars, and T is absolute temperature (in degrees Kelvin). The maximum error in the estimate of N introduced by using this approximation for saturation vapor pressure is less than 0.1% when used over the entire range of temperature expected in the atmosphere, $-70-+40°C$.

If the dew-point temperature is given, the substitution of the dew-point temperature into (2.5.3) yields the vapor pressure for use in (2.5.2). The dew point is used to describe the vapor pressure in terms of saturation vapor pressure with respect to water. Meteorologists generally use saturation with respect to water, not with respect to ice, when reporting dew point or relative humidity for temperatures less than 0°C. For conversion from frost point (temperature for saturation with respect to ice) to dew point see the meteorological tables.[5] Psychrometer data are generally reported as wet-bulb temperature and the ambient (dry-bulb) temperature by

$$e = e_s(T) - [0.00066(1 + 0.00115w)]P(t - w), \qquad (2.5.4)$$

where t is dry-bulb temperature (degrees centigrade), w the wet-bulb temperature (degrees centigrade), p the pressure (mbars), and e and $e_s(T)$ are in millibars as before. Hygrometer data are generally reported in terms of relative humidity expressed in percent. The vapor pressure is related to relative humidity by

$$e = e_s(T) \frac{U}{100} \left[1 - \left(1 - \frac{U}{100}\right) \frac{e_s(T)}{P} \right]^{-1} \qquad (2.5.5)$$

$$\simeq e_s(T) \frac{U}{100}, \qquad (2.5.6)$$

where U is the relative humidity (percent) and P, e, $e_s(T)$ are measured in millibars. Approximation (2.5.6) is in error by 8% at 40°C, 2% at 20°C, and by less than 0.6% at 0°C.

Profiles of radio refractivity may be prepared from profiles of temperature, pressure, or dry-air density and from profiles of one of the water-vapor parameters. In Chapter 2.2 a temperature profile was presented for the 45°N latitude (July) U.S. Standard atmosphere together with a mixing-ratio model to be used with any standard midlatitude temperature profile. Using both profiles and the hydrostatic equation to relate pressure and height [see (2.2.1)], we calculated the radio refractivity profiles presented in Figs. 1 and 2. To indicate the effect of water vapor, two additional profiles were computed, one for saturation below the tropopause (100% humidity) and one for no water vapor below the tropopause (the dry term only). Due to the lapse (decrease with height) of temperature in the troposphere and the relative dryness of the atmosphere above the tropopause, water-vapor effects

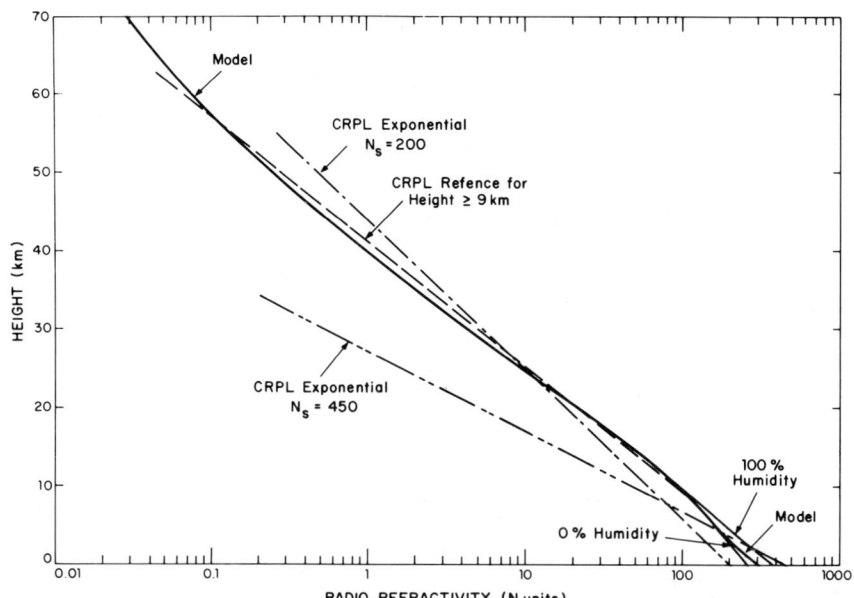

FIG. 1. Radio-refractivity profiles for model atmospheres.

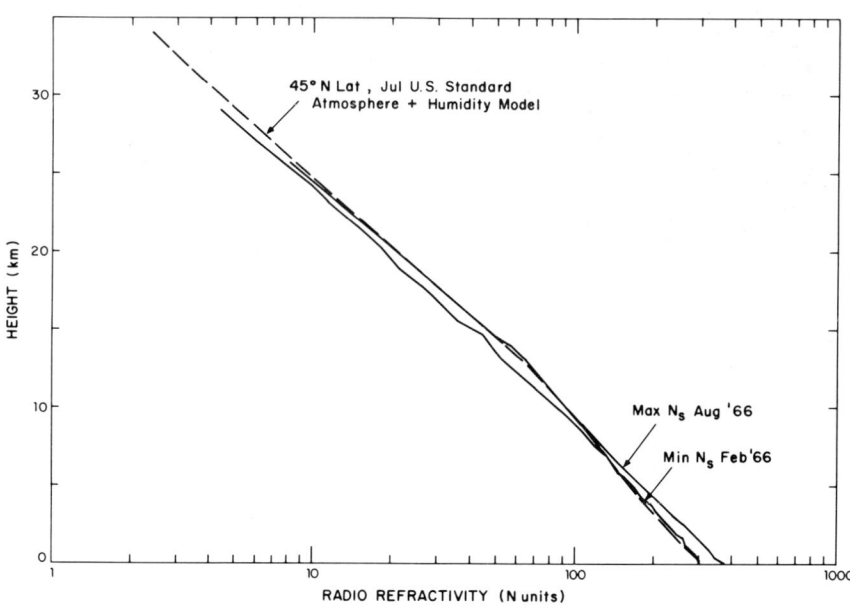

FIG. 2. Radio-refractivity profiles (Albany, New York). The profile for a model atmosphere is shown for comparison.

are only important near the surface. The variation in water vapor near the surface and in mesoscale volumes above the surface (for instance, in clouds) contribute to the major short time-period changes in the radio-refractivity profile.

The radio-refractivity profiles were generally observed to have an exponential decrease with height although the best fit required at least two exponential curves, one above the tropopause and one below. Since the major changes in bending or path length occur near the surface only the exponential representing the behavior near the surface is required to model refraction effects. The exponential model atmospheres prepared by the Central Radio Propagation Laboratory of the National Bureau of Standards (CRPL)[4] are also shown on Fig. 1 for the extremes of surface radio refractivity ($N_s = 200$, 450 N units).

Individual radio-refractivity profiles may show significant departures from the model profiles. Figure 2 shows two profiles calculated from routine National Weather Service (NWS) radiosonde observations made in Albany, New York during 1966. These profiles were selected from a large number of profiles used for a statistical analysis of bending and path length change in the sections that follow. For comparison, the profile for 45°N latitude U.S. Standard atmosphere for July and the humidity model is also included in Fig. 2. The variability of the measured profiles with height, the alternating regions or layers of large and small radio-refractivity gradients are typical of all measured profiles. These layers are smoothed in the process of generating model profiles and are generally of little importance either to bending or path-length computations. They may, however, be of importance to bending computations at low elevation angles (less than 5°). The layers with large gradients are indicative of stable regions in the atmosphere where turbulence-caused radio-refractivity fluctuations will occur. The fluctuations give rise to scattering phenomena.

2.5.3. Bending

An accurate estimate of the pointing error caused by the curvature of a ray trajectory is required to point an antenna at a radio source. For a horizontally homogeneous (stratified) atmosphere, bending occurs only in the great circle plane, and azimuth-angle correction is not required. The elevation-angle correction or elevation-angle error depends on ray curvature. For astronomical sources, the true direction to the source (direction in the absence of an atmosphere) at the antenna and the direction to the source (tangent to the ray) at a point on the ray above the atmosphere are identical. The elevation-angle error for astronomical sources, therefore, is equal to the bending angle.

2.5. REFRACTION EFFECTS IN THE NEUTRAL ATMOSPHERE

The bending angle (or bending) may be computed for a known radio-refractivity profile using the ray-tracing techniques. As an example, we have computed ray position and bending for several rays using a model-refractivity profile. The results of the computations are presented in Table I, which lists

TABLE I. Ray Parameters for a Standard Atmosphere[a,b]

Initial elevation angle (deg)	Height (km)	Range (km)	Bending (mdeg)	Elevation-angle error (mdeg)	Range error (m)
0.0	0.1	41.2	97.0	48.5	12.63
	1.0	131.1	297.8	152.8	38.79
	5.0	289.3	551.2	310.1	74.17
	25.0	623.2	719.5	498.4	101.0
	80.0	1081.1	725.4	594.2	103.8
5.0	0.1	1.1	2.6	1.3	0.34
	1.0	11.4	25.1	12.9	3.28
	5.0	55.2	91.7	52.4	12.51
	25.0	241.1	176.7	126.3	24.41
	80.0	609.0	181.0	159.0	24.96
50.0	0.1	0.1	0.2	0.1	0.04
	1.0	1.3	1.9	1.0	0.38
	5.0	6.5	7.0	4.0	1.47
	25.0	32.6	14.3	10.3	3.05
	80.0	104.0	14.8	13.4	3.13

[a] U.S. Standard Atmosphere Supplements, 1966. Environmental Sci. Serv. Administration. Dept. of Commerce, Washington, D.C. (1966).
[b] N. Sissenwine, D. D. Grantham, and H. A. Salmela, AFCRL-68-0556, Air Force Cambridge Res. Lab., Bedford, Massachusetts (October 1968).

the initial elevation angle (direction of tangent to the ray at the starting point) for each of the rays, positions of points along the ray, and bending for each of the points. The range values and heights specify the positions along the ray. The bending values are reported in millidegrees (1 mdeg = 3.6 arcsec). Table I also contains other entries which will be discussed later in this section.

The model profile for these computations was obtained from the U.S. Standard Atmosphere for 45°N latitude July,[6] and the humidity-profile

[6] U.S. Standard Atmosphere Supplements, 1966, Environmental Sci. Serv. Administration, U.S. Dept. of Commerce, Washington, D.C. (1966).

model developed by Sissenwine et al.[7] The bending computations show that most of the bending takes place near the surface; above 10 km little additional bending occurs. For the 0° initial elevation-angle ray, 41% of the bending was accomplished within the first kilometer of the atmosphere, 56% within the first two, and 89% by a height of 10 km. At a 25-km height 99% of the bending is reported and by 50 km essentially all the bending. Ionospheric bending may cause additional ray curvature for frequencies below 2 GHz as discussed in Section 2.1. The total bending at 2 GHz for a typical daytime ionosphere is within 1% of the value for the troposphere alone for all initial elevation angles. For higher initial elevation angles, the bending values do not increase as rapidly with height. The 5° initial elevation-angle ray attains 14% of the bending by a height of 1 km and 99% by 25 km; the 50° initial elevation-angle ray attains 13% by 1 km and 97% by 25 km.

The bending values for the rays reported in Table I are typical of those that occur, but actual radio-refractivity profiles change from hour to hour and from one location to another. In order to explore the variability we made additional computations using a large number of profiles obtained at a single location to investigate the statistical variation of bending. Figure 3 shows curves representing average bending and root mean square (rms) bending as a function of *initial elevation angle*, that is, antenna-pointing angle *not* true elevation. The atmospheric profiles here were obtained from the routine, twice-daily radiosonde soundings made by the National Weather Service in Albany, New York. Data for each day in February and August, 1966–1968 were used, subject to the condition that the radiosondes reached a height of 25 km. This requirement was met by 77% of the profiles. The computed values gave 97–99% of the total bending.

For comparison Fig. 3 also shows values computed for the 45°N. latitude model[6,7] (Table I), the CRPL statistical model,[8] and the CRPL exponential model.[4] The CRPL statistical model includes 77 profiles chosen to represent a wide range of locations and extreme weather conditions. A straight line segment shows the slope of the function representing cosecant of the elevation angle csc(α_0). At the higher elevation angles the rms variations follow this dependence, but the bending of course does not. Dotted lines in Fig. 3 show bending computed for the extreme values of the CRPL exponential atmosphere shown in Fig. 1. These curves, corresponding to surface refractivity values of $N_s = 200$ and 450, indicate the extent of possible variations in bending values. Surface refractivity for the Albany data ranged from $N_s = 296$–360, well within the extremes of the CRPL exponential model.

[7] N. Sissenwine, D. D. Grantham, and H. A. Salmela, AFCRL-68-0550, Air Force Cambridge Res. Lab., Bedford, Massachusetts (October 1968).
[8] B. R. Bean, B. A. Cahoon, and G. D. Thayer, Nat. Bur. Std. Tech. Note 44, Central Radio Propagat. Lab., Nat. Bur. Std., Boulder, Colorado (1960).

2.5. REFRACTION EFFECTS IN THE NEUTRAL ATMOSPHERE

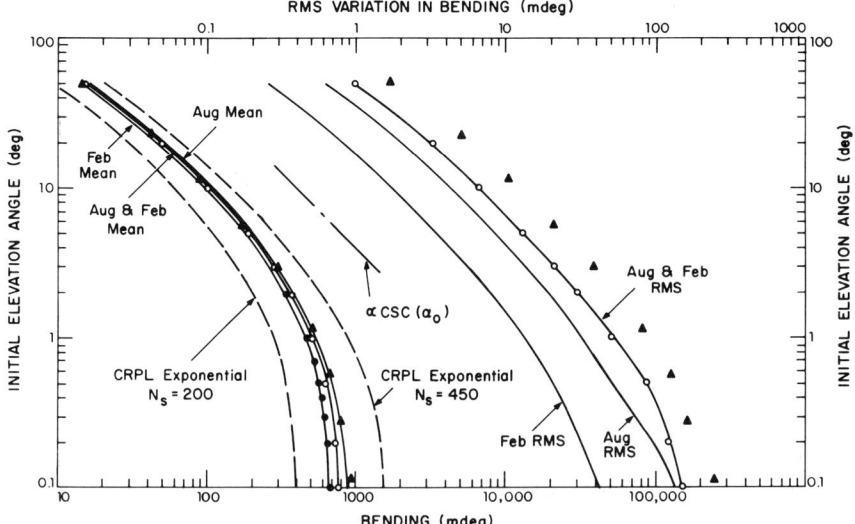

FIG. 3. Mean bending values and rms deviation from the mean for the Albany, New York data set. All solid curves represent Albany data, while values computed for the 45° N latitude model (●) and the CRPL statistical model (▲) are also shown.

The Albany radiosonde soundings were made routinely at 0000 and 1200 UT. We expect that our mean values are within 10% of the true mean values, allowing for the fact that the sounding times do not coincide generally with periods of maximum or minimum bending. Radiosonde data also tend to underestimate bending errors due to the smoothing of the profiles caused by sensor lag.

Let us consider now the antenna-pointing problem which requires the estimation of bending values in order to compensate for atmospheric refraction in pointing toward astronomical radio sources. Since a significant part of the total bending occurs in the first kilometer of atmosphere (see Table I), we would expect a correlation between total bending and the surface value of the index of refraction, a quantity which is relatively easy to measure. Figure 4 shows an example of the statistical relationship between these two quantities. The bending values for an initial elevation angle of 5° are plotted here against the surface refractivity for the entire Albany data set. The solid line on the figure represents a straight-line fit by the method of least squares. Surface refractivity N_s is clearly a useful predictor of the total bending. For various elevation angles, Table II, based on further analysis of the Albany data, gives the regression coefficients a and b, where the estimated total bending ζ is computed from $\zeta = a + bN_s$. Table II also lists the correlation

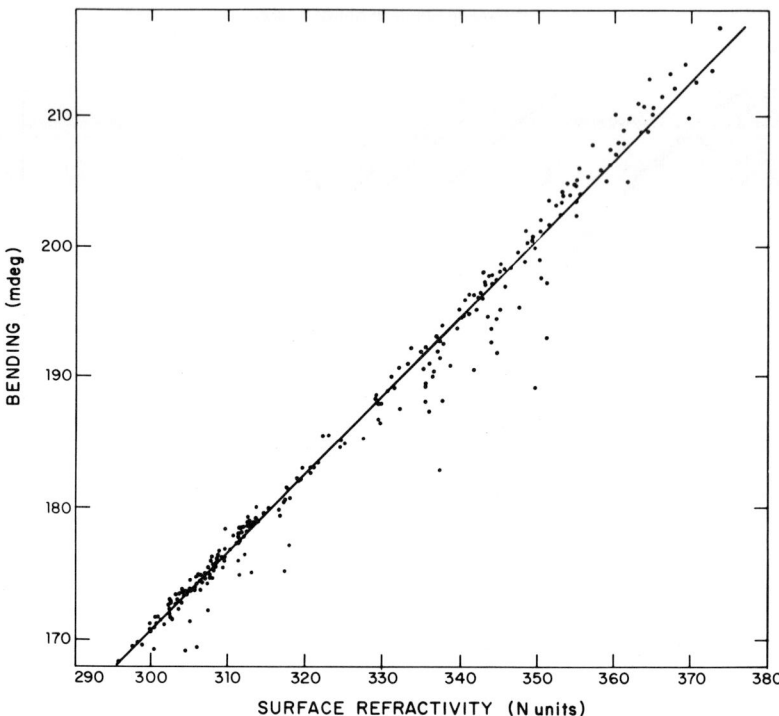

FIG. 4. Scatter diagram of bending versus surface radio refractivity for an initial elevation angle of 5° for the Albany, New York data set at a height of 25 km.

TABLE II. Regression Parameters for Estimating Bending Angle, Given Surface Refractivity

Elevation angle (deg)	a (mdeg)	b (mdeg/N unit)	Correlation coefficient	rms error (mdeg)	95% deviation (mdeg)
0.1	−1112.8	5.778	0.81	89.0	151.0
0.2	−889.2	4.951	0.85	64.0	119.0
0.5	−512.3	3.473	0.94	26.0	58.0
1.0	−268.3	2.372	0.97	12.0	28.0
2.0	−95.9	1.409	0.99	5.0	11.0
3.0	−41.0	0.985	0.99	3.1	6.6
5.0	−10.2	0.610	0.99	1.9	3.8
10.0	−0.3	0.309	0.99	0.99	1.8
20.0	+0.6	0.151	0.99	0.49	0.88
50.0	+0.2	0.046	0.99	0.15	0.27

2.5. REFRACTION EFFECTS IN THE NEUTRAL ATMOSPHERE

coefficients, the rms errors in the estimates, and the 95% confidence values. Ten values of the initial elevation angle are included in this table. Bending varies smoothly as a function of elevation angle as Fig. 3 shows, and the tabular values can be interpolated as a function of elevation angle to obtain values of a and b for arbitrary elevation angles.

Figure 5 shows interpolated values of the regression coefficients plotted

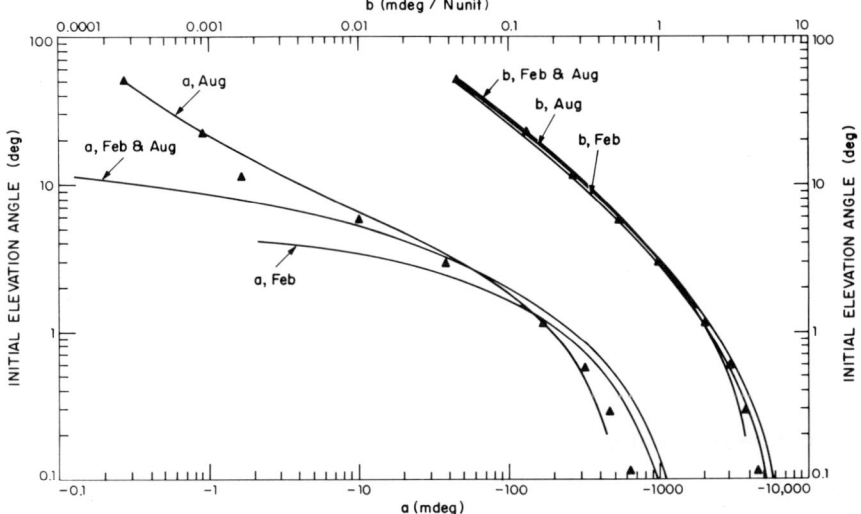

FIG. 5. Regression coefficients for the prediction of bending values given a surface radio-refractivity measurement. The coefficients for the Albany data set are based on $\zeta = a + bN_s$. Coefficients for the CRPL statistical model (▲) are also plotted.

against initial elevation angle. The regression coefficients for the CRPL statistical model are also plotted for comparison. All the data sets appear to give nearly the same values for the slope coefficient b. The constant coefficient a however, appears to depend somewhat on the choice of data set. This variability may to some degree be a consequence of the limited nature of each data set. The Albany data and the CRPL statistical model proposed by Bean et al.[8] lead to nearly the same regression coefficients. For example, consider a surface refractivity of 375 N units and an initial elevation angle of 10°. The difference between predictions based on the CRPL statistical model and on the Albany data (February and August) is 3.3 mdeg. Considered separately the February and August data sets lead to prediction's that differ by 1.8 mdeg. Although a correction procedure tailored for a particular climate may be somewhat more accurate, the Albany data (Table II) or the CRPL model[8] provide a practical basis for refraction correction.

2.5.4. Path Length

The propagation time of an electromagnetic wave through an inhomogeneous, nonionized medium differs from the transit time in free space (vacuum) because the ray path is curved and the propagation velocity is reduced (index of refraction greater than one). The changes in propagation time are of little importance to single-antenna observations but are of great importance to interferometric measurements particularly where long base lines are used. When the propagation time is used to define an electrical path length or range along a ray, the electrical path length is longer than the straight-line distance in the absence of an atmosphere, and the difference is called the *range error*. For interferometers the difference between the range errors on the two or more paths is of importance. The difference between the range errors for a pair of paths when multiplied by $2\pi/\lambda$, where λ is the wavelength of observation, provides an estimate of the differential phase shift caused by the atmosphere.

The range error ΔR may be expressed as

$$\Delta R = \int_{\text{ray}} n \, ds - D = \int_{\text{ray}} (n - 1) \, ds + S - D, \qquad (2.5.7)$$

where n is the index of refraction, D the straight-line distance, and S the distance measured along the ray path. At high elevation angles the bending is small, and the ray trajectory approximately follows the straight-line path. Then $(S - D) \to 0$ and

$$\Delta R \simeq \int_{\text{ray}} (n - 1) \, ds = 10^{-6} \int_{\text{ray}} N \, ds. \qquad (2.5.8)$$

For the extreme cases of bending, the term $S - D$ is still less than 1 cm for elevation angles above $50°$. For high elevation angles and small amounts of bending, the approximately straight-line path tranverses the important portions of the atmosphere below 10 km within a relatively short horizontal distance. For these distances, the spherical constant-height surfaces may be approximated as plane surfaces:

$$\Delta R \simeq 10^{-6} \csc(\alpha_0) \int_{h_0}^{h} N \, dh, \qquad (2.5.9)$$

and the range error is approximately proportional to the cosecant of the initial elevation angle α_0.

The range error may be computed numerically for a given profile. Table I in Section 2.5.3 includes the results of calculations of range error for a horizontally homogeneous atmosphere specified by the model U.S. Standard Atmosphere for $45°$N. latitude, July,[6] and by the humidity profile proposed

2.5. REFRACTION EFFECTS IN THE NEUTRAL ATMOSPHERE

by Sissenwine et al.[7] For a 90° elevation angle this profile gives a range error of 2.40 m, computed for a height up to 80 km. About 12% of this range error occurs in the first kilometer of height, 73% by 10 km, and 98% by 25 km. We have also calculated from the Albany data set the mean range error, averaged over all seasons, and the rms deviation of these data from the average. Figure 6 shows the results of these computations. For the zenith ray the

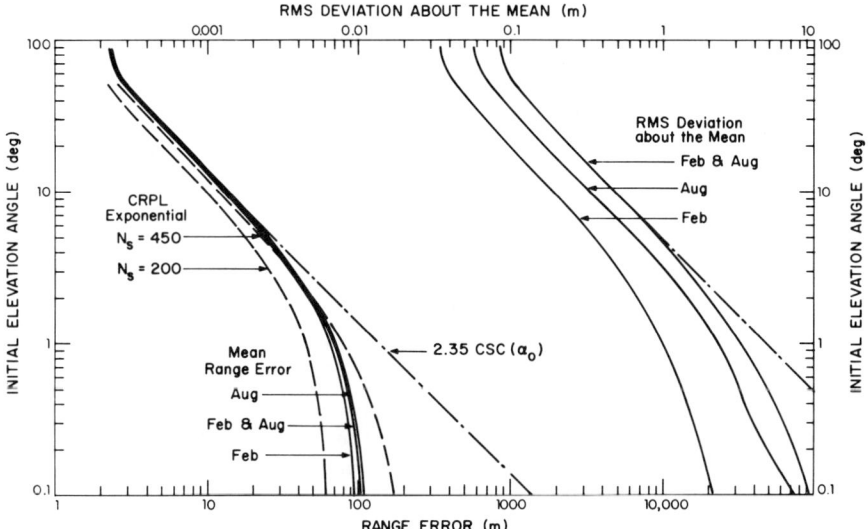

FIG. 6. Mean range-error values and rms deviation from the mean for the Albany, New York data set computed to a height of 25 km.

magnitude of the range error was 2.35 m and the maximum deviation from the average was 21 cm. We show in Fig. 6 dashed curves representing the CRPL model atmosphere[4] for extreme values. At elevations above 5° both these curves lie below the range of possible values based on the Albany data. Ionospheric contributions have not been included in these computations, but for frequencies higher than 20 GHz the ionospheric contribution is less than 1% of the range error at all elevation angles.

The range of validity of the cosecant approximation in (2.5.9) can be observed by noting the dot–dashed lines in Fig. 6. For example, at elevation angles above 20° this approximation deviates from the mean range error by no more than 5 cm.

Measurements of surface refractivity may be used to estimate the range error by means of a procedure analogous to that used in the estimation of atmospheric bending in Section 2.5.3. Table III shows the results of a regres-

TABLE III. Regression Parameters for Estimating Zenith Range Error, Given the Surface Refractivity

Data set	Sample size	Regression parameters[a]		rms deviation (m)	Correlation coefficient
		a(m)	b(m/N unit)		
Albany, N.Y.[b]					
February & August	273	1.16	0.00364	0.033	0.92
February	133	0.906	0.00446	0.024	0.71
August	140	1.32	0.00319	0.039	0.73
CRPL statistical Model[c]	77	1.46	0.00296	0.037	0.98

[a] The parameters are defined so that the range error estimate $\Delta R = a + bN_s$.
[b] See the text for discussion of these data.
[c] B. R. Bean, B. A. Cahoon, and G. D. Thayer, Nat. Bur. Std. Tech. Note 44, Central Radio Propagation Lab., Nat. Bur. Std. Boulder, Colorado (1960).

sion analysis of the Albany data. Estimates of range error obtained in this way were smaller by roughly a factor 2.5 than estimates based on the mean value of the zenith range error for the Albany data. These data suggest that the zenith range error may be estimated with an rms error of perhaps 6 cm if one uses regression coefficients tailored to the local climate. This technique, however, does not take into account clouds and inhomogeneties in the water-vapor distribution.

Equation (2.5.9) provides a connection between the range error and refractivity-profile values in the approximation for a horizontally stratified atmosphere. The index of refraction may be expressed in terms of dry-gas and water-vapor densities ρ_D and ρ_v, respectively, so that (2.5.9) becomes

$$\Delta R \simeq 10^{-6} \csc(\alpha_0) \left\{ \left[\int_{h_0}^{h} 0.223 \rho_D \, dh \right] + \left[\int_{h_0}^{h} 0.332 \rho_v \, dh + \int_{h_0}^{h} 1.72 \times 10^3 \frac{\rho_v}{T} \, dh \right] \right\}$$
$$= \Delta R_D + \Delta R_v \qquad (2.5.10)$$

Here we have separated the dry-gas and water-vapor contributions, and the range error is measured in meters with height in meters and densities in grams per meter cubed.

The first term ΔR_D is called the *dry term* and ΔR_v the *wet term*. Using the hydrostatic equation, we find that the integral of ρ_D with height is approximately proportional to the surface pressure

$$P_0 - P_h = \int_{h_0}^{h} g\rho \, dh \simeq g \int_{h_0}^{h} \rho_D \, dh \qquad (2.5.11)$$

Hence the dry term gives a range-error contribution in meters of

$$\Delta R_D \simeq 2.27 \times 10^{-3} P_0. \qquad (2.5.12)$$

2.5. REFRACTION EFFECTS IN THE NEUTRAL ATMOSPHERE

In the preceding equations P_0 is the surface pressure (mbars) and g the acceleration due to gravity. The connection between surface pressure and the dry term is therefore one-to-one, provided that the hydrostatic and dry-gas density approximations are adequate for the weather conditions and the required precision of estimate of ΔR_D. The dry term consititutes approximately 90% of the total zenith range error and may be estimated with an accuracy of 0.2% or better,[9] or an error of approximately 0.5 cm.

The wet term ΔR_v is the source of variability in the estimate of zenith range error. The wet term depends on conditions along the path which are imperfectly correlated with surface conditions. For example, the changes in refractivity associated with the changes in humidity in clouds are shown in the cross sections in Fig. 7. The data shown in the figure were obtained from a series of aircraft refractometer soundings made along an East–West path just south of Cape Kennedy in Florida.[10,11] Range errors were calculated for rays passing through this distribution from various positions along the ground. Zenith angles of 0 and 70° were assumed. Figure 7 shows the relative range error as a function of horizontal position. The figure can be interpreted as the time history of observations if the clouds drifted overhead. The variation in N units ranged up to 30 in the larger cloud, leading to a change in relative range error of 9 cm for vertical incidence. Note that in Fig. 7 the changes observed at one elevation angle are not simply related to those observed at another.

Basart et al.[12] reported relative changes in range error for paths separated horizontally by 11.3 km at the National Radio Astronomy Observatory, Green Bank, West Virginia. This path separation is large enough for clouds to affect each path separately. For a series of 30-min periods the rms difference between range errors on the two paths varied from about 0.03 to 1 cm. The variations reported by Basart et al. appear consistent with our computations shown in Fig. 7.

A method has been proposed for estimating the wet-term contribution by measuring thermal emission from the atmosphere along a ray path at two frequencies. One frequency must be near that of the water-vapor transition at 22.235 GHz to obtain a measure of the integrated water-vapor emission. The other frequency should be chosen nearby to determine the nonresonant

---

[9] H. S. Hopfield, *Radio Sci.* **6**, 357 (1971).

[10] R. M. Cunningham, Scale and type of atmospheric refractive index anomalies, *Proc. Tropospheric Refraction Effects Meeting*, 3rd **1**, 33-43. ESD-TDR-64-148, Electron. Syst. Div., AFSC, USAF, Bedford, Massachusetts (1964).

[11] J. H. Meyer, Digital atmospheric profile generation, *Proc. Tropospheric Refraction Effects Meeting*, 3rd **I**, 43-57. ESD-TRD-64-148, Electron. Syst. Div. AFSC, USAF, Bedford, Massachusetts (1964).

[12] J. P. Basart, G. K. Miley, and B. G. Clark, *IEEE Trans. Antennas Propagat.* **AP-18**, 375 (1970).

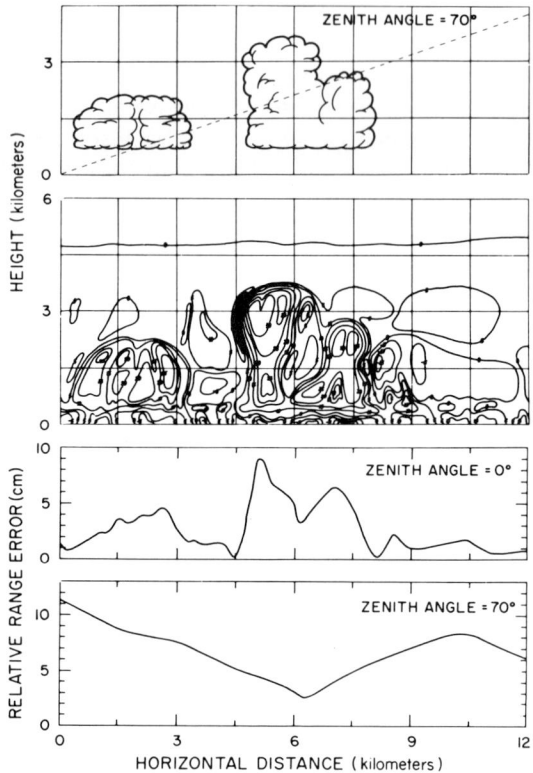

FIG. 7. Radio-refractivity and range-error cross sections for cumulus clouds near Valkaria, Florida, August 7, 1963 [R. M. Cunningham, *Proc. Tropospheric Refraction Effects Meeting*, 3rd **1**, 33–43. ESD-TRD-64-148, Electron. Syst. Div., AFSC, USAF, Bedford, Massachusetts (1964); J. H. Meyer, *ibid* **1**, 43–57.]

emission from clouds, etc. Although the difference in thermal emission at two such frequencies would be related to integrated water-vapor content, the thermal emission and the range error have different origins. The weighting of each elemental contribution of water vapor along the path, as evidenced by expressions (2.3.4) and (2.5.7), is quite different. Such radiometric measurements may, however, be of some use in refining estimates of path length. Numerical experiments by Schaper *et al.*[13] indicate that in the absence of clouds radiometric measurements may improve estimates of range error by a factor of two or three as compared to estimates based of surface-refractivity measurements alone.

[13] L. W. Schaper, Jr., D. H. Staelin, and J. W. Waters, *Proc. IEEE* **58**, 272 (1970).

3. RADIOMETERS

3.1. Radiometer Fundamentals*

3.1.1. Introduction

In a radiometer system, the purpose of the receiver is to select and amplify the signal received by the antenna and to provide an output signal to a chart, digital recorder, or other display/processing unit. Of prime importance is the accurate reproduction of the amplitude and spectral characteristics of the input signal. The receiver must be linear in output even when it operates over a large dynamic range of amplitude and should introduce a minimum of noise to the signal being amplified.

Typical power gains of receiver systems are 120 dB. Bandwidths used in observations vary in width from a few kilohertz (the frequency resolution required for precision radio spectral line measurements) up to a few gigahertz (for a wide bandwidth continuum receiver at centimeter or millimeter wavelengths). Output integration times might vary from 10 μsec for pulsar measurements up to tens of minutes for spectral line observations.

The signals of interest are at very low power levels. For a spectral line observation the power received from a given spectral feature might be of the order 10^{-20} W. More typically, for a continuum observation a source might give 10^{-16} W at the receiver input (approximately the same as the noise power observed through a 100-kHz bandwidth from a resistor at liquid nitrogen temperature).

Given the wide range of signal amplitudes and spectral characteristics from different sources, it follows that a receiver system is generally designed with particular measurements in mind, and therefore system specifications will vary widely.

3.1.2. Types of Signals in Radio Astronomy

In its initial development, radio astronomy dealt primarily with continuum radiation. Most observations employed wide bandwidths (greater than 1 MHz) and long time constants. One exception to this general rule was observation of the sun, where more detailed studies of the time and frequency

---

* Chapter 3.1 is by R. M. Price.

structure of the radiation received were carried out. The discovery of spectral line emission from neutral hydrogen atoms created much interest in spectral line observations. This interest has been greatly enhanced by the subsequent discovery of spectral lines from a number of different molecules in interstellar space. In addition, the discovery of pulsars has led to renewed interest in high-resolution (in both time and frequency) studies of received signals.

3.1.2.1. Continuum Signals. The radiation received from continuum radio sources has a constant spectral density over the observation bandwidths. Such signals are noiselike and cause a voltage at the receiver input which can be represented, to a very close approximation, as white Gaussian noise. The input signal cannot, therefore, be distinguished from noise generated within the receiver itself.

3.1.2.2. Spectral Line Signals. Transitions between various discrete energy levels in atoms and molecules in interstellar regions give rise to a number of spectral lines that lie in the radio portion of the spectrum. Both emission and absorption lines have been observed. Widths of individual spectral features vary from 1 kHz up to hundreds of kilohertz. For a given transition, the spectrum observed in the direction of a given object might cover a total frequency range of 10 to 20 MHz because of superposition of components with different Doppler velocities in the line of sight.

3.1.2.3. Special Types of Signals. Emission from the sun or from pulsars has a complex time–frequency structure. In the case of pulsars, the emission comes at very regular intervals of the order of 1 sec. The pulse width is typically 5% of the pulse period and in some cases has time-variable substructure at least as fine as 10 μsec. The signals are also dispersed in time, and lower frequencies arrive slightly later than the high-frequency radiation from the same pulse.

Emission from the sun takes on a number of forms, some rather complex. Characteristic time scales for events can vary from milliseconds up to hours. Time-dependent spectral features are often observed in solar events.

3.1.3. Measurement of Radio Astronomy Signals

The basic experimental problem of radio astronomy is the detection of low-level noiselike signals in the presence of the irrelevant noise power that enters the feed or is generated within the receiver system. Most of the techniques described in Section 3.1.5 deal with this problem. First, however, it is necessary to define some of the terms that will be useful in this discussion.

3.1. RADIOMETER FUNDAMENTALS

3.1.3.1. Power in Terms of Equivalent Noise Temperature. It is convenient to represent the power in the system in terms of an equivalent temperature by the use of Nyquist's theorem.[1] This is written as $P = kTB$, where P is the power delivered from a resistor at absolute temperature T through a rectangular bandwidth B, and k the Boltzmann's constant. It is then possible to consider all power in the receiver system as coming from a hypothetical resistor, at the required temperature, placed across the receiver input.

3.1.3.2. System Temperature. The system temperature is a figure of merit applied to a radiometer and represents the total output of the system as coming from a resistor at temperature T_{sys} placed across the input to an ideal receiver. The system temperature is thus composed of the antenna temperature, a correction for losses in the feed line from the feed to the receiver input, and the equivalent noise temperature of the receiver itself. All of the temperatures are referred to the receiver input.

The antenna temperature is the equivalent temperature of the radiation resistance of the antenna feed. It is the sum of the following contributions:

(1) the antenna temperature attributable to the observed source,
(2) the antenna temperature attributable to the radio background on which the source is measured,
(3) power received by the collecting antenna from outside its primary reception beam (i.e., from sidelobes) (This includes contributions from both sky and ground radiation.), and
(4) the antenna temperature caused by radiation produced in the atmosphere

$$T_a = T_{source} + T_{background} + T_{atmosphere} + T_{sidelobes}.$$

It is necessary to correct T_a for losses in the transmission line between the feed terminals and the receiver input. The antenna temperature, as measured at the receiver input, is

$$(1 - \alpha)T_a + \alpha T_0,$$

where T_a is the antenna temperature at the feed terminals, T_0 the ambient temperature, and α the fractional loss (0–1) in the transmission line.

The receiver temperature T_r is the equivalent noise temperature of the receiver. Noise generated within the receiver is nearly constant and comes mostly from thermal noise in components and discrete charge carrier effects in transistors or tubes (shot effect).

[1] H. Nyquist, *Phys. Rev.* **32**, 110 (1928).

TABLE I. System Noise Temperature Contributions in rf Amplifiers and Radiometric Systems

Receiver configuration	Cooled (°K)	Operating frequency (GHz)	Bandwidth (MHz)	rf amplifier type and gain (dB)	T_{sys} (°K)	T_{amp} (°K)	T_{loss} (°K)	T_{feed} (°K)	$T_{2nd\,stage}$ (°K)	Remarks
Total power[a]	No	0.150	4	Transistor 20	400	200		200	1	Tunable over 230 MHz
Total power[b]	No	0.4	—	Degenerate parametric gB = 25[hh] MHz	170					
Dicke modulation[c]	No	0.4	10	Mixer	320[gg]			~40		
Dicke modulation[d]	77	1.4	15	Parametric 16/stage	120	34			6	Two-stage
Dicke modulation[e]	77	1.41	10	Degenerate parametric 20	105	35	~40	~20	10	Gain servo-controlled
Dicke modulation[b]	No	1.42 2.40	—	Up-converter (1.1) and degenerate parametric gB = 60 MHz	160	70 44				
Dicke modulation[f]	1.9	1.65	6.5	T. W. Maser 30	37	0	7	20	1	Tunable over 230 MHz
Total power[g]	4.4	2.295	15	T. W. Maser 47	16.3	4.6	2.22	9.3	.24	
Noise-adding modulation[h]	4.2	2.295	12	T. W. Maser	40					Computer for backend
Dicke modulation[i]	No	2.65	100	Degenerate parametric 17	180	60	46	20	30	

Dicke modulation[j]	16	2.695	70	Parametric 30	60	24	23	12	1.5	Two-stage
Correlation[k]	No	2.7	400	Parametric 15	80	25	21	15	2	Two-stage degenerate
Dicke modulation[l]	No	5.0	200	Parametric ~17	215[gg]	96			~23	100-MHz bandwidth (i.f.)
Dicke modulation[m]	No	8.0	1000	Tunnel diode 10.8	1300	630				Three-stage
Dicke modulation[n]	No	19.0	1600	Tunnel diode 42	1340	1165				
Dicke modulation[o]	No	110.0	40	Mixer	1200[gg]					
[p]	No	1.8	50	Parametric 10		170				Integrated stripline construction
[q]	No	3.52	600	Parametric 20		120				Balanced configuration
[r]	No	4–8	~120	Parametric 17		100–240				Balanced configuration
[s]	4.2	4.0	500	Parametric 16		14				Two-stage, second stage at 77°, $T_{amp} = 83°$
[t]	16	4.0	150	Parametric 30		25				Two-stage tunable 3.7–4.26 GHz
[u]	4.2	4.1	240	Parametric 24		13				Two-stage
[v]	No	7.5	500	Parametric 15		63				Balanced configuration

TABLE I (cont.)

Receiver configuration	Cooled (°K)	Operating frequency (GHz)	Bandwidth (MHz)	rf amplifier type and gain (dB)	T_{sys} (°K)	T_{amp} (°K)	T_{loss} (°K)	T_{feed} (°K)	$T_{2nd\ stage}$ (°K)	Remarks
w	No	8–11.6	3600	Tunnel diode 9		750				
x	4.4	8.448	17	T. W. Maser 41.5		18				Tunable over 140 MHz
y	No	18	550	Parametric 20		280				Double tuned
z	77	18.7	600	Parametric 20		100				Two-stage
aa	20	24	85	Degenerate parametric 15		150				2-GHz electronic tuning range
bb	No	26	1000	Mixer −9		2600				Integrated
cc	No	33	800	Degenerate parametric 15		140				
cc	No	35	500	Mixer		750				
dd	20	45.8	390	Degenerate parametric 23		∼40[gg]				
ee	No	46.0	160	Degenerate parametric 20		354[gg]				
ff	No	94	200	Mixer −10.2		11,000				Integrated mixer i.f., $F_{dB} = 16$
cc	No	94	500	Mixer		1550				

[a] R. A. Batchelor, *Proc. IRE (Aust.)* **30**, 99 (1969).
[b] B. J. Robinson, *Proc. IRE (Aust.)* **24**, 119 (1963).
[c] M. B. Mackey, *Proc. IRE (Aust.)* **25**, 515 (1964).
[e] M. Uenohara and J. P. Elward, Jr., *IEEE Trans. Antennas Propagat.* **AP-12**, 939 (1964).
[f] F. F. Gardner and D. K. Milne, *Proc. IRE (Aust.)* **24**, 127 (1963).
[g] O. E. H. Rydbeck and E. Kollberg, *IEEE Trans. Microwave Theory Tech.* **MTT-16**, 799 (1968).
[g] G. S. Levy, D. A. Bathker, W. Higa, and C. T. Stelzried, *IEEE Trans. Microwave Theory Tech.* **MTT-16**, 596 (1968).
[h] G. D. Nicolson, *IEEE Trans. Microwave Theory Tech.* **MTT-18**, 169 (1970).
[i] B. F. C. Cooper, T. E. Cousins, and L. Gruner, *Proc. IRE (Aust.)* **25**, 221 (1964).
[j] N. J. Keen, *Nachr. Tech. Z.* **3**, 168 (1971).
[k] R. A. Batchelor, J. W. Brooks, and B. F. C. Cooper, *IEEE Trans. Antennas Propagat.* **AP-16**, 228 (1968).
[l] M. P. Hughes, E. Moley, D. R. Parenti, and J. J. Whelehan, *IEEE Trans. Antennas Propagat.* **AP-13**, 423 (1963).
[m] T. V. Seling, *Proc. IEEE* **52**, 423 (1964).
[n] H. G. Pascalar, P. R. Jordan, *Proc. IEEE* **54**, 442 (1966).
[o] P. M. Solomon, K. B. Jefferts, A. A. Penzias, and R. W. Wilson, *Astrophys. J.* **168**, L107 (1971).
[p] P. Bura, R. Camisa, W. Y. Pan, S. Yuan, and A. Block, *IEEE Trans. Microwave Theory Tech.* **MTT-16**, 424 (1968).
[q] H. C. Okean and H. Weingart, *IEEE Trans. Microwave Theory Tech.* **MTT-16**, 1057 (1968).
[r] J. Edrich, *IEEE J. Solid State Circuits* **SC-7**, 32 (1972).
[s] J. G. Josenhans, *IEEE Trans. Microwave Theory Tech.* **MTT-16**, 791 (1968).
[t] R. Damino and J. Kliphuis, *Proc. IEEE* **54**, 1618 (1966).
[u] J. A. Stoveman, *Proc. IEEE* **54**, 1500 (1966).
[v] L. E. Dickens, *Proc. IEEE* **60**, 328 (1972).
[w] H. C. Okean and P. J. Meier, *1971 IEEE-GMTT Int. Microwave Symp. Digest, Washington, D.C.* 186 (1971).
[x] S. M. Petty and R. C. Clauss, *IEEE Trans. Microwave Theory Tech.* **MTT-16**, 47 (1968).
[y] Y. Kinoshita and M. Maeda, *1970 IEEE-GMTT Int. Microwave Symp. Digest, Newport Beach, California* 95 (1970).
[z] S. Takahashi, M. Nojima, T. Fukada, and A. Yamada, *1970 IEEE-GMTT Int. Symp. Digest* 100 (1970).
[aa] J. Edrich, *1970 IEEE-GMTT Int. Microwave Symp. Digest, Newport Beach, California* 104 (1970).
[bb] D. Dobramysl, *1971 IEEE-GMTT Int. Microwave Symp. Digest, Washington, D.C.* 18 (1971).
[cc] M. Cohn, L. E. Dickens, and J. W. Dozier, *1969 IEEE-GMTT Int. Microwave Symp. Digest, Dallas* 225 (1969).
[dd] J. Edrich, *IEEE Trans. Microwave Theory Tech.* **MTT-22**, 581 (1974).
[ee] J. Edrich, *Proc. IEEE* **59**, 1125 (1971).
[ff] J. J. Kirwan and C. J. Abronson, *Proc. IEEE* **54**, 809 (1966).
[gg] Double-sideband value of noise temperature.
[hh] gB is gain-bandwith factor in MHz.

Thus the system temperature is given by

$$T_{sys} = T_a(1 - \alpha) + \alpha T_0 + T_r.\dagger$$

Table I shows the contributions to total system temperature for several operational radiometer systems. In many systems most of the power received at the feed has its origin in ground spillover radiation. Another large contribution often comes from the switch at the receiver input (placed there to operate the system in the modulated mode for stabilization of the gain). Within the receiver itself, the main contribution comes from the first stage of amplification. For an amplifier with cascaded stages, the second-stage contribution T_2 is less important. It is given by $T_2 = T_{amp2}/G_{amp1}$, where T_{amp2} is the noise temperature of the second stage, and G_{amp1} the power gain of the first-stage amplification. This holds for individual stages or for separate amplifiers. Thus the contribution of later stages is of decreasing importance in low-noise systems.

3.1.3.3. Minimum Detectable Signal: Statistical Considerations. In order to predict the performance of a system, and compare different types of systems, the concept of a minimum detectable signal (MDS) for a radiometer system has been established. Simply stated, the MDS is the signal that causes deflection at the receiver output equal to the standard deviation (rms) of the output fluctuations about the mean.

It can be shown[3,4] that the standard deviation ΔT of the output about the mean for a radiometer system is

$$\Delta T = T_{sys}/(B\tau)^{1/2}, \qquad (3.1.1)$$

where T_{sys} is the total system temperature (including the contribution due to the observed source), B the noise equivalent (rectangular) bandpass, and τ the true integration time at the receiver output. Ideally then, the value of ΔT

---

[2] M. E. Tiuri, *in* "Radio Astronomy" (J. D. Kraus, ed.), p. 236. McGraw-Hill, New York, 1966, and references therein.

[3] E. J. Kelly, D. H. Lyons, and W. L. Root, *J. Soc. Ind. Appl. Math.* **11**, 235 (1963).

[4] A. van der Ziel, "Noise: Sources, Characterization, Measurement." Prentice-Hall, Englewood Cliffs, New Jersey, 1970.

---

† Another figure of merit for a receiver system is noise figure F. This quantity is related to the receiver temperature by the relationship

$$T_r = (F - 1)290°.$$

The above noise figure is defined for a single sideband system.[2] For a double sideband system,

$$T_r = \tfrac{1}{2}(F - 1)290°.$$

The noise figure is also sometimes expressed in decibels, where $F_{dB} = 10 \log_{10} F$.

is dependent on (1) the total system temperature, (2) the width and shape[5,6] of the input bandpass accepted at the receiver, and (3) the types of smoothing applied.[6] Tables II and III give factors to be applied for several representative

TABLE II. Input Bandpass Filters and Their Widths

Input bandpass filter	To obtain relative sensitivity multiply B in Eq. (3.1.1) by
Rectangular	1.00
Gaussian[a]	0.94
Single-tuned circuit[a]	2.00
Two synchronously-tuned stages[a]	1.29

[a] The filter bandwidth B represents the width between half-power points.

TABLE III. Output Smoothing Filters and Their Equivalent Time Constant

Output smoothing filter	Value of τ in Eq. (3.1.1)
Single section RC low pass	$2RC$
Critically damped LRC	$4(LC)^{1/2} = 8L/R$
Ideal low pass $<f_0$	$1/f_0$
Ideal running mean over T	T

types of input bandpasses and output filters used in radiometer systems. These factors are discussed in greater detail by Bracewell,[5] Colvin,[6] and Cooper.[7] See Section 3.5.4 for a discussion of ΔT for autocorrelation spectrometers.

Although the concept of MDS is quite valid as a theoretical quantitative measure of system performance, in practice a signal three to five times this level is considered more reliable for detection of a weak source. Positive detection of a source also depends on the confusion limit[8] for the antenna at the observing frequency and other factors such as the amount of radio-frequency interference at the time of the observation and the receiver-system stability.

[5] R. N. Bracewell, in "Handbuch der Physik" (S. Flugge, ed.), Vol. LIV, p. 42, Springer-Verlag, Berlin and New York, 1962.
[6] R. S. Colvin, Stanford Radio Astronomy Inst. Publ. 18A, Stanford Univ, Stanford, California, 1961.
[7] B. F. C. Cooper, *Proc. IRE (Aust.)* **31**, 41 (1970).
[8] S. Von Hoerner, *Publ. Nat. Radio Astron. Observ.* **1**, 19 (1962).

3.1.3.4. Minimum Detectable Signal: Practical Limitations. In an ideal system, the statistical rms output fluctuations caused by the system noise temperature set the limit to the sensitivity of the system. In practice this is often not the case. For instance, a reasonable system temperature might be 100°. This means that an overall gain change of 1% or a variation in bandpass characteristics that permits passage of 1% more power will cause a change in output level equal to that of a radio source that would give an antenna temperature of 1°. Such changes in gain and bandpass in a receiver system are referred to as "instabilities" and generally set the limit on the system sensitivity (minimum detectable signal).

In the simplest case, such instabilities can be expressed as $\Delta G/G$ and $\Delta B/B$ and can be considered as random variables. Then the rms fluctuations at the receiver output are proportional to

$$T_{\text{sys}} \left[\frac{1}{B\tau} + \left(\frac{\Delta G}{G}\right)^2 + \left(\frac{\Delta B}{B}\right)^2 \right]^{1/2}.$$

Only a few studies have been made of the characteristics of instabilities. It is generally considered[6] that they exhibit a concentration of spectral density at very low frequencies (much less than 1 Hz), that their occurrence rapidly decreases as shorter time intervals are considered, and that they exhibit possible peaks associated with frequencies common in systems elements (e.g., 60 Hz from power lines), or microphonics and mechanical excitation.

The problems caused by such instabilities depend greatly on their spectra, and the type of observations for which the receiver system is used. For instance, slow drifts are of little consequence in a source survey or measurement of a single source where the drift can easily be removed from the baseline. Such a drift might, however, render a scan of the galactic continuum background useless.

3.1.4. The Basic Receiver System

Within the receiver, the signal undergoes a number of processes: filtering, amplification, frequency changes (mixing or heterodyning), detection, and integration. The components or devices that carry out each of these functions are, in general, imperfect. They add noise to the signal, have nonlinear characteristics, or exhibit instabilities. We will now consider a receiver system and the functions that take place in each part of the system.

3.1.4.1. The Ideal Receiver. We often define an "ideal" receiver system which we can describe in functional and mathematical terms. It is then possible to compare actual receiver systems with this ideal to determine the relative merits of different types of receivers.

3.1. RADIOMETER FUNDAMENTALS

We will assume that our ideal system has the following characteristics: filters with rectangular bandpasses, absolute gain and bandpass stability, perfect square-law detection, and true integration at its output.

Figure 1 shows a total-power receiver. The signal and its spectrum at each stage of the receiver system are shown. In this system, a band of frequencies of bandwidth B (or Δf) is centered about the signal frequency (or observing

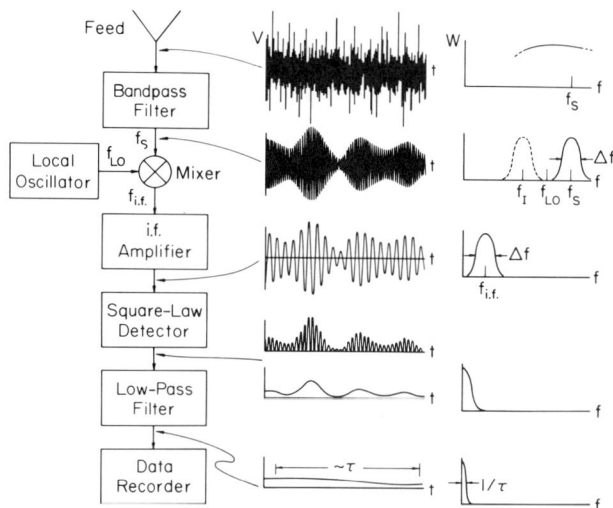

FIG. 1. Superheterodyne total-power receiver system showing a received signal and its spectrum as it passes through the receiver. (See the discussion in Section 3.1.4.)

frequency) f_s. This signal is then mixed with the output of a local oscillator at f_{LO}. This shifts the band down to an intermediate frequency ($f_{i.f.}$), often 10 or 30 MHz.

The i.f. signal which has the same spectral and intensity information as the original rf band is then amplified. In most receivers the bulk of the amplification occurs at the intermediate frequency. It is generally the bandpass characteristics of the i.f. amplifier that determine the bandpass for the receiver. (In Fig. 1 the bandpass filter is indicated separately for clarity.)

The signal then passes through a square-law detector, which detects the envelope of the i.f. waveform. The output of a square-law detector is proportional to the square of the applied voltage amplitude. Thus the detector output is proportional to the power incident on the receiver input.

Finally the detector output is integrated, or passed through a low-pass filter to remove its high-frequency components. This output signal goes to the output recording device.

3.1.4.2. Single-Sideband and Double-Sideband Systems.

In a superheterodyne receiver the centers of the bands accepted at

$$f_{LO} + f_{i.f.} = f_S, \qquad f_{LO} - f_{i.f.} = f_I.$$

where S is signal and I image. A system in which both of these frequency bands are retained is called a "double-sideband system." In many cases the image frequency is rejected by a filter preceding the mixer resulting in single-sideband operation.

Double-sideband systems are most commonly used for continuum observations where rf amplification is not used. Since the signal is accepted in both sidebands this reduces the effective noise temperature of the system by a factor of 2 (see Section 3.1.3).

Image-frequency rejection or single-sideband operation is generally used in observations of spectral features (where the image band contributes only noise and no signal). Single-sideband receivers are also necessary where the first-stage rf amplifier has a bandpass too narrow to pass both mixer sidebands, or when interfering signals might be present within the image sideband.

3.1.5. Practical Receiver Configurations

In discussions of receiver systems, reference is sometimes made to the "front end" or the "back end" of the receiver. In general, the front end is the equipment normally found in the receiver box mounted at the focus of the telescope structure. This might typically include the rf components and amplifiers, the mixer, i.f. amplifier, and supporting power supplies and control equipment. The back end, generally in an equipment room distant from the focus, perhaps in the telescope base, includes an i.f. amplifier, detector, synchronous demodulator, integrator, and perhaps the data recorder. Both terms are rather loose and no strict definitions apply to them. For instance, the local oscillator might be found in either section.

3.1.5.1. Total-Power Receivers.

The total power receiver is the simplest receiver system used in radio astronomy. It can take a number of forms. Several of the most common are described here. The primary disadvantage of the total power system is that instabilities caused by changes in gain and bandwidth in the amplifiers affect the output level.

3.1.5.1.1. SUPERHETERODYNE RECEIVERS. A superheterodyne receiver is shown in Fig. 1. The input signal (with or without rf amplification) is mixed with the signal from a local oscillator. The resulting i.f. signal is further amplified and then detected by a square-law detector. The output, after suitable integration, or smoothing, is displayed or recorded. The bandpass of the receiver is generally determined by the i.f. bandpass characteristics or by a bandpass filter either before or after the i.f. amplifiers.

3.1. RADIOMETER FUNDAMENTALS

A superheterodyne receiver can either be a single or double sideband. Most systems presently in use are single sideband. This allows for more flexibility in the choice of the intermediate frequency, allows use of the system for spectral line measurements, and simplifies calibration procedures.

3.1.5.1.2. TUNED RADIO-FREQUENCY RECEIVERS. The other type of total-power system currently in use is the tuned radio-frequency (TRF) receiver[9] shown in Fig. 2. In this system, the radio-frequency signal is

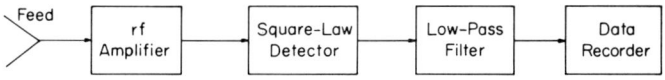

FIG. 2. Tuned radio-frequency, total-power receiver system.

amplified directly to a level where it can be detected. Such receivers usually have very large bandwidth to obtain the required sensitivity, and thus are applicable mainly in the gigahertz range. Tunnel diode amplifiers are commonly used in such receivers. After detection, the signal is smoothed and recorded.

3.1.5.1.3. SELOVE RECEIVERS. The Selove receiver[10] has two separate but identical total-power channels. The signal is amplified in one channel, and a constant-reference signal (e.g., from a resistive termination at a known temperature) is processed in the second. The two outputs are smoothed and their difference taken. This difference in signal is the output of the system. To the extent that the instabilities in one channel match those of the other, the receiver output is stabilized.

Selove receivers as such have not found wide use in radio astronomy, but a variation of this technique has been used for a spectral line receiver. A wideband system is used to observe the relatively narrowband spectral feature of interest. A portion of the bandpass in which the signal does not lie is detected and used as a comparison channel. This method was used by Kerr et al.[11] on a multichannel hydrogen line spectrometer.

3.1.5.1.4. STABILIZATION OF TOTAL-POWER RECEIVERS. Total-power systems exhibit instabilities that limit their wide use in radio astronomy. These systems are sensitive to variations in amplifier characteristics such as gain, bandwidth, and noise temperature. These characteristics, particularly gain, are dependent on power-supply variations, temperature changes, mechanical stresses, and aging. Many of the special techniques applied to radio-astronomy receiver systems seek to eliminate or decrease the effects of instabilities.

[9] H. G. Pascalar and P. R. Jordan, *Proc. IEEE* **54**, 442 (1966).
[10] W. Selove, *Rev. Sci. Instrum.* **25**, 120 (1954).
[11] F. J. Kerr, J. V. Hindman, and C. S. Gum, *Aust. J. Phys.* **12**, 270 (1959).

To stabilize total-power systems we most commonly

(1) regulate power supplies to one part in 10^4 or better;

(2) maintain the receiver in a constant-temperature environment (with changes less than $1°K$);

(3) ensure mechanical rigidity of all components and connections in the system;

(4) maintain a constant local oscillator power level in mixer systems, or a constant bias level in the case of tunnel diode and parametric amplifiers.[12,13] For example, in a parametric amplifier the dependence of gain G on pump power P is given by $\Delta G/G = \sqrt{G}\,\Delta P/P$, where ΔG and ΔP are small changes in gain and pump power, respectively.

3.1.5.2. Modulated Receivers. Receivers in which the gain stabilization is accomplished by modulation techniques are used most commonly in radio astronomy. Such receivers are based on the evidence that the spectrum of gain fluctuations in the amplifiers decreases rapidly with frequency to nearly zero above 1 kHz. If the gain can be monitored at a rate of the order of the highest frequency of gain instability, then corrections for the majority of the gain instabilities are possible by an automatic gain-control circuit.

3.1.5.2.1. DICKE SYSTEM RECEIVER. In this widely used receiver technique[14] (see Fig. 3) the receiver input is switched many times per second

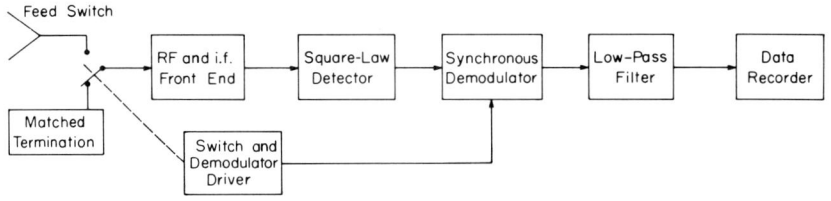

FIG. 3. Dicke modulated receiver system.

between the signal to be measured and a constant reference source. After conversion, amplification, and detection, the modulated signal is passed through a synchronous demodulator that produces an output voltage proportional to the difference in power (or temperature) between the reference and signal inputs. If this difference is kept small, say a few percent of the

[12] P. Penfield, Jr., and R. P. Rafuse, "Varactor Applications," p. 497. M.I.T. Press, Cambridge, Massachusetts, 1962.

[13] B. J. Robinson, B. J. Seegar, C. L. van Damme, and J. T. DeJager, *Proc. IRE* **48**, 1648 (1960).

[14] R. H. Dicke, *Rev. Sci. Instrum.* **17**, 268 (1946).

system temperature, the effect of gain changes on the output will be correspondingly reduced.

In a switched or modulated system, several questions immediately arise. What is the best switching rate, and what is the optimum switching waveform? In answer to the latter question, the intuitive notion that square-wave switching is best is correct. It has also been shown, however, that the demodulation must also be square wave and the bandpass between the detector and the synchronous demodulator be sufficiently broadband to allow passage of the higher harmonics of the switching frequency that contain information from the square wave modulated signal. Table IV, adapted from Colvin,[6]

TABLE IV. Relative Minimum Detectable Signal Depending on Modulation and Demodulation Waveform in Modulated Receiver Systems[a]

Demodulation waveform	Modulating waveform	
	Square wave	Sinusoidal
Square wave	1.00	$\pi/2 = 1.57$
Sinusoidal	$\pi/2\sqrt{2} = 1.11$	$\sqrt{2} = 1.41$

[a] Adapted from Colvin.[6]

shows the relative values of the minimum detectable signal obtained by use of different modulation and demodulation waveforms.

The theoretical sensitivity of a modulated system is only half that of a total-power system (if we assume optimum switching and demodulation waveforms). A factor of root two is caused by the fact that only half of the time is spent looking at the signal. Another factor of root two arises from the comparison of the two noiselike signals in the synchronous demodulator.

In practice, the sensitivity of the Dicke modulated system also suffers from two other factors. First, the modulating switch in the signal line at the receiver input introduces losses that increase the effective system noise temperature. Second, because of the response of the receiver system to the transients during the switch operation (during which there is likely to be a large mismatch at the receiver input), it is necessary to blank the receiver gain for a short time during the operation of the switch. This time characteristically varies from less than 100 μsec for a diode switch up to 500 μsec for a switched circulator. In the case of a switched circulator and a switch rate of 10 Hz, the receiver is in the blanked state for 10 msec out of each second, which reduces sensitivity less than 1%. When the circulator switches at 100 Hz with the same blanking time, the loss in sensitivity becomes approximately 5%.

In a modulated system it is desirable to modulate the signal at a rate faster than the fastest characteristic time scale for amplifier variations. Tests by Steinberg[15] and Yaroshenko[16] indicate that gain variations drop to nearly zero above a rate of 1 kHz. In most systems now in use switching rates of a few tens of hertz, and in some cases hundreds of hertz, are used. The practical limitations of the switching system, particularly in ferrite devices, however, has limited switching frequencies in most of the present systems to a few tens of hertz.

3.1.5.2.2. GAIN-MODULATION RECEIVER. In the Dicke system the immunity to variations in gain depends on the system being balanced, i.e., having nearly the same temperatures for the signal and the reference positions of the switch. This is often difficult to attain, especially at frequencies above 1 GHz where the antenna temperature might be 20°K and the reference temperature 80°K. In such situations, a gain-modulation receiver can be used.[17] In this system, shown in Fig. 4, an adjustable passive attenuator

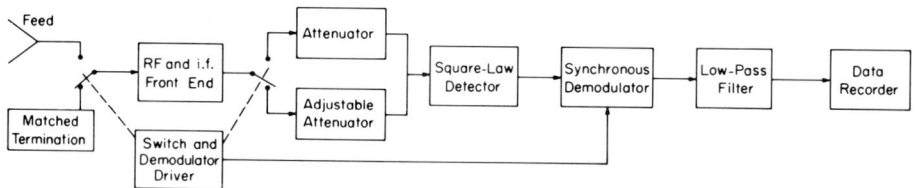

FIG. 4. Gain-modulated receiver system.

is switched in phase with the input switch so as to keep equal the power received from the signal and reference portions of the switch cycle. This method can be applied with the gain modulation before or after the mixer in a superheterodyne system.

3.1.5.2.3. NOISE-ADDING RECEIVER. The noise-adding receiver (NAR) first described by Ohm and Snell[18] is a special type of modulated receiver in which a fixed amount of noise is added periodically (50% duty cycle) to the input of a total-power receiver. The synchronously detected output then contains the information necessary to correct for gain fluctuations in the receiver. Figure 5 shows the block diagram for such a receiver system.

[15] J. L. Steinberg, *Onde Elec.* **32**, 519 (1952).
[16] V. Yaroshenko, *Radiotechnica* **7**, 749 (1964).
[17] T. Orhaug and W. Waltman, *Publ. Nat. Radio Astron. Observ.* **1**, 179 (1962).
[18] E. A. Ohm and W. W. Snell, *Bell Syst. Tech. J.* **42**, 2047 (1963).

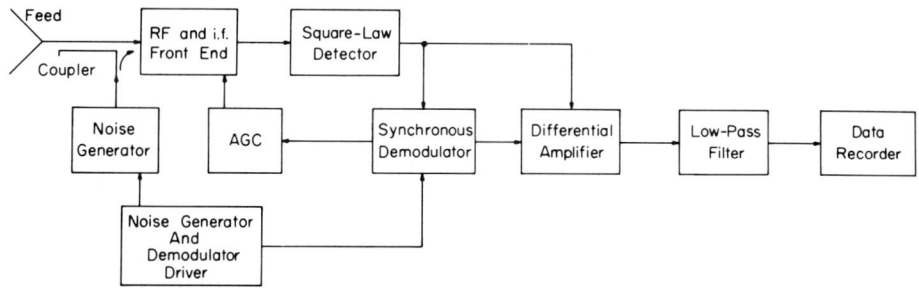

FIG. 5. Noise-adding receiver system.

The NAR is particularly attractive for very low-noise systems ($<20°K$) where the additional contribution to the system temperature of a conventional Dicke switch ($\sim 7°K$) would cause an appreciable fractional increase in system temperature. A version of this type of radiometer which has been developed for use on the ultralow-noise receivers of the Deep Space Network of NASA has been described in the literature.[19,20]

In the NAR the amplitude of the added noise is important, as is the stability of the noise source. Instabilities in the noise source have the same effect as gain variations in a total-power system.

3.1.5.3. Servo Control Receivers. As noted, a simple switched receiver is insensitive to gain variations insofar as it is balanced, i.e., the noise power from the signal and reference sides of the switch are equal. Ryle and Vonberg[21,22] suggested a system, shown in Fig. 6, in which this balance was maintained at all times. A continuously variable noise source injects noise into the reference side of the switch. The level of this signal is controlled by a servo system. In this case the receiver is part of the servo loop. The output signal of the receiver system is simply the value of the level of the noise required to balance the system.

Until recently, such systems have been used only for observations at decimeter and meter wavelengths where current controlled noise diode sources were available. With the development of PIN diode attenuators[23,24]

[19] G. D. Nicolson, *IEEE Trans. Microwave Theory Tech.* **MTT-18**, 169 (1970).
[20] P. D. Batelaan, R. M. Goldstein, and C. T. Stelzried, Space Programs Summary No. 37-65, Vol. II, p. 66, Jet Propulsion Lab. (1970).
[21] M. Ryle and D. D. Vonberg, *Proc. Roy. Soc. London Ser. A* **193**, 98 (1948).
[22] K. E. Machin, M. Ryle, and D. D. Vonberg, *Proc. Inst. Elec. Eng. Part 3*, **99**, 127 (1952).
[23] J. K. Hunton and A. G. Ryals, *IEEE Trans. Microwave Theory Tech.* **MTT-10**, 262 (1962).
[24] H. C. Okean and R. Pflieger, *1971 IEEE-GMTT Intern. Microwave Symp. Digest, Washington, D.C.* 180 (1971).

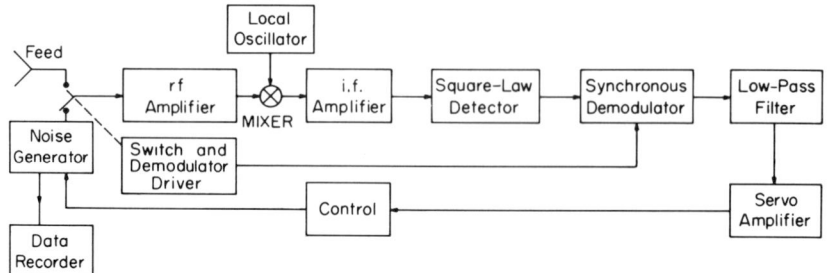

FIG. 6. Servo control (Ryle–Vonberg) receiver system.

and solid-state noise sources[25] this method may be extended into the microwave region.

3.1.5.4. Correlation Receivers

3.1.5.4.1. TWO-CHANNEL CORRELATION RECEIVERS. The application of correlation techniques to a receiver for use on a single-dish radio telescope was first analyzed by Blum.[26] The design and operation of such a two-channel system is described by Batchelor et al.[27] A correlation system is shown in Fig. 7.

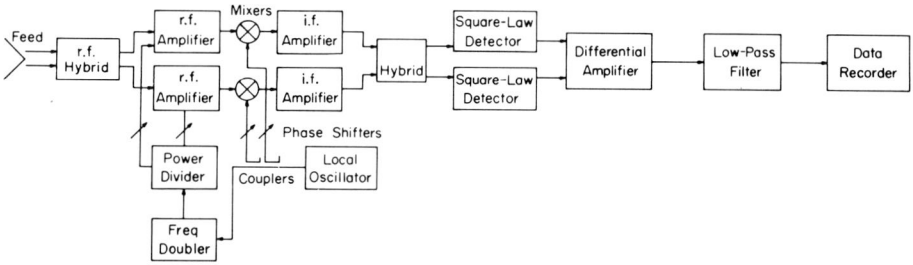

FIG. 7. Two-channel correlation receiver system.

In a correlation receiver, the input signal and a reference signal are combined in a 3-dB hybrid and the outputs fed into two identical receivers. After subsequent amplification and conversion, the signals are passed through another 3-dB hybrid. The sum and difference outputs of this hybrid are detected and fed into a differential dc amplifier. The output of this amplifier is proportional to the difference in the effective noise temperatures of the two original input signals. The dc component from the internal noise contributions of the two receiver channels is cancelled out.

[25] N. J. Keen, *Radio Electron. Eng.* **41**, 133 (1971).
[26] E. J. Blum, *Ann. Astrophys.* **22**, 139 (1959).
[27] R. A. Batchelor, J. W. Brooks, and B. F. C. Cooper, *IEEE Trans. Antennas Propagat.* **AP-16**, 228 (1968).

In the system described by Batchelor *et al.*, two degenerate parametric amplifiers are used in each channel. Since the pump power and local oscillator are derived from the same source, the phase relationships between the pump and local oscillator signals are carefully adjusted[28] to ensure a minimum of correlated noise between the two channels.

A serious problem in correlation systems is cross-coupling between the two channels through the hybrid at the input, or because of mismatch in the feed. Any such noise coupled from one channel to the other is correlated, and appears at the output as a signal. A major problem is that it is often difficult to obtain good isolation between the inputs over a broad bandwidth.

3.1.5.4.2. MANY-CHANNEL CORRELATION RECEIVERS. Aitken[29] has suggested an n-channel correlation receiver where the input power is divided equally into n channels. In each of these channels the signals are amplified and converted to i.f. The output of each channel is then multiplied with the output of every other channel [requiring $n(n-1)/2$ multipliers]. These products are then added to give the final output. Such a system has stability equal to that of the two-channel correlation system, and sensitivity proportional to the number of channels used. For instance, when $n = 6$, the sensitivity is 1.66 times better than for the two-channel case. The primary difficulty in realizing such a system is to build a low-loss n-way power splitter with high isolation between the outputs.

3.1.6. Special-Purpose Receivers

The receiver systems described in the preceding sections are those commonly used for continuum measurements. A number of special techniques are required for adapting such systems to special purposes such as spectral-line observations. In the following sections we describe receivers for measurements involving radio spectral lines, solar radio emissions, pulsar emissions, and very long baseline interferometry.

3.1.6.1. Spectral-Line Receivers. A spectral-line receiver requires facilities often not contained in a continuum receiver. In the "front end" or rf section the principal differences are that (1) the system must be single-sideband (which also excludes the use of degenerate parametric amplifiers), and (2) the local oscillator often has the facility for frequency switching (rather than load switching). In the "back end" or i.f. section, the receiver must have sufficient frequency resolution or selectivity to measure accurately the profile

---

[28] A. S. Van der Vorst and R. S. Colvin, *IEEE Trans. Antennas Propagat.* **AP-14**, 667 (1966).
[29] G. J. M. Aitken, *IEEE Trans. Antennas Propagat.* **AP-16**, 224 (1968).

of the spectral line of interest. Such resolution can be obtained by three different methods.

3.1.6.1.1. MULTICHANNEL SPECTROMETERS. The most straightforward method for obtaining high-frequency resolution over a broadband is to use a multichannel receiver. A number of such multichannel systems have been built and described.[11,30,31] In these systems, the i.f. bandwidth is divided into a number of separate, narrowband channels, generally spaced in frequency by an amount equal to their bandwidths. Individual channel bandwidths typically take on values of 1, 10, or 100 kHz. These systems are described further in Chapter 3.4.

3.1.6.1.2. SINGLE-CHANNEL TUNABLE SPECTROMETERS. Another possible approach to a spectrometer is to tune a single narrowband channel over the frequency band of interest.[32] Such a system has been built for observations of line radiation from hydrogen atoms in external galaxies. Problems arise with calibration of the gain and bandwidth responses at the various frequencies.

3.1.6.1.3. AUTOCORRELATION SPECTROMETERS. Another approach to a spectrometer uses autocorrelation techniques. Because information about the spectrum of the received signal is contained in the autocorrelation func-

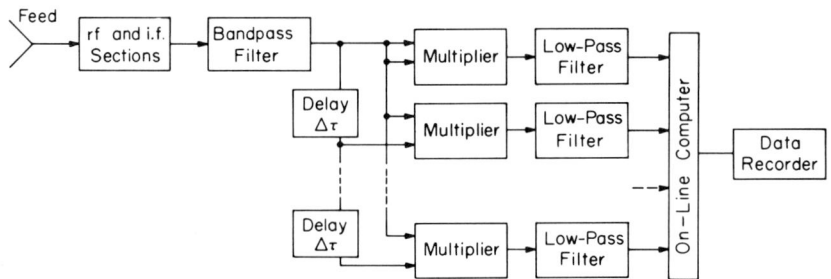

FIG. 8. Block diagram of autocorrelation spectrometer.

tion (ACF) of this signal, it becomes possible to autocorrelate the wideband signal at i.f. and then perform a Fourier transform on the autocorrelation function to obtain the spectrum.[33] The basic configuration of such a system is shown in Fig. 8.

The ACF can be obtained by either analog or digital techniques. Digital techniques are generally preferred because of the stability offered. The application of many-bit (full amplitude) sampling to such a system has been limited

---

[30] R. X. McGee and J. D. Murray, *Proc. IRE (Aust.)* **24**, 191 (1963).
[31] R. A. Batchelor, J. W. Brooks, and M. W. Sinclair, *Proc. IRE (Aust.)* **30**, 39 (1969).
[32] C. A. Muller, *Philips Tech. Rev.* **17**, 351 (1956).
[33] J. Blum, *C. R. Acad. Sci. Paris* **250**, 3279 (1960).

by the time required to perform the necessary multiplications. To avoid this excessive processing time and to obtain broad bandwidths (requiring high sampling rates) a one-bit technique developed by Weinreb[34] is widely used. The sensitivity of such a spectrometer is only 64% of that of a many-bit correlation system.

With the development of faster digital logic units, a two-bit correlation system has been suggested.[35,36] Such a system would have the same gain stability as a one-bit system while the sensitivity increases to 88% of that of a many-bit correlation system. Autocorrelation spectrometer systems are described in detail in Chapter 3.5.

3.1.6.1.4. LOCAL OSCILLATOR SYSTEMS FOR SPECTRAL-LINE RECEIVERS. Special consideration must be given to the local oscillator for a spectral-line receiver. The frequency stability required is higher than that in a continuum receiver. High spectral purity is necessary to avoid any spurious responses to signals outside the band of interest. Such oscillator systems also provide for frequency switching.

3.1.6.2. Interferometer Receivers.[37-39] In principle, the receivers used for interferometer or synthesis measurements are similar to those described elsewhere in this chapter. Interferometer systems are discussed in detail in Part 5.

3.1.6.3. Very Long Baseline Interferometry Receivers. Very long baseline interferometer (VLBI) measurements of continuum sources require no special techniques in the rf and i.f. sections of the receiver except that the local oscillator be controlled by an extremely accurate frequency source such as an atomic clock or hydrogen maser. In such a system, the i.f. signal is processed in a special "back end" for subsequent computer analysis. A complete description of such systems is given in Chapter 5.3. of this volume.

3.1.6.4. Special Pulsar Receivers. Several special equipmental techniques have been developed for observations of pulsars to ensure adequate time and frequency resoution.[40-42] Such systems are described in detail in Chapter 4.5 of this volume.

3.1.6.5. Solar Radio-Astronomy Receivers. Observations of solar radio

[34] S. Weinreb, Tech. Rep. 412, Res. Lab. of Electron. M.I.T. (August 30, 1963).
[35] B. F. C. Cooper, *Aust. J. Phys.* **23**, 521 (1970).
[36] J. G. Ables, B. F. C. Cooper, A. J. Hunt, G. G. Moorey, and J. W. Brooks, *Rev. Sci. Instrum.* **46**, 284 (1975).
[37] B. J. Robinson, *Ann. Rev. Astron. Astrophys.* **2**, 401 (1964).
[38] B. G. Clark, *IEEE Trans. Antennas Propagat.* **AP-16**, 143 (1968).
[39] K. H. Wesseling, *IEEE Trans. Antennas Propagat.* **AP-15**, 332 (1967).
[40] J. M. Sutton, D. H. Staelin, R. M. Price, and R. Weimer, *Astrophys. J.* **159**, L89 (1970).
[41] M. S. Ewing, R. A. Batchelor, R. D. Friefeld, R. M. Price. and D. H. Staelin, *Astrophys. J.* **162**, L169 (1970).
[42] J. H. Taylor and G. R. Huguenin, *Astrophys. J.* **167**, 273 (1971).

events often require receiver systems with very different frequency and time resolution than those used in continuum observations. The signals of interest are generally 1000 times stronger than typical continuum sources. A large variety of systems have been developed and described in the literature. These include a three-octave wideband spectrometer,[43,44] image-forming synthesis arrays,[45] and a number of systems to record the total and polarized radiation from a variety of different types of solar events.[46,47]

3.1.6.6. Polarization Receivers The radiation of interest in radio astronomy is often polarized. For many observations we wish to determine the polarization state of the received radiation.† There are a number of techniques for obtaining this information,[48] some of which require a special receiver system (see Section 5.1.3).

The fundamental method of observing polarized emission is to rotate a linearly polarized feed about its forward beam axis. Any linear polarization will then appear as a sinusoidal variation in the output of the receiver.[50] No special receiver techniques are required for this method except that the feed must be mechanically rotatable and so constructed to minimize changes in mismatch which might arise due to the rotation.

Wielebinski et al.[51] have used an extension of this technique in which the receiver input is switched between two orthogonal dipole feeds. The receiver output is sensitive to power differences between the two probes and thus measures only the linearly polarized component of the radiation. The feed is rotated or set at several fixed positions to obtain all of the linear polarization information.

The preceding methods have the disadvantage that any change in the match of the feeds as they rotate causes a variation in the receiver gain which appears at the receiver output as spurious polarization. This effect can be

---

[43] S. Suzuki, C. F. Attwood, and K. V. Sheridan, *IEEE Trans. Antennas Propagat.* **AP-14**, 91 (1966).

[44] J. Hanasz, U. V. G. Rao, and K. V. Sheridan, *IEEE Trans. Antennas Propagat.* **AP-14**, 804 (1966).

[45] J. P. Wild, *Proc. IRE (Aust.)* **28**, 279 (1967).

[46] K. V. Sheridan, *Proc. IRE (Aust.)* **24**, 174 (1963).

[47] N. Fourikis, *Proc. IRE (Aust.)* **32**, 361 (1971).

[48] M. Born and E. Wolf, "Principles of Optics." Pergamon, Oxford, 1959.

[49] M. H. Cohen, *Proc. IRE* **46**, 172 (1958).

[50] F. F. Gardner and J. B. Whiteoak, *Phys. Rev. Lett.* **9**, 197 (1962).

[51] R. Wielebinski, J. R. Shakeshaft, and I. I. K. Pauliny-Toth, *Observatory* **82**, 158 (1962).

---

† The polarization state of the radiation is completely specified by the Stokes parameters. A description of the definition of Stokes parameters is given in standard optics texts.[48] Their use in radio astronomy has been discussed by Cohen.[49]

minimized by correlating the voltages from the two orthogonal feeds.[52] Correlation techniques are also commonly used in interferometer systems. An anlysis of these techniques and the required receiver system is given by Morris, Radhakrishnan, and Seielstadt.[53]

More recently precision measurements of discrete source polarization have been performed with a correlation receiver of the type described by Batchelor et al.[27] and shown in Fig. 7. In this system the two receiver inputs are provided by the input junction hybrid in the feed. Hybrids following orthogonal feed probes are also used to detect circularly polarized radiation.[54] Receivers designed for particular applications such as solar or pulsar observations where it is desirable to measure all of the Stokes parameters almost simultaneously have been described by Manchester,[54] Suzuki and Tanishi,[55] and Cohen.[56]

3.1.7. Present Trends in Receiver Systems

The preceding sections have been devoted to the various aspects of operational radio astronomy receiver systems. Many of the future developments in such systems can be expected in the improvement of low-noise amplifiers. Several recently developed amplifiers are listed in Table I in Section 3.1.3. Another trend that can be expected is more extensive use of special-purpose computers in the signal processing. Systems have already been built that use a computer for the detection and synchronous demodulation stages of processing.[20] Such digital systems overcome many of the problems of instability such as gain variations, zero drifts, and bandwidth changes. The detector law of such a system is strictly linear.

The use of hybrid-integrated systems[57,58] in the receiver front ends also should reduce problems in cooling these components and reduce losses from long waveguides or cable lengths.

3.1.8. Considerations in Radiometer System Design

To achieve maximum sensitivity from a radiometer system, it is not sufficient to consider only the sensitivity and stability of the receiver system. From Table I it is evident that the present limits to system performance in many

---

[52] G. Westerhout, C. L. Seeger, W. N. Brouw, and J. Tinbergen, *Bull Astron. Inst. Neth.* **16**, 187 (1962).

[53] D. Morris, V. Radhakrishnan, and G. A. Seielstadt, *Astrophys. J.* **139**, 551 (1964).

[54] R. N. Manchester, *Astrophys. J. Suppl. Ser.* **23**, 283 (1971).

[55] S. Suzuki and A. Tsuchiya, *Proc. IRE* **46**, 190 (1958).

[56] M. H. Cohen, *Proc. IRE* **46**, 183 (1958).

[57] P. Bura, R. Camisa, W. Y. Pan, S. Yuan, and A. Block, *IEEE Trans. Microwave Theory Tech.* **MTT-16**, 424 (1968).

[58] R. Damino and J. Kliphuis, *Proc. IEEE* **54**, 1618 (1966).

cases are not caused by receiver noise temperatures. With more common usage of very low-noise amplifiers, system performance depends more on other aspects of the overall system design. A few of the factors that should be considered in designing a radiometer system are:

(1) The choice of radio frequency and bandwidth used in observations must be commensurate with the radio-frequency interference environment.

(2) Post-detection signal processing must be used to remove effects of man-made or natural interference, e.g., to remove lightning from meter-wavelength observations.

(3) The design of collecting surfaces must minimize leakage.

(4) Feeds must be designed to optimize reception of signals from the collecting surface and minimize spillover and distant sidelobes.

(5) Adequate electrical shielding of on-line computers should be used to prevent interference with the receiver system.

(6) Collecting antennas should be designed to allow receiver "front ends" to be as close as possible to feeds to reduce cable and waveguide losses.

(7) In interferometer systems or arrays, the use of rf preamplifiers at each feed point helps overcome the effects of line losses (gain and phase variations of such amplifiers must be checked closely).

(8) Cable attenuation, phase characteristics, and resonances must be considered when long cable lengths are used between "front ends" and "back ends."

ADDENDUM[†]

The literature search for this introduction was concluded in October, 1972. Since that time several publications of interest in this area have appeared. An overview of millimeter wavelength receiver systems has been given by Penzias and Burrus.[59] A millimeter wavelength cooled parametric amplifier has been reported by Edrich.[60] A number of articles of interest are included in the *Proceedings of the IEEE* special issue on radio and radar astronomy.[61]

[59] A. A. Penzias and C. A. Burrus, *Ann. Rev. Astron. Astrophys.* **11**, 51 (1973).
[60] J. Edrich, *IEEE Trans. Microwave Theory Tech.* **MTT-22**, 581 (1974).
[61] *Proc. IEEE* **61**, 1169-1376 (1973).

† October 1974

3.2. Parametric Amplifiers*

3.2.1. Fundamentals of Nonlinear Reactances

Parametric amplification of signals using nonlinear reactances was predicted as early as 1931 by Lord Rayleigh. It was not until 1957, however, that Weiss built the first experimental parametric amplifiers based on the nonlinear inductance of ferrite material.[1] At present the variable-capacitance diode or varactor is the only device used in practice.[2,3]

The varactor is a *p–n* or Schottky barrier (metal–semiconductor) junction diode with a voltage-dependent capacitance

$$C_j = \frac{C_j^0}{(1 - V/\phi)^n}, \tag{3.2.1}$$

where C_j^0 is the junction capacitance at zero bias, V the bias potential, ϕ the contact or barrier potential, $n = 0.5$ for an abrupt junction (ideal Schottky barrier type), and, $n = 0.333$ for a linearly graded diffused junction.

In amplifiers an rf pump current may be applied to the diode. The pump current is much larger than the signal current; this results in a variation of the elastance $S_j(V) = 1/C_j(V)$ between a minimum value $S_{\min}$, as determined by the onset of significant forward current (0.3–3 µA), and a maximum value $S_{\max}$, as determined by the reverse breakdown voltage of the diode. Neglecting higher-order terms the elastance can be written as:

$$S_j(t) = S_m(1 + 2\gamma \cos 2\pi f_2 t), \tag{3.2.2}$$

where f_2 is the pump frequency, S_m the mean value of $S_j(t)$, and γ the capacitance modulation factor which can assume a maximum value

$$\gamma_{\max} = \frac{(S_{\max} - S_{\min})}{2(S_{\max} + S_{\min})}. \tag{3.2.3}$$

In practice the maximum value of γ lies below 0.25.

[1] M. T. Weiss, *Phys. Rev.* **107**, 317 (1957).
[2] L. A. Blackwell and K. L. Kotzebue, "Semiconductor–Diode Parametric Amplifiers." Prentice-Hall, Englewood Cliffs, New Jersey, 1961.
[3] P. Penfield and R. P. Rafuse, "Varactor Applications." MIT Press, Cambridge, Massachusetts, 1962.

---

* Chapter 3.2 is by Jochen Edrich.

When only the pump and the signal frequencies f_2 and f_1, respectively, are present, transfer of power from the pump to the signal source i.e., gain, is possible only if the pump frequency is equal to an even multiple of the signal frequency (degenerate case, parametron). Otherwise, a power flow at a third frequency, the "idler" frequency must be permitted. Therefore most parametric amplifiers or converters contain at least three frequency-selective circuits which are coupled to the nonlinear reactance: the signal, the pump and the idler circuits. In the very general case there are two driving circuits for the frequencies f_1 and f_2, and many dissipative circuits for the frequencies $|mf_1 + nf_2|$, where m and n are integers. The power into and out of a nonlinear reactance which is coupled to these circuits as shown in Fig. 1 can be

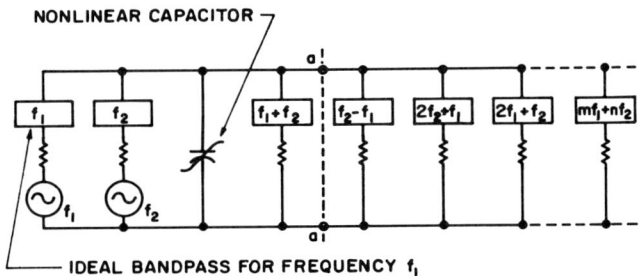

FIG. 1. Equivalent circuit illustrating the Manley–Rowe relations.

described by the Manley–Rowe equations[4]:

$$\sum_{m=0}^{\infty} \sum_{n=-\infty}^{\infty} \frac{mP_{mn}}{mf_1 + nf_2} = 0, \quad (3.2.4a)$$

$$\sum_{m=-\infty}^{\infty} \sum_{n=0}^{\infty} \frac{nP_{mn}}{mf_1 + nf_2} = 0. \quad (3.2.4b)$$

Referring to Fig. 1, P_{mn} represents the power flowing into the ideal lossless nonlinear capacitor at the frequencies $|mf_1 + nf_2|$. Here, f_1 and f_2 are the frequencies of the two driving sources. In most practical parametric amplifier and converter applications three frequency-selective circuits with midband frequencies f_1, and f_2 and f_3 are employed. There are two cases each of which represents several modes of operation:

3.2.1.1. Noninverting Case: $f_3 = f_1 + f_2$. Let us first assume that the nonlinear capacitor in Fig. 1 is excited at f_1 and f_2, and a resistive load is provided for the power at a third frequency $f_3 = f_1 + f_2$. We also assume the small-signal case where the signal power at frequency f_1 is much smaller than

[4] J. M. Manley and H. E. Rowe, *Proc. IRE* **44**, 904 (1956).

the pump power at frequency f_2. With the assumption $f_3 = f_1 + f_2$ the Manley–Rowe equations (3.2.4) can be simplified to

$$\frac{P_1}{f_1} + \frac{P_3}{f_3} = \frac{P_2}{f_2} + \frac{P_3}{f_3} = 0. \tag{3.2.5}$$

If the pump power P_2 is taken to be positive, then it is evident from Eq. (3.2.5) that the power P_3 at the frequency f_3 is negative, i.e. power output is supplied by the capacitor at frequency f_3. Therefore the input power P_1 must be positive. This indicates absolute stability for the device. Using Eq. (3.2.5) the maximum gain becomes

$$G_{13} = \frac{P_3}{P_1} = \frac{f_3}{f_1} \quad \text{for} \quad f_3 > f_1. \tag{3.2.6}$$

This amplifying converter is called the "noninverting up-converter" or "upper-sideband up-converter" (modulator).

Let us now assume that a small signal power P_3 is applied to the capacitor at the frequency f_3 and a resistive load instead of a generator is provided at the frequency f_1. Equation (3.2.5) then shows that the powers P_1 and P_2 are negative. Since P_2 is negative, the nonlinear capacitor exhibits a negative resistance to the pump circuit. Therefore this device, called the "noninverting down-converter" (demodulator), is potentially unstable.

3.2.1.2. Inverting Case: $f_3 = f_2 - f_1$. We first assume that the difference frequency f_3 lies between the pump frequency f_2 and the signal frequency f_1. Thus the Manley–Rowe equations (3.2.4) simplify to

$$\frac{P_1}{f_1} - \frac{P_3}{f_3} = \frac{P_2}{f_2} + \frac{P_3}{f_3} = 0. \tag{3.2.7}$$

If we take the pump power P_2 to be positive, then the powers P_1 and P_3 must be negative. Therefore the capacitor presents a negative resistance to both the frequencies f_1 and f_3. Three modes of operation can be obtained in this case: The input power is applied at f_1 and the output power is withdrawn at $f_3 > f_1$; this device is called the "lower-sideband or inverting up-converter" (modulator). When $f_3 < f_1$ the device is called the "inverting down-converter" (demodulator).

By far the most important mode is when input and output signal frequencies are the same and equal to f_1, the pump frequency is equal to f_2, and the power at the "idler" frequency f_3 is simply dissipated in a circuit, called the "idler circuit." If the pump frequency f_2 is greater than twice the signal frequency, the mode of operation is called *nondegenerate*; if f_2 is exactly twice f_1 it is called *degenerate*. The negative resistance appearing in the signal circuit at f_1 can result in parametric amplification of the signal power as will be discussed in more detail in the following section.

3.2.2. Fundamentals of Parametric Amplifiers

Figure 2 shows the equivalent circuit of a parametric amplifier using series-resonant idling and signal circuits. These circuits are coupled by the time-varying capacitance $C_A = (2\gamma S_m)^{-1}$ whose amplitude represents the ac component of the elastance S_j according to Eq. (3.2.2). The average capacitance $C_m = 1/S_m$ of the junction and its spreading resistance r are shown in both the signal and the idling circuits, since the currents at both frequencies flow through these elements. The other elements L_1, C_1 and L_3, C_3 represent the signal and idler circuit tuning reactances. The idler circuit (in most modern amplifiers) is resistively terminated with the spreading resistance r of its varactor. The signal circuit is driven by a signal generator with a generator resistance R_1. The time-varying capacitance C_A in Fig. 2 can be described by its Z-matrix representation:

$$\begin{bmatrix} V_1 \\ V_3{}^* \end{bmatrix} = \begin{bmatrix} 0 & \frac{-1}{2j\omega_3 C_A} \\ \frac{1}{2j\omega_3 C_A} & 0 \end{bmatrix} \begin{bmatrix} I_1 \\ I_3{}^* \end{bmatrix}. \quad (3.2.8)$$

Here ω_1 and ω_3 are the angular frequencies, 2π times f_1 or f_3, respectively. From Eq. (3.2.8) and Fig. 2 the impedance of the amplifier referred to the capacitance C_A can be derived as

$$Z_N = R_N + jX_N = -\frac{\gamma^2}{\omega_1 \omega_3 C_m{}^2 Z_3{}^*}. \quad (3.2.9)$$

At the midband frequency, where $X_1 = X_N = 0$ and $Z_3{}^* = r$, this impedance Z_N becomes a negative real resistance

$$R_N = -\frac{\gamma^2}{\omega_1 \omega_3 C_m{}^2 r}. \quad (3.2.10)$$

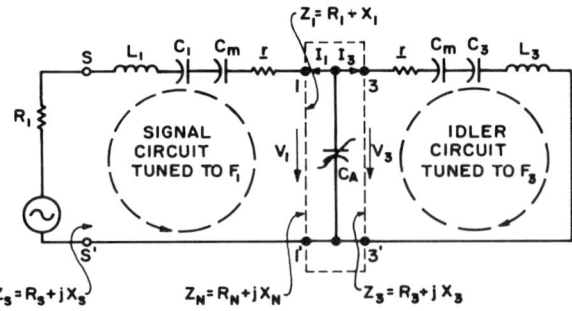

FIG. 2. Equivalent circuit of a parametric amplifier with series-resonant circuits and no external idler load. ($C_m = 1/S_m$; $C_A = 1/2\gamma S_m$).

Similarly the input impedance Z_s at the signal generator terminals S–S' becomes a real resistance

$$R_s = R_N + r. \qquad (3.2.11)$$

This leads to the midband power gain G of the reflection amplifier which is equal to the square of the voltage reflection coefficient at the terminals S–S':

$$G = \left[\frac{(R_1 - R_s)}{(R_1 + R_s)}\right]^2. \qquad (3.2.12)$$

In order to achieve a stable gain ($1 \ll G < \infty$) two requirements have to be met: The first requirement, $G \gg 1$, says that the gain of the amplifier has to be sufficiently high; this is the case if R_s is negative and its absolute value is sufficiently large. This, in turn requires that the amplifier be pumped ($\gamma > 0$) and the spreading resistance r be low enough ($r \ll |R_N|$). The second requirement, $G < \infty$, is met if the sum of R_1 and R_s is positive; this can be assured by proper transformation of R_1.

As Eq. (3.2.12) shows, negative resistance amplifiers are very sensitive to changes in circuit impedance. In particular, changes of the negative resistance caused by fluctuations of the pump-power level and frequency will affect the gain strongly through the γ-factor [Eq. (3.2.10)]. Therefore, most practical amplifiers contain automatic gain-control circuits; these circuits amplify fluctuations of the detected varactor current and use them for leveling of the pump power via ferrite or pin-diode modulators.[5]

In the past, various devices including hybrids were used to convert the parametric one-port device into a two-port amplifier. Today only circulators are employed for this purpose. The reason lies in the low insertion loss and the broad bandwidth that can now be achieved with junction-type circulators up to about 60 GHz and Faraday-type circulators up to more than 100 GHz.[6–10] Figure 3 shows a typical configuration using two circulator junctions which form the so-called *four-port circulator*. The first junction is terminated and therefore acts as an isolator which improves the input match and helps to protect the amplifier from input impedance changes. The second junction provides the input–output separation. An arrow indicates the direction of power flow in the circulators. The block with input impedance Z_s represents the circuit containing the varactor circuits.

---

[5] W. Heinlein and P. G. Metzger, *Frequenz* **16**, 320 (1962).

[6] H. Bosma, *IEEE Trans. Magn.* **MAG-4**, 587 (1968).

[7] G. P. Rodigue, *IEEE Trans. Microwave Theory Tech.* **MTT-11**, 351 (1963).

[8] J. Edrich and R. G. West, *1969 INTERMAG Conf. Proc.* 481 (1969); *IEEE Trans. Magn.* **MAG-5**, 481 (1969).

[9] J. B. Bastillo and L. E. Davis, *IEEE Trans. Microwave Theory Tech.* **MTT-18**, 25 (1970).

[10] J. Edrich, P. Hardee, and R. G. West, *Proc. 1973 Eur. Microwave Conf.*, Brussels (September 1973).

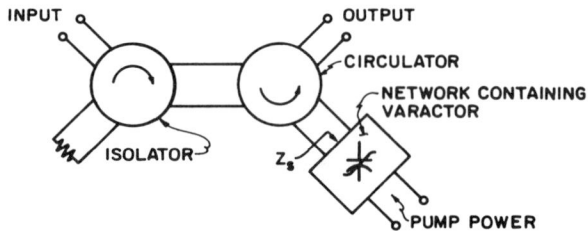

FIG. 3. Typical configuration of a parametric amplifier using a four-port circulator.

An important performance parameter is the bandwidth. It is greatly influenced by the varactor. Practical varactors for the microwave frequency range are mostly encapsulated in pill or micropill packages; they can be characterized up to about 20 GHz by the equivalent circuit shown in Fig. 4.

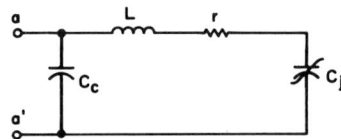

FIG. 4. Equivalent circuit of a pill-type varactor diode in the microwave frequency range.

Here, the elements C_j, C_c, L, and r represent the junction capacitance, the case capacitance, the lead inductance, and the combined spreading and lead resistance, respectively. When such a varactor is connected with its signal and idler circuits many more elements have to be added for tuning, broad-banding, and matching purposes. Various authors have discussed the general gain-bandwidth limitations.[11-16]

Modern nondegenerate amplifiers often employ one of the varactor self-resonances for their idler circuit in order to maximize the gain-bandwidth product.[14,17,18] The first resonance—called *series* resonance—occurs at the frequency

$$f_s = \frac{1}{2\pi(LC_m)^{1/2}}, \qquad (3.2.13)$$

[11] J. C. Greene and E. W. Sard, *Proc. IRE* **48**, 1583 (1960).
[12] E. S. Kuh and M. Fukada, *IRE Trans. Circuit Theory* **CT-9**, 410 (1961).
[13] G. L. Matthaei, *IRE Trans. Microwave Theory Tech.* **MTT-9**, 23 (1961).
[14] W. Heinlein and P. G. Metzger, *Frequenz* **16**, 391 (1962).
[15] B. T. Henoch, *IRE Trans. Microwave Theory Tech.* **MTT-11**, 62 (1963).
[16] J. T. de Jager, *IEEE Trans. Microwave Theory Tech.* **MTT-12**, 459 (1964).
[17] J. Edrich, *Frequenz* **20**, 337 (1966).
[18] C. S. Aitchison, R. Davies and C. D. Payne, *IEEE Trans. Microwave Theory Tech.* **MTT-16**, 46 (1968).

3.2. PARAMETRIC AMPLIFIERS

where the inductance L and the average value C_m of the junction capacitance C_j in Fig. 4 are in resonance. The second self-resonance frequency, called *parallel-resonance frequency* f_p, occurs when the input impedance of the varactor at terminals a–a' in Fig. 4 becomes a pure resistance. The best bandwidth results can be achieved if the series-resonance frequency is equal to the idler frequency. The upper limit for the gain-bandwidth product becomes then[17]

$$G^{1/2}B = \frac{(2f_3{}^2\gamma/f_1\tilde{Q})(\tilde{Q}^2 - f_3/f_1)}{\tilde{Q}^2 + (f_1 C_c/f_3 C_m)[(f_3/f_1)^4 - 2(f_3/f_1)^2 + 1 + C_m f_3{}^4/(C_c f_1{}^4)]}.$$

(3.2.14)

Here the standard definitions of the dynamic $\tilde{Q}$

$$\tilde{Q} = \gamma \frac{f_c}{f_1} = \frac{\gamma}{2\pi f_1 C_m r} \qquad (3.2.15)$$

and the cutoff frequency

$$f_c = \frac{1}{2\pi C_m r} \qquad (3.2.16)$$

at the operating point of the varactor are used. Inspection of Eq. (3.2.14) reveals that a small case capacitance C_c is desirable; in addition a high idler-frequency f_3 and a high dynamic Q are needed throughout the microwave frequency range in order to achieve an optimum gain-bandwidth product. The theory predicts relative gain-bandwidth products $\sqrt{G}\,B/f_1$ of up to 50% for single-tuned signal and idler circuits. In order to obtain even larger bandwidths it is common practice to multitune the signal circuit. This way one can obtain a maximally flat or a Chebyshev response with a certain passband ripple which results in an increase of the bandwidth by another factor 2 or 3.[13,16,19–21]

The noise of modern parametric amplifiers for the microwave region originates primarily from the loss resistance r of the varactor and is therefore of thermal nature. The equivalent noise temperature T of such nondegenerate parametric amplifiers can be described by[22–24]

$$T = T_A\left[1 + \frac{\Delta T_0}{T_A}\left(\frac{f_3}{f_1} + 1\right)^2\right]\left[\frac{f_3}{f_1} + \frac{1 + (f_3/f_1)}{\tilde{Q}^2 - (f_3/f_1)}\right]. \qquad (3.2.17)$$

[19] W. J. Getsinger, *IEEE Trans. Microwave Theory Tech.* **MTT-11**, 486 (1963).
[20] V. Porra and P. Sommervuo, *IEEE Trans. Microwave Theory Tech.* **MTT-16**, 880 (1968).
[21] K. Garbrecht and W. Heinlein, *Microwave J.* **13**, 77 (1970).
[22] H. Hefner and G. Wade, *J. Appl. Phys.* **29**, 1323 (1958).
[23] R. C. Knechtli and R. D. Weglein, *Proc. IRE* **48** 1218 (1960).
[24] K. Garbrecht, *1965 Int. Solid-State Circuits Conf. Digest*, Philadelphia 22 (1965).

Since T is proportional to the ambient temperature T_A, the noise temperature can be substantially reduced by cooling, a technique that is applied in many low-noise systems. The term $\Delta T^0/T_A$ in Eq. (3.2.17), which is due to pump heating, is negligibly small at room temperature ($T_A = 290°K$); it can, however, become noticeable at temperatures T_A which are below $10°K$.[24] The plot of the relative noise temperature T/T_A versus the ratio of idling to signal frequency in Fig. 5 shows that there are minima for the noise temperature.

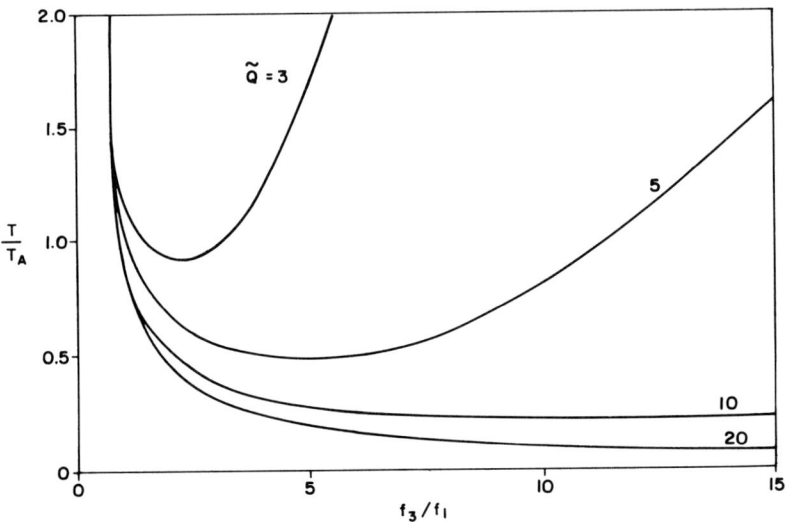

FIG. 5. Noise temperature T/T_A of a nondegenerate parametric amplifier as a function of the idler-to-signal frequency ratio f_3/f_1. Parameters: dynamic $\tilde{Q}$, T_A is the physical temperature of the varactor, and $\Delta T_0 = 0$ (negligible pump heating).

These minima depend on the dynamic $\tilde{Q}$ of the diode and they occur at a ratio of f_3/f_1, which increases when raising the dynamic $\tilde{Q}$ of the varactor.

In the degenerate case ($f_1 = f_3$) the idler and signal channels coincide. The output-noise contributions from the two sidebands, above and below one-half of the pump frequency, are therefore correlated with each other through the amplification and conversion process. Unless the input-signal statistics are known, the type of detector specified, and the interpretation of the detector output given, it is not possible to compare the noise of degenerate amplifiers with the noise of nondegenerate ones. There are, however, some applications where these two types can be compared. One of them is the single-sideband operation which is particularly important for spectral line observations; here, the signal originating from a single spectral line is contained solely in one sideband, called the *signal band*. For the practical case of small input-noise powers in the other band, called the *idler band*, the single-sideband noise temperature is given by[2]

$$T_{SSB} = 2T_A[(1/\tilde{Q}) + (1/\tilde{Q}^2)]. \qquad (3.2.18)$$

This noise temperature T_{SSB} is always larger than the minimum noise temperature of a nondegenerate amplifier. A far more important application of degenerate amplifiers is in the field of broadband, radiometric observations of continuum sources. Here, the signal is contained in both sidebands. The best amplifier scheme is one which uses as second stage a mixer with the local-oscillator frequency equal to one-half the pump frequency. This operation is commonly referred to as the "synchronously pumped" mode. Generally, a single source is used to provide both local-oscillator power for the mixer following the amplifier and pump for the amplifier through a doubler. A phase shifter in one of these two feeding lines allows adjustment of the phase so that the two sidebands of the amplifier are folded in-phase into the two sidebands of the mixer. The noise temperature of this mode is often referred to as "double-sideband" noise temperature T_{DSB}. It is half the value of the single-sideband noise temperature

$$T_{DSB} = T_A[(1/\tilde{Q}) + (1/\tilde{Q}^2)], \qquad (3.2.19)$$

and therefore can, under optimized conditions, be lower than the noise temperature of a nondegenerate amplifier using the same diode.

3.2.3. Design Considerations and Practical Parametric Amplifiers

Before designing a parametric amplifier it must be cleared whether the system under consideration requires a very large bandwidth, extremely low noise, a high degree of stability, or a very good linearity. In contrast to communications and radar applications, radio-astronomy receivers usually require a high degree of stability and are rarely affected by nonlinearities because of the low signal-power levels involved. Low noise is a prime requisite for spectral-line receivers, whereas continuum receivers are designed so that the ratio

$$K = T/\sqrt{B} \qquad (3.2.20)$$

of the noise temperature T and the square root of the bandwidth B are minimized in order to minimize the rms noise fluctuation ΔT_{rms}; this requirement often leads to cryogenically cooled amplifiers because cooling improves the noise temperature T substantially. In addition, it generally broadens the bandwidth because of the larger capacitance modulation factor γ which can be achieved at cryogenic temperatures. These advantages can be utilized from approximately 1.4 up to at least 50 GHz because coolable circulators with broad bandwidths have been developed throughout this frequency range.[10,25–29]

Turning to specific designs of parametric amplifiers we will find two structures which are most frequently used. The first one is the single-varactor or single-ended version. A schematic cross section is shown in Fig. 6. The

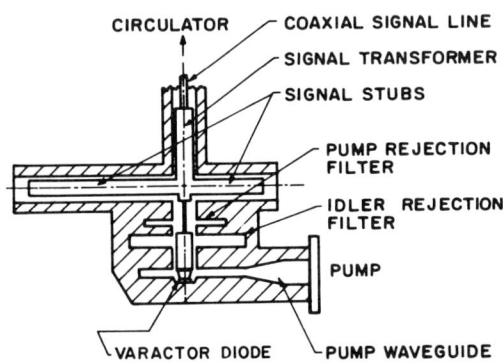

FIG. 6. Cross-sectional view of a broadband parametric amplifier with one varactor, single-tuned idling circuit, and double-tuned signal circuit.

varactor is mounted across the middle of a reduced-height pump waveguide; idler energy cannot propagate in it since its cutoff frequency is higher than the idler frequency. A radial choke is placed into the signal line which produces a short at the diode for the idler frequency. This method yields the advantages of broad bandwidth as previously discussed. Open-circuited stub lines can provide the desired double tuning of the signal circuit. Theoretical analysis shows that the gain-bandwidth product is of the form $G^{1/4}B = $ const for the double-tuned case (one additional circuit) and $G^{1/6}B = $ const for the triple-tuned case (two additional circuits). For a maximally flat response, the limiting bandwidth is $B \log G = $ const.[20,30] In practice it is difficult to add more than two circuits.

[25] R. C. Comstock and C. E. Fay, *J. Appl. Phys.* **36**, 1253 (1965).
[26] D. H. Roth, W. W. Sshedelbeck and G. H. Schollmeier, *IEEE Trans. Magn.* **MAG-2**, 256 (1966).
[27] J. Edrich and R. G. West, *IEEE Trans. Microwave Theory Tech.* **MTT-18**, 743 (1970).
[28] J. Edrich and R. G. West, *1972 INTERMAG Conf. Proc.*, Kyoto, 508 (1972); *IEEE Trans. Magn.* **MAG-8**, 508 (1972).
[29] T. Okajima, M. Kudo, K. Shirahata, and D. Taketomi, *IEEE Trans. Microwave Theory Tech.* **MTT-20**, 812 (1972).
[30] R. M. Fano, *J. Franklin Inst.* **249**, 57, 139 (1950).

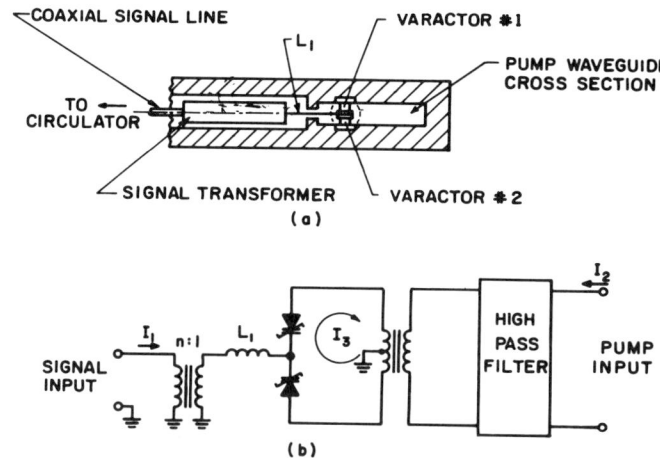

FIG. 7. (a) Cross-sectional view of a balanced parametric amplifier with single-tuned signal and idler circuits; (b) Equivalent circuit of (a).

Figure 7 shows a schematic cross section of a balanced configuration using two varactor diodes. Again, the idler frequency f_3 is close to the series resonance frequency given by Eq. (3.2.13). A shorting lace across both diodes, as indicated by a dashed line in Fig. 7a, helps to improve the desired short at the idler frequency. This bridge-type configuration can be described by the equivalent circuit in Fig. 7b. As compared to the single-ended version it has the advantages of eliminating the need for idler- and pump-rejection filters in the signal line. This should, under idealized conditions, lead to a limiting value of the bandwidth that is twice the one achievable with single-ended amplifiers. Under realistic conditions, however, this ideal factor cannot be approached very closely.[17] Nevertheless, balanced configurations have certain advantages that led to designs with the largest instantaneous bandwidth[18,31–34] (up to 40%) and the widest tuning range[35–38] (up to one octave).

[31] J. Kliphuis, *Proc. IRE* **48**, 1583 (1960).
[32] D. Neuf, R. Holifield, and E. Barnell, *Proc. IEEE* **51**, 1365 (1965).
[33] H. Bruntrup, J. Edrich, and K. Garbrecht, *Proc. NTG Conf. Satellite Commun., Berlin, 1966* 62 (1966).
[34] H. C. Okean and H. Weingart, *IEEE Trans. Microwave Theory Tech.* **MTT-16**, 1057 (1968).
[35] K. L. Kotzebue, and L. B. Fletcher, *1965 G-MTT Symp. Digest, Clearwater, Florida* 101 (1965).
[36] R. L. Sleven, *Microwave J.* **5**, 167 (1962).
[37] J. Edrich, *IEEE J. Solid-State Circuits* **SC-7**, 32 (1972).
[38] J. Edrich, *IEEE Trans. Microwave Theory Tech.* **MTT-18**, 1173 (1970).

TABLE I. Bandwidth of Single-Varactor Parametric Amplifiers[a]

Signal frequency f_1 (GHz)	Pump frequency f_2 (GHz)	Gain G(dB)	Bandwidth 3-dB points B (%)	Description
1	10	19	20	Two-stage, double-tuned signal circuits[b]
3	15	12	14	Single-stage, quadruple-tuned signal circuits[c]
4	~25	12	13–15	Single-stage, triple-tuned signal circuits[d,e]
9.2	25.8	10	10	Single-stage, multituned signal circuit[f]
18	53	10	7	Single-stage, triple-tuned signal circuit[g]

[a] Bandwidths cited are for maximally flat characteristics (+0.2 dB). The amplifiers are in a nondegenerate mode.
[b] C. E. Barnes, W. J. Bertram, and M. J. Cowan, *1964 Int. Solid State Circuits Conf. Digest, Philadelphia, Pennsylvania* 24 (1964).
[c] K. M. Johnson, *Proc. IRE* **49**, 1943 (1961).
[d] H. Bruntrup, J. Edrich, and K. Garbrecht, *Proc. NTG Conf. Satellite Commun., Berlin, 1966* 62 (1966).
[e] L. A. Blackwell and J. L. Halpain, in "Low Temperature Refrigeration of Microwave Systems" (W. H. Hogan and G. Klipping, ed.), p. 151. Boston Tech. Publ., Boston, Massachusetts, 1967.
[f] K. M. Johnson, *Proc. IRE* **50**, 332 (1962).
[g] T. Okajima, M. Kudo, K. Shirahata, and D. Taketomi, *IEEE Trans. Microwave Theory Tech.* **MTT-20**, 812 (1972).

TABLE II. Bandwidth of Balanced, Nondegenerate, Parametric Amplifiers

Signal frequency f_1 (GHz)	Pump frequency f_2 (GHz)	Gain G (dB)	Bandwidth 3-dB points B(%)	Description
2.4	—	20	40	Four-stage, compensated signal circuits[a]
3.5	20	20	15	Two-stage, double-tuned signal circuit[b]
7.5	70	15	7	Single-stage, double-tuned signal circuit[c]

[a] D. Neuf, R. Holifield and E. Barnell, *Proc. IEEE* **51**, 1365 (1965).
[b] H. C. Okean, and H. Weingart, *IEEE Trans. Microwave Theory Tech.* **MTT-16**, 1057 (1968).
[c] L. E. Dickens, *Proc. IEEE* **60**, 328 (1972).

3.2. PARAMETRIC AMPLIFIERS

Tables I and II show the bandwidth of practical single-ended and balanced configurations which were designed for applications requiring a large instantaneous bandwidth. In some applications a wide tuning range is preferred over a large instantaneous bandwidth. Such tuning is usually accomplished by variation of one or more of the following parameters: varactor bias voltage, pump power, pump frequency, mechanical or electrical tuning of the signal circuit. As Table III shows one can, in practice, tune up to one octave of frequency.

TABLE III. Wide-band Tunable Parametric Amplifiers

Tuning range Δf (GHz)	Pump frequency f_2 (GHz)	Gain G (dB)	Instantaneous bandwidth B (%)	Noise temperature T (°K)	Description
1–2	7–8	20	0.8–2.1	233–259	Signal-, bias, and pump-tuned, single ended[a]
1.8–4	10–13	20	0.1–0.6	233–285	Signal- and bias-tuned, single-ended[a]
2.2–4	17	25	0.2–0.5	150–400	YIG-tuned, single-ended[b]
4–8	22.6	17	1.2–2.1	100–250	Signal- and bias-tuned, balanced[c]
22–24	44–48	17	0.5–1.0	290–450[d]	Bias- and pump-tuned, degenerate, single-ended[e]

[a] R. L. Sleven, *Microwave J.* **5**, 167 (1962).
[b] K. L. Kotzebue and L. B. Fletcher, *1965 G-MTT Symp. Digest, Clearwater, Florida* 101 (1965).
[c] J. Edrich, *IEEE J. Solid-State Circuits* **SC-9**, 32 (1972).
[d] Single-sideband noise temperature.
[e] J. Edrich, *IEEE Trans. Microwave Theory Tech.* **MTT-18**, 1173 (1970).

The use of two idler circuits rather than conventional single-idler circuit configurations has been shown to improve the stability of nondegenerate amplifiers[39,40]; however, it normally degrades the bandwidth and always degrades the noise performance. More than two idler circuits are hard to realize at microwave frequencies. For these reasons multiple-idler configurations have not found applications yet.

[39] R. L. Ernst, *IEEE Trans. Microwave Theory Tech.* **MTT-15**, 9 (1967).
[40] A. R. Kerr, *IEEE Trans. Microwave Theory Tech.* **MTT-18**, 277 (1970).

Traveling wave structures using varactors have been investigated by many authors.[41-46] Recently it has been shown that, theoretically, amplification can be achieved even for greater mismatches (between the signal, idler, and pump phase constants), than the critical value;[45] this value can give a product of voltage gain and relative bandwidth of about 2 dB. Practically, values up to 1.35 dB have been achieved at low microwave frequencies. The requirements on the uniformity of the varactor array at microwave frequencies are, however, so stringent that it has, up to the author's knowledge, not yet been possible to construct practically useful traveling wave parametric amplifiers.

There are a number of operation modes possible when using parametric converters, as was pointed out when discussing the Manley–Rowe equations. The only type of practical importance, however, is the upper-sideband up-converter. According to Eq. (3.2.6) it can provide a stable gain which is given by the ratio of output to input frequency. In practice, circuit as well as diode losses will reduce the gain substantially. Two other features, however, make this converter a very useful device. First, the limit of its noise temperature is set by the thermal noise in the series resistance of the varactor. Therefore, the minimum noise temperature of upper-sideband up-converters without additional idlers is equal to the minimum noise temperature of nondegenerate amplifiers. An idler in the lower sideband of the converter can increase the gain and decrease the noise temperature. This converter is thus intrinsically a very low-noise device. Because of its limited gain, however, it is useful in a low-noise application only if it is followed by another low-noise device like a parametric amplifier or a maser. Sometimes, this is a very good combination. For example, one can build extremely wide-band tunable low-noise converter-type receivers by using a variable pump frequency. An excellent example for the low noise performance is a cooled converter followed by a maser as reported by Sard et al.[47] which achieved a noise temperature of less than 30°K in S-band. The wide tuning range of such a device which can extend over several octaves in the lower microwave frequency region, makes it very attractive for spectral-line receivers.

The bandwidth, noise, and stability of a negative-resistance-type parametric amplifier is greatly influenced by its circulator if the amplifier is cryogenically cooled. At room temperature, the performance of junction

---

[41] A. L. Cullen, *Nature (London)* **181**, 332 (1958).
[42] R. S. Engelbrecht, 1959 *Solid-State Circuits Conf. Digest, Philadelphia* 8 (1959).
[43] P. P. Lombardo and E. W. Sard, *Proc. IRE* **49**, 995 (1959).
[44] W. Heinlein, *Arch. Elektrischen Uebertragung* **15**, 547 (1961).
[45] M. J. Colles, R. C. Smith, and C. R. Stanley, *Proc. IEEE* **57**, 73 (1969).
[46] D. Raicu, *Proc. IEEE* **58**, 1149 (1970).
[47] E. Sard, B. Peyton, and S. Okwit, *IEEE Trans. Microwave Theory Tech.* **MTT-14**, 608 (1966).

circulators in stripline or waveguide is not a limiting factor in regard to loss and bandwidth. Up to one octave of frequency range can be achieved with more than 20-dB isolation per junction in the conventional microwave bands. Even at 45 GHz one can still obtain up to 15%. The insertion loss increases from about 0.1 dB at 1.4 GHz to only 0.35 dB at 45 GHz. Cooling of circulators leads to a slight deterioration of the bandwidth and insertion loss in the frequency range from 3 to about 18 GHz.[25,26] Below 3 GHz changes of existing aluminum-doped yttrium–iron garnets due to temperature effects lead to an increase of the insertion loss by a factor of 2 to 3 and a decrease of the isolation bandwidth by a factor of 3 to 5 as compared to room temperature.[27,28] Very recent results on coolable circulators for the millimeter range indicate that it is now possible to achieve sufficiently low insertion loss (<0.8 dB) and broad isolation bandwidths (>13%) up to at least 46 GHz.[10,29] This progress in the area of cryogenically cooled circulators has greatly aided the development of low-noise cooled parametric amplifiers. Its significance can best be understood by looking at the noise budgets of three low-noise amplifiers for 1.4, 1.66, and 46 GHz (Table IV). The noise contribution of the uncooled circulator in the 1.66-GHz system constitutes more than 40% of the system noise temperature. At 46 GHz the use of an uncooled 4-port circulator instead of the cooled one would raise the system noise temperature by more than 100%. As can be seen from Table IV the input line contributions are another significant noise source. Noise contributions of mounts and varactors of cooled amplifiers are in general small, provided diffused or Schottky barrier varactors with gallium arsenide (GaAs) are used in conjunction with a pump frequency that is low enough to prevent pump heating.[24] Silicon (Si) yields better bandwidth than GaAs, and noise performance that is comparable to GaAs at room temperature. Si, however, has been found unsuitable for cooling below 77°K because of a drastic increase of its resistance in Si diodes at low temperatures.[24,48]

The development of GaAs Schottky barrier varactors with cutoff frequencies well beyond 600 GHz made it possible to develop parametric amplifiers for the millimeter wave range.[49-54] The required small junction capacitances ($C_m = 0.01$–0.1 pF) can be achieved with planar honeycomb structures on

[48] M. Uenohara, in "Advances in Microwaves" (L. Young, ed.), Vol. 2, p. 89. Academic Press, New York, 1967.
[49] M. W. Sharpless, *Bell Syst. Tech. J.* **35**, 1385 (1956).
[50] D. Kahng, *Bell Syst. Tech. J.* **44**, 215 (1964).
[51] D. T. Young and J. C. Irvin, *Proc. IEEE* **53**, 2130 (1965).
[52] T. P. Lee and C. A. Burrus, *IEEE Trans. Microwave Theory Tech.* **MTT-16**, 289 (1968).
[53] M. Cohn, L. E. Dickens, and J. W. Dozier, *IEEE 1969 Int. Microwave Symp. Digest, Dallas* 225 (1969).
[54] J. Edrich, *IEEE Trans. Microwave Theory Tech.* **MTT-18**, 1173 (1970).

TABLE IV. Noise Budgets of Four Parametric Amplifier Systems
Developed for Radio Astronomy Applications

Midband frequency	1.4 GHz[a] (°K)	1.66 GHz[c] (°K)	23 GHz[e] (°K)	46 GHz[g] (°K)
Varactor mount	~3	~3	304	<10
Circulator	3[b]	10[d]	88	20[h]
Input line	4	6	—	6
Input switch	—	—	70[f]	—
Second stage	3	3	12	17
Unknown contribution	4	2	20	6
Total system (excluding feed and sky contributions)	17 ± 3 (SSB)	24 ± 3 (SSB)	494 ± 30 (SSB)	59 ± 20 (DSB)

[a] 20°K cooled, two-channel, nondegenerate, parametric receiver with 20°K cooled circulator. Line receiver being used on NRAO 140-ft and 300-ft telescopes in Greenbank, West Virginia, description by D. L. Thacker, NRAO Electron. Div. Int. Rep. No. 124 (Sept. 1972).

[b] J. Edrich and R. G. West, *1972 INTERMAG Conf. Proc.*, *Kyoto* 508 (1972); *IEEE Trans. Magn.* **MAG-8**, 508 (1972).

[c] 20°K cooled, two-channel, nondegenerate, parametric receiver with uncooled circulator. Line receiver tunable from 1540 to 1750 MHz, used on NRAO 42.7-m (140-ft) and 91.4-m (300-ft) telescopes in Greenbank, West Virginia. Report (Dig. of 1969 IEEE Int. Conf. on Communications, Digest, p. 49 (June 1969)).

[d] J. Edrich and R. G. West, 1969 *INTERMAG Conf. Proc.* 481 (1969); *IEEE Trans. Magn.* **MAG-5**, 481 (1969).

[e] Uncooled spectral line receiver using single-stage, synchronously pumped, degenerate amplifier. Tuning range 21.5–24.0 GHz. Used on NRAO 36-ft telescope on Kitt Peak, NRAO 140-ft telescope in Greenbank, West Virginia and NRL 85-ft telescope in Washington, D.C., described by J. Edrich, *IEEE Trans. Microwave Theory Tech.* **MTT-18**, 1172 (1970); NRAO Electron. Div. Int. Rep. No. 114 (December 1971).

[f] Including contribution of a calibration signal coupler.

[g] 20°K cooled, continuum and spectral line receiver using single-stage, degenerate amplifier with 20°K cooled circulator. To be used on NRAO 36-ft telescope on Kitt-Peak, Arizona; described by J. Edrich, *Digest IEEE Int. Microwave Symp.*, 1973 paper No. III-5 (June 1973).

[h] J. Edrich, P. G. Hardee, ad R. G. West, *Proc. 1973 Eur. Microwave Conf.* (September 1973).

very thin epitaxial silicon or GaAs layers. The small junction areas are obtained by cutting circular holes (1–10 μm diameter) through insulating silicon oxide (SiO) layers using conventional photoetching or—more recently—ion-beam machines.[55] Deposition of suitable metals such as gold or

[55] H. L. Stover, H. M. Leedy, R. P. Bryan, H. G. Moorehead, and T. G. Armstrong, *Int. Electron. Devices Meeting, Washington, D.C., 1972*, 174 (1972).

platinum on top of the epitaxial semiconductor dots in these holes creates metal–semiconductor junctions which are called *Schottky barrier junctions*.[50] Ion implantation into the epitaxial layer of the holes yields high-quality *p–n* junctions.[55] All of these methods provide a very reproducible way to build junctions with cutoff frequencies up to 2000 GHz. Special encapsulations such as the "Sharpless wafer" are used for these junctions in order to reduce stray reactances.[49–54] Practically useful uncooled amplifiers have been reported up to 60 GHz[56] and work is in progress around 94 GHz[57]. Extremely low noise temperatures of less than 60°K have recently been

TABLE V. Important Characteristics of Various Millimeter Wave Parametric Amplifiers

Midband frequency (GHz)	17.9[a]	18.4[b]	23[c]	24[d]	33.6[e]	46[f]	60[g]
Physical temperature (°K)	300	40	300	20	300	18	300
Gain (dB)	20	10	18	15	15	22	18
3-dB bandwidth (MHz)	560	900	150/250	85/210	900	200	200
Tuning range (GHz)	—	—	2.5	2	—	—	—
Noise temperature excl. second stage (°K)	300 (SSB)	100 (SSB)	390/470 (SSB)	71 (DSB)	144 (DSB)	40 (DSB)	605 (SSB)
Pump frequency (GHz)	53	52.5	43–48	46–50	67.2	92	134
Pump power (mW)	—	44	12/22	12	20	30	20

[a] Double-tuned, uncooled, nondegenerate amplifier, described by Y. Kinoshita and M. Maeda, *IEEE Trans. Microwave Theory Tech.* **MTT-18**, 1114 (1970).
[b] Triple-tuned, 4°K cooled, nondegenerate amplifier with 4°K cooled circulator, described by T. Okajima, M. Kudo, K. Shirahata, and D. Taketomi, *IEEE Trans. Microwave Theory Tech.* **MTT-20**, 812 (1972).
[c] Single-tuned, single-stage, uncooled, electronically tuned, synchronously pumped degenerate amplifier, described in NRAO Electron. Div. Int. Rep. No. 114, Dec. 1971.
[d] Single-tuned, single-stage, 20°K cooled, degenerate amplifier with uncooled circulator, described by J. Edrich, *IEEE Trans. Microwave Theory Tech.* **MTT-18**, 1173 (1970).
[e] Triple-tuned, single-stage, uncooled, degenerate amplifier, described by M. Cohn, L. E. Dickens, and J. W. Dozier, *IEEE 1969 G-MTT Int. Microwave Symp. Digest, Dallas* 225 (1969).
[f] Single-tuned, 20°K cooled, degenerate amplifier with 20°K circulator; description by J. Edrich, *1973 Int. Microwave Symp. Digest* No. III-5 (June 1973).
[g] Uncooled, nondegenerate amplifier; description by H. L. Stover, H. M. Leedy, R. P. Bryan, H. G. Moorehead, T. G. Armstrong, P. V. Cooper, R. E. Bryan, and H. C. Bell, Jr., *1973 IEEE Int. Solid-State Circuits Conf. Digest, Philadelphia* 80 (1973).
[h] Including contributions of input waveguide switch and second-stage uncooled parametric amplifier.

[56] H. L. Stover, H. M. Leedy, R. P. Bryan, H. G. Moorehead, and T. G. Armstrong, P. V. Cooper, R. E. Bryan, and H. C. Bell, Jr., *1973 IEEE Trans. Solid-State Circuits Conf. Digest, Philadelphia* 80 (1973).
[57] H. C. Okean, J. R. Asmus, and L. J. Steffek, *1973 IEEE-G-MTT Int. Microwave Symp. Digest, Boulder, Colorado* 78 (1973).

achieved with 20°K cooled amplifiers at frequencies up to 50 GHz.[58] Bandwidths of such millimeter wave systems are presently limited to approximately 1 GHz by their i.f. amplifiers. Table V lists important characteristics of some reported amplifiers in the frequency range from 18 to 60 GHz. The upper signal-frequency limit of parametric amplifiers is set by the availability of reliable pump sources. The present progress in the area of millimeter wave solid-state and tube sources and multipliers on the one side, and high-powered laser sources in the submillimeter region on the other side offer promising avenues for raising this upper frequency limit beyond 100 GHz.

In Figures 8–11 photographs of various cooled and uncooled amplifiers are shown which are spanning the signal-frequency range from 1.6 up to

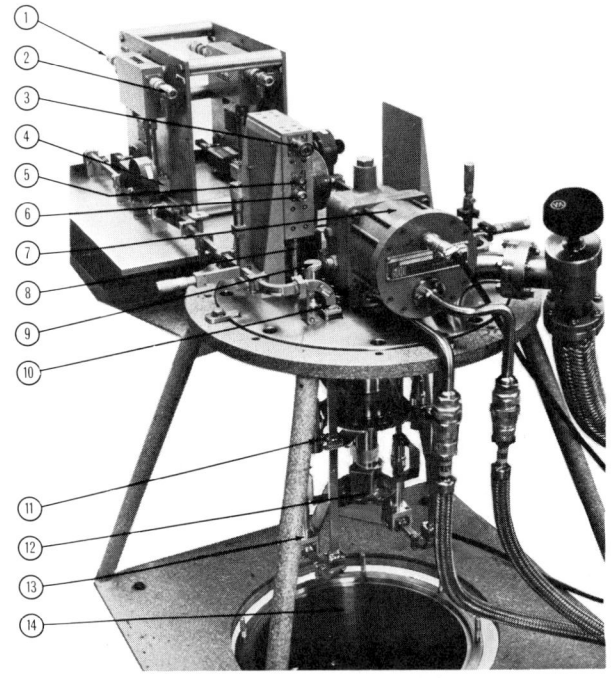

FIG. 8. Exposed view of a 20°K cooled, two-stage, two-channel, nondegenerate parametric amplifier receiver for 1660 MHz. [J. Edrich; *1969 IEEE Int. Conf. Commun. Digest, Boulder, Colorado* 49 (1969)]. This receiver is being used on the NRAO 42.7-m (140-ft) and 91.4-m (300-ft) telescopes in Greenbank, West Virginia. Its noise performance is given in Table IV. (1) Second stage output, (2) second stage input, (3) first stage input, (4) second stage paramp mount, (5) circulator bias connector, (6) varactor bias connector, (7) refrigerator head, (8) stainless steel input line, (9) Input to 77°K cold reference load, (10) pump input for first stage, (11) 77°K cold reference load, (12) 20°K station, (13) 20°K cooled first stage paramp mount, (14) Vacuum dewar.

[58] J. Edrich, *1973 G-MTT Symp. Digest, Boulder, Colorado* 72 (1973).

3.2. PARAMETRIC AMPLIFIERS

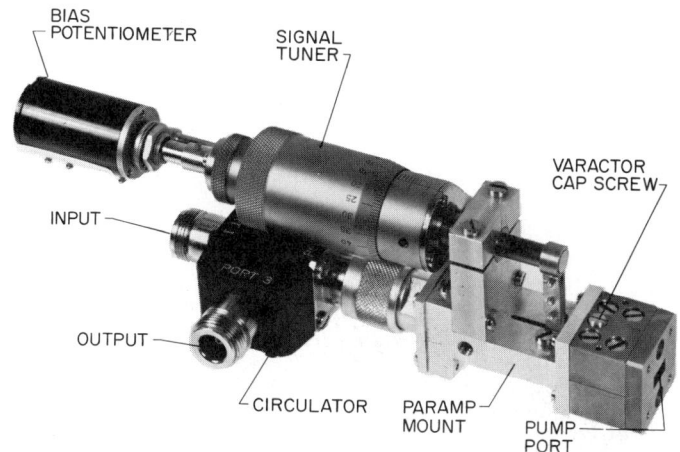

Fig. 9. The 4-8-GHz tunable parametric amplifier [J. Edrich, *IEEE J. Solid State Circuits* **SC-7**, 32 (1972)]; performance data in Table III.

46 GHz. Figure 12 is a "state of the art" plot of noise temperature versus signal frequency. The table at the end of Chapter 3.3 summarizes important performance characteristics of parametric amplifiers and compares them with masers. As is evident from this table, parametric amplifiers are being used in most sensitive continuum-receivers and are gaining importance as preamplifiers in spectral-line receivers because of their broad bandwidth and low noise temperature.

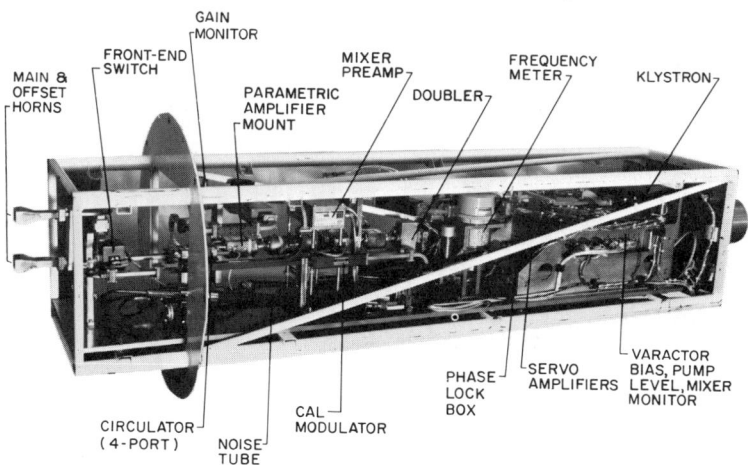

Fig. 10. The 22-24-GHz remotely tunable front-end which uses the synchronously pumped degenerate parametric described in Tables IV and V [J. Edrich, *IEEE Trans. Microwave Theory Tech.* **MTT-18**, 1173 (1970)].

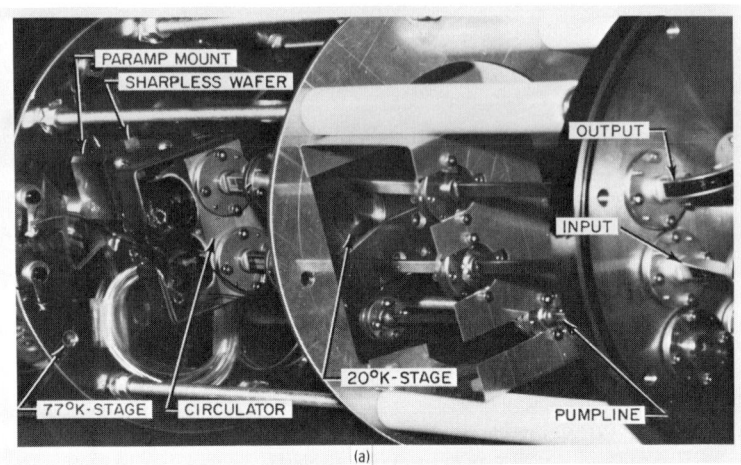

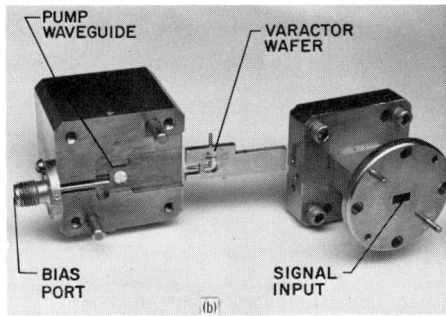

Fig. 11. 20°K cooled, degenerate, parametric, amplifier receiver described in Tables IV and V. [J. Edrich, *1973 G-MTT Symp. Digest, Boulder, Colorado* 72 (1973)]. (a) Exposed view with the vacuum dewar removed; (b) exposed view of the paramp mount.

3.2. PARAMETRIC AMPLIFIERS

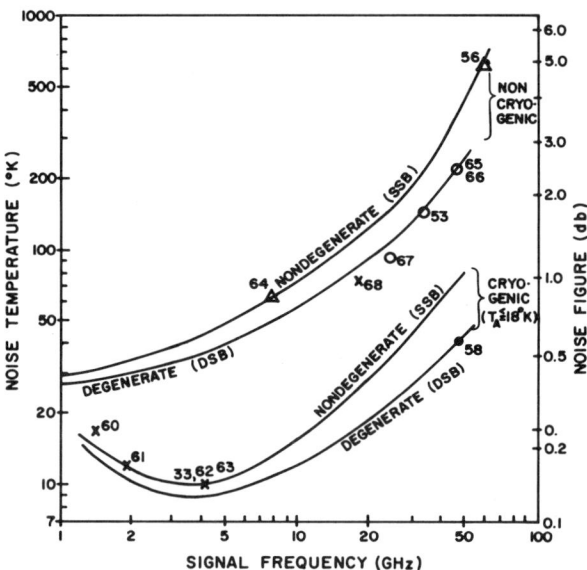

FIG. 12. Noise performance of parametric amplifiers versus signal frequency—state of the art, SSB is the single-sideband noise; DSB the double sideband-noise. Numbers at measurement points refer to reference list.

3.3. Maser Amplifiers*

3.3.1. Basic Properties of Maser Amplifiers

The word MASER (an acronym for "*M*icrowave *A*mplification by *S*timulated *E*mission of *R*adiation") was coined by Charles H. Townes and co-workers at Columbia University after they successfully operated the ammonia maser oscillator,[1] the first electronic device making direct use of stimulated emission, in 1954. A similar device, the hydrogen maser oscillator, emitting at the hydrogen hyperfine-splitting frequency (1420 MHz), is now in frequent use by radio astronomers as the most accurate frequency standard available (see Chapter 5.4). Maser oscillators and amplifiers operate according to the same fundamental principles of stimulated emission, but we will here only treat the maser amplifier. Rather than a gas, practical maser amplifiers use solid-state single crystals as the active medium, thereby taking advantage of the much higher densities of active particles available in a solid.

As a further distinction, the original ammonia maser made use of only two energy levels of the ammonia molecules, whereas practical maser amplifiers, without exception, utilize at least three discrete energy levels, as first proposed by N. Bloembergen.[2] A typical energy-level diagram is shown in Fig. 1, which also illustrates the three-level principle. The example chosen is ruby, one of the most widely used active materials. The type of ruby employed in maser amplifiers, so-called "pink ruby" is single-crystal Al_2O_3 with roughly 0.05% of its Al^{3+} ions substituted with Cr^{3+}. The ground state of the Cr^{3+} ion is fourfold degenerate due to its spin ($S = 3/2$). Half of this degeneracy is lifted by the crystal field, which the Cr^{3+} ion (generally referred to as the "spin") is exposed to in the Al_2O_3 lattice, giving rise to the zero-field splitting corresponding to a frequency of 11.46 GHz. The rest of the degeneracy is removed on application of a magnetic field, and the energy levels shown in Fig. 1 result from competition between crystal field and magnetic field effects. Over the range of magnetic fields for which these two effects are of comparable importance, all four levels can in general be connected via magnetic dipole-type transitions of good strength. Several such transitions are illustrated. If the crystal is in thermal equilibrium, the populations n_i on the various levels

---

[1] J. P. Gordon, H. J. Zeiger, and C. H. Townes, *Phys. Rev.* **99**, 1264 (1955).
[2] N. Bloembergen, *Phys. Rev.* **104**, 323 (1956).

---

*Chapter 3.3 is by K. Sigfrid Yngvesson.

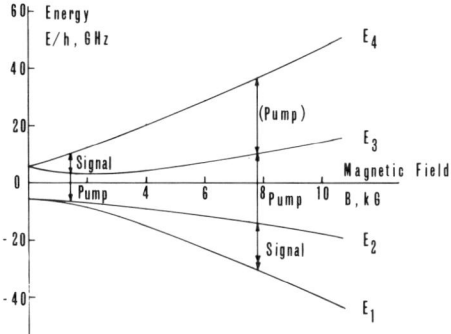

FIG. 1. Energy levels versus magnetic field for ruby, with the magnetic field perpendicular to the c-axis ($\theta = 90°$). The low-field operating point marked is used at low frequencies. The higher-field push–push operating point is also indicated.

will follow the Boltzmann distribution $n_i \sim e^{-E_i/kT}$, where E_i is the energy level i, k Boltzmann's constant, and T the absolute temperature. If a sample of ruby is placed in a resonant cavity, the power absorbed from an incident wave at frequency $f_{34} = (E_4 - E_3)/h$ (transition marked "signal" to the left in Fig. 1), will be

$$P_{34} = hf_{34}(n_3 - n_4)\bar{w}_{34}, \qquad (3.3.1)$$

where $\bar{w}_{34}$ is the transition probability caused by the incident wave and averaged over the ruby sample. Conversely, if we have a nonequilibrium situation in which $n_4 > n_3$, the sample will show a net emission of power due to stimulated emission being dominant over absorption. The incident signal will thus be reflected from the cavity with gain. To maximize the gain, according to (3.3.1) we want to maximize $n_4 - n_3$ and the transition probability $\bar{w}_{34}$. If $n_4 > n_3$, transition $3 \to 4$ is said to be *inverted*. According to the three-level idea, one makes use of (at least) one of the remaining levels to accomplish this inversion in a process referred to as *pumping*. For example, we may apply enough power to the ruby at a frequency corresponding to transition $2 \to 4$ in order to cause a transition rate which is much larger than that induced by lattice vibrations, which act so as to retain the populations at the Boltzmann distribution values. Eventually, populations n_2 and n_4 will then be nearly equalized, and depending on the particular energy-level spacings in the configuration chosen, transition $3 \to 4$ may become inverted (see Fig. 2). A typical level of pump power required is 100 mW if the maser is operated at liquid helium temperatures (4.2°K or below). Operation at temperatures higher than this is seldom attempted, since (a) much higher pump power is required due to stronger spin–lattice interaction and (b) population differences are smaller, decreasing the gain below usable levels.

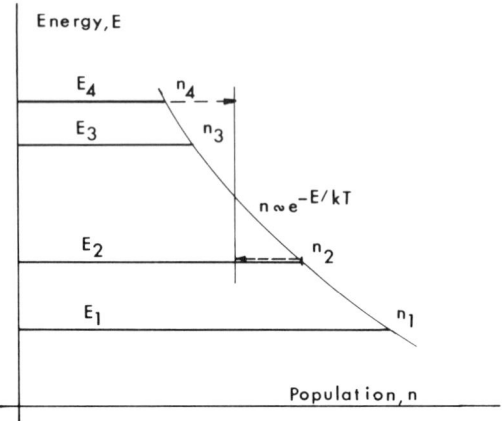

FIG. 2. Populations in a simple four-level system.

It is convenient to define the inversion ratio for transition $3 \to 4$

$$I_{34} = \frac{n_{3p} - n_{4p}}{n_{30} - n_{40}} \qquad (3.3.2)$$

Here, subscript 0 indicates thermal equilibrium, whereas p is used for the populations in the pumped state, assuming that the pump is saturated, i.e., the populations of the transition being pumped are equalized. As defined, I is negative in the inverted state. Simple expressions can be derived for the inversion ratio, if we make the approximation $hf \ll kT$. Note that this approximation is usually good for masers amplifying in the centimeter range, but not for millimeter wave masers. Introducing f_s for the signal frequency, and f_p for the pump frequency,[3]

$$-I \simeq (f_p/2f_s) - 1 \qquad (3.3.3)$$

for the case when the lattice vibration-induced (relaxation) probabilities are equal for all transitions. For the optimum relaxation probabilities

$$-I \simeq (f_p/f_s) - 1. \qquad (3.3.4)$$

These expressions can only be regarded as a rough guide to the expected inversion ratios and reliable values require a direct experimental measurement. Relaxation probabilities may be calculated using the theory due to Van Vleck and Kronig[4] and in one case with very low spin concentration of Cr^{3+} in TiO_2, good agreement between predicted and measured inversion

---

[3] A. E. Siegman, "Microwave Solid State Masers," McGraw-Hill, New York, 1964.
[4] R. deL. Kronig, *Physica* **6**, 33 (1939); J. H. Van Vleck, *J. Chem. Phys.* **7**, 72 (1939); *Phys. Rev.* **57**, 426 (1940).

3.3. MASER AMPLIFIERS

ratios were obtained.[5] At the higher concentrations used in actual masers, however, relaxation is generally speeded up by processes that are presently incompletely understood, and calculation of the inversion ratio is not possible. Thus one must resort to experimental determination for every new case.

In order to evaluate fully the gain and bandwidth obtainable from a maser amplifier we need to explore further the properties of the active material as well as the circuits used to couple the electromagnetic wave to the active material. The circuit may be of the cavity-type or a traveling-wave type structure. In either case we may characterize the active material by its (resonant) paramagnetic susceptibility χ''. If we assume a Lorentzian line shape, it is given by[3]

$$\chi'' = \left(\frac{2g^2\beta^2\mu_0}{h}\right) NI(\Delta n_s)_0 \,\bar{\sigma}\bar{\sigma}^* \left(\frac{1}{\Delta f_L}\right). \quad (3.3.5)$$

The constant $2g^2\beta^2\mu_0/h = 1.3 \times 10^{-18}$ m^3 sec^{-1}, where g is the spectroscopic splitting factor ($\simeq 2$), β the Bohr magneton, μ_0 the permeability of vacuum, and h Planck's constant. In practice a convenient expression is obtained by using the same numerical value for the constant while assuming the following units for the variables in (3.3.5): N is the concentration of spins per centimeter cubed, $(\Delta n_s)_0$ the fractional population difference at the signal transition in thermal equilibrium, Δf_L the linewidth of the signal transition in megahertz, and $\bar{\sigma}\bar{\sigma}^*$ the normalized transition probability for the particular polarization of the microwave fields. The maximum value of $\bar{\sigma}\bar{\sigma}^*$ (for circularly polarized fields) is equal to $\frac{1}{2}$ for a pure Kramers doublet transition, 2 for $S = \frac{3}{2}$ (such as Cr^{3+}) and $\frac{9}{2}$ for $S = \frac{5}{2}$ (such as Fe^{3+}). Further, χ'' is related to the magnetic Q-value Q_m and the filling factor η by

$$1/Q_m = \chi''_{opt} \frac{\int \mathbf{H}^* \cdot \sigma\sigma^* \mathbf{H}\, dV}{\sigma^2_{opt} \int \mathbf{H}^* \cdot \mathbf{H}\, dV} = \chi''_{opt}\eta. \quad (3.3.6)$$

Here, $\mathbf{H}$ is the rf magnetic field vector and χ''_{opt} and σ^2_{opt} are the values of $\bar{\sigma}\bar{\sigma}^*$ for optimum polarization of $\mathbf{H}$. The integrals may be performed over the entire volume of the cavity for a cavity maser and for one period of the structure for a traveling-wave maser. They take into account both the nonoptimum polarization of the fields in a particular structure–crystal combination (usually circular polarization is close to optimum) and the fact that the active material does not in general fill the entire region of the structure which contains rf magnetic fields. The filling factor may differ substantially from structure to structure.

---

[5] K. S. Yngvesson, Res. Rep. No. 84, Res. Lab. Electron., Chalmers Univ. Techn., Sweden (1968).

If a single cavity maser (SCM) is operated at voltage gain $g \gg 1$, the voltage gain-bandwidth product is approximately constant, and given by

$$gB \simeq \frac{2f_s}{|Q_m| + (f_s/\Delta f_L)}. \tag{3.3.7}$$

Use of special techniques, such as extra coupling cavities, will result in substantially wider bandwidths[3,6] and Fig. 3 shows the relation of gain and

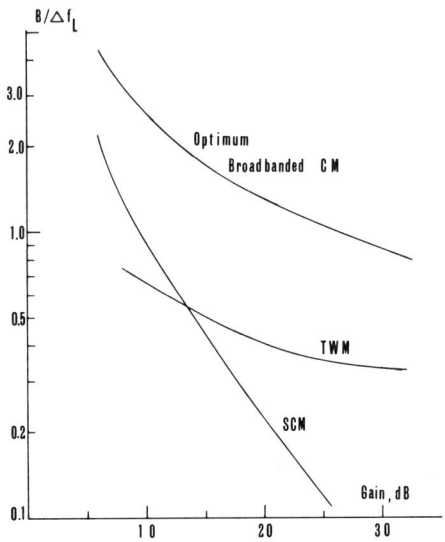

FIG. 3. Normalized bandwidth versus gain for cavity masers (two different cases) and traveling-wave masers. For the cavity masers, it is assumed that $|Q_m| \ll f_s/\Delta f_L$.

bandwidth in this case, and compares it to Eq. (3.3.7). Cavity masers were used at several radio-astronomy observatories in the late 1950s and early 1960s.[7-9] Data on these masers have been summarized by Jelley.[10] At the present (1974), only traveling-wave masers seem to be in active use, and we will return to the reasons for this switch when discussing systems considerations.

[6] R. L. Kyhl, R. A. McFarlane, and M. W. P. Strandberg, *Proc. IRE* **50**, 742 (1962).
[7] J. A. Giordemaine, L. E. Alsop, C. H. Mayer, and C. H. Townes, *Proc. IRE* **47**, 1062 (1959).
[8] J. J. Cook, L. G. Cross, M. E. Bair, and R. W. Terhune, *Proc. IRE* **49**, 768 (1961).
[9] J. V. Jelley and B. F. C. Cooper, *Rev. Sci. Instrum.* **32**, 166 (1961).
[10] J. V. Jelley, *Proc. IEEE*, **51**, 30 (1963).

In a traveling-wave maser (TWM), the signal power grows exponentially with distance from the input, such that the gain in decibels contributed by the active material, the electronic gain, is

$$G_{dB}^e = 27.3 S \frac{1}{|Q_m|} \frac{L}{\lambda_0}. \qquad (3.3.8)$$

Here S is the slowing factor equal to c/v_g where c is the speed of light in vacuum and v_g the group velocity with which the wave propagates through the structure, and L the length of active material from input to output and λ_0 the free-space wavelength at the signal frequency.

A TWM bandwidth decreases much more slowly when the gain increases than a SCM bandwidth. It is given by

$$B = \Delta f_L [3/(G_{dB}^e - 3)]^{1/2}. \qquad (3.3.9)$$

At typical operational gains, the bandwidth is therefore about one-third Δf_L. Equation (3.3.9) has also been plotted in Fig. 3, for comparison with the two cavity-maser cases. One should note that it is possible to "stagger tune" the magnetic field, and thereby increase the TWM bandwidth at the expense of gain.

We can now predict the performance of a maser design, given the required material parameters to substitute in the preceding expressions. Fairly complete material data are available regarding three crystals: ruby,[3,11–14] Cr^{3+} in TiO_2 (rutile),[5,15,16] and Fe^{3+} in TiO_2.[16] These data cover the signal frequency range 1–20 GHz, while above 20 GHz material data are very scarce. If we restrict ourselves to considering TWM's, a useful figure of merit for the active material is $m = \chi'' \Delta f_L$. Optimizing m optimizes the gain while maintaining a fixed bandwidth (i.e., for materials with small Δf_L, magnetic-field stagger-tuning would be used to boost the bandwidth to the desired value).

In Fig. 4 we have used available data to plot the figure of merit versus signal frequency, using the three different materials mentioned. One should note that the approximate optimum concentration of spins has been empirically determined for these crystals. Very high concentrations give lower $|\chi''|$ because cross-relaxation effects prevent large inversion. The energy levels of a

[11] W. S. Chang and A. E. Siegman, reprinted in J. Weber, *Rev. Mod. Phys.* **31**, 7 (1959).

[12] G. I. Haddad and D. H. Paxman, *IEEE Trans. Microwave Theory Tech.* **MTT-12**, 406, (1964).

[13] E. O. Schulz-DuBois, *Bell Syst. Tech. J.* **38**, 271 (1959).

[14] R. C. Clauss, JPL Space Programs Summary 37-61, Vol. III, p. 90 (1970).

[15] K. S. Yngvesson, *IEEE J. Quantum Electron.* **QE-2**, 165 (1966).

[16] P. O. Anderson, A. Jelenski, and E. Kollberg, Res. Rep. No. 107, Res. Lab. of Electron. Chalmers Univ. of Tech. (1972).

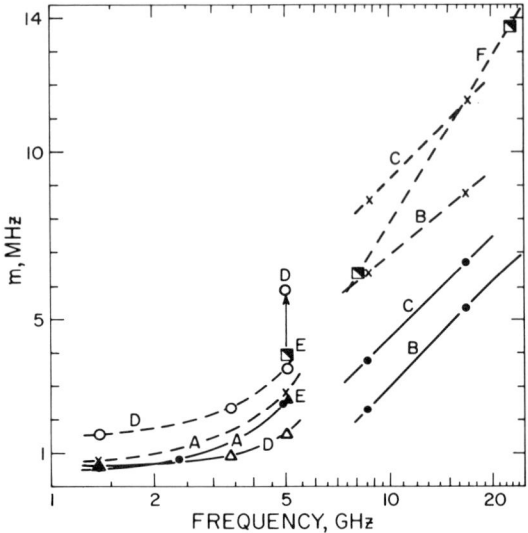

FIG. 4. The figure of merit ($m = \chi'' \Delta f_L$) versus frequency for important active maser-materials. Solid lines are for data at 4.2°K, dashed lines for data at 2°K. The upward arrow indicates the improvement accomplished in the figure of merit of Cr–TiO$_2$ by heat treatment and subsequent quenching. The notation A–F, as explained in the text, is used for the following operating points. A—ruby, $\theta = 90°$, low frequencies; B—ruby, $\theta = 54.7°$, push–pull; C—ruby, $\theta = 90°$, push–push; D—Cr–TiO$_2$, $\theta = 0°$ (c-axis); E—Fe–TiO$_2$, $\theta = 0°$ (c-axis), and F—Fe–TiO$_2$, $\theta = 90°$, $\psi = 45°$ (a-axis). The points plotted are Cr–TiO$_2$ at 4.2°K ($\triangle$) and 2°K ($\circ$), Fe–TiO$_2$ at 4.2°K ($\blacktriangle$) and 2°K ($\blacksquare$), and ruby at 4.2°K ($\bullet$) and 2°K ($\times$).

particular material also depend on the angle between the magnetic field and the crystal symmetry axes. For example, at low frequencies with ruby (curve A in Fig. 4) the magnetic field is aligned perpendicular to the c-axis ($\theta = 90°$), and the energy levels are the ones already shown in Fig. 1. At frequencies around 8 GHz and higher, the so-called push–pull† point, for which the angle θ between the magnetic field and the c-axis is 54.7°, has been commonly used. The energy level diagram for this operating point is shown in Fig. 5a and the figure of merit as curve B in Fig. 4. As can be seen from the diagram, transitions $1 \rightarrow 3$ and $2 \rightarrow 4$ coincide in frequency, since the energy levels are symmetric with respect to the zero of energy, as defined in Fig. 5a. If pump power is applied at the proper frequency, both transitions will be pumped,

---

† The terminology refers to pump action in "pushing" spins out of the lower signal-transition state and "pulling" spins into the upper state.

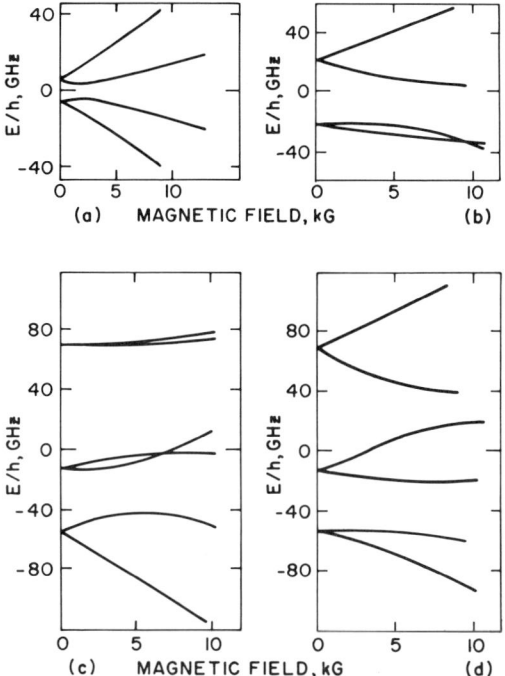

FIG. 5. Energy levels versus magnetic field for (a) curve B, ruby, $\theta = 54.7°$, push–pull; (b) curve D, Cr–TiO$_2$, $\theta = 0°$ (c-axis); (c) curve E, Fe–TiO$_2$, $\theta = 0°$ (c-axis); and (d) curve F, Fe–TiO$_2$, $\theta = 0°$, $\psi = 45°$ (a-axis), where the curves are those in Fig. 4.

and both will contribute to inversion of transition $2 \rightarrow 3$. If all relaxation rates are equal, the inversion ratio for push–pull pumping is approximately $(f_p/f_s) - 1$ (compare Eq. (3.3.3) for the single pump-transition case). R. Clauss[14] has recently shown that a "push–push" pumping scheme at $\theta = 90°$ (indicated to the right in Fig. 1) is superior to the push–pull scheme, and the figure of merit for the push–push scheme is shown by curve C in Fig. 4. It has the further advantage that particular orientation of the magnetic field is not required, also simplifying assembly of the maser. The push–pull scheme, on the other hand, puts stringent demands on exact orientation. Another important factor, particularly in a long maser, is the crystal quality available. The direction of the c-axis in ruby, for example, shows random variations and also a continuous trend as one proceeds along the length direction of the crystal. As a consequence, doubling the length of the maser may not double the gain in decibels as Eq. (3.3.8) predicts if the operating point used is sensitive to variations in orientation. The push–push point is much to be preferred in this respect. Obviously, the push–push scheme does require two pump sources instead of one, though. It should also be added that the figure of merit for presently available (Czochralski-grown) ruby is superior by at least a factor of two compared to that which could be obtained with the Verneuil-grown type, which was the only one available in the early 1960s; this must be kept in mind when comparing data.

In presenting data for TiO_2 (rutile), we have the choice of two dopants, Cr^{3+} or Fe^{3+}. Up to 3 or 4 GHz, the best material is $Cr-TiO_2$ [5,15,16] (see curve D in Fig. 4 and the energy level diagram in Fig. 5b). The magnetic field should be parallel to the c-axis. $Cr-TiO_2$ has the advantage of a larger zero-field splitting, 43 GHz, which results in a larger inversion ratio since higher pump frequencies are used [compare Eqs. (3.3.3) and (3.3.4)].

At frequencies from 4 to 7.5 GHz, the best operating point is obtained by using $Fe-TiO_2$ with the magnetic field parallel to the c-axis (energy levels in Fig. 5c; for figure of merit, see curve E, Fig. 4). Above 7.5 GHz less data are available, but good results at 8 and 22 GHz have been obtained by orienting the magnetic field parallel to the a-axis, using $Fe-TiO_2$, as shown in curve F in Fig. 4, and Fig. 5d.

In general, the quality of rutile which is available is inferior to that of ruby, so that one has to take greater care in avoiding degradation of the performance due to bad crystal quality. Considerable improvement in performance has been obtained by Anderson et al.[16] by quenching the crystals in air after a heat treatment in oxygen at 1200°C. The curves drawn in Fig. 4 refer to the material as supplied by the manufacturer, whereas the arrows indicate the improvement that has been obtained by the heat treatment. If the heat-treated crystals are used, TiO_2 has a significantly larger figure of merit than ruby for frequencies from 1 to 5 GHz. At frequencies higher than 5 GHz comparable figures of merit are obtained with some advantage for ruby around 8 GHz. Higher gain results if the pump is frequency modulated in both ruby and rutile. This may indicate some inhomogeneous broadening of the pump transition, i.e., it does not saturate uniformly.

A good way of making a stable cavity circuit is to coat a piece of ruby with silver and cut coupling holes for the signal and pump, respectively, in the silver layer.[8] The silvered ruby cavity is then clamped to the signal and pump waveguides; the signal may be coupled through a cavity for increased bandwidth. Cavity masers require a ferrite circulator to separate the input and output waves as shown in the block diagram of Fig. 6.

The basic design used in most TWM's was developed by DeGrasse et al.[17] and is shown in Fig. 7a. The periodic arrangement of copper pins, parallel to the broad wall of the waveguide, the "comb," allows slow wave propagation

---

[17] R. W. DeGrasse, E. O. Schulz-DuBois, and H. E. D. Scovil, *Bell Syst. Tech. J.* **38**, 305 (1959).

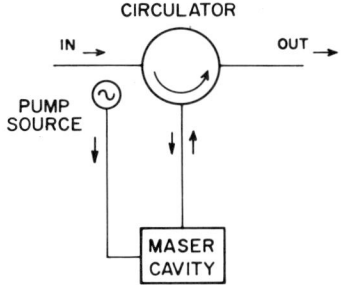

FIG. 6. Block diagram of a cavity maser.

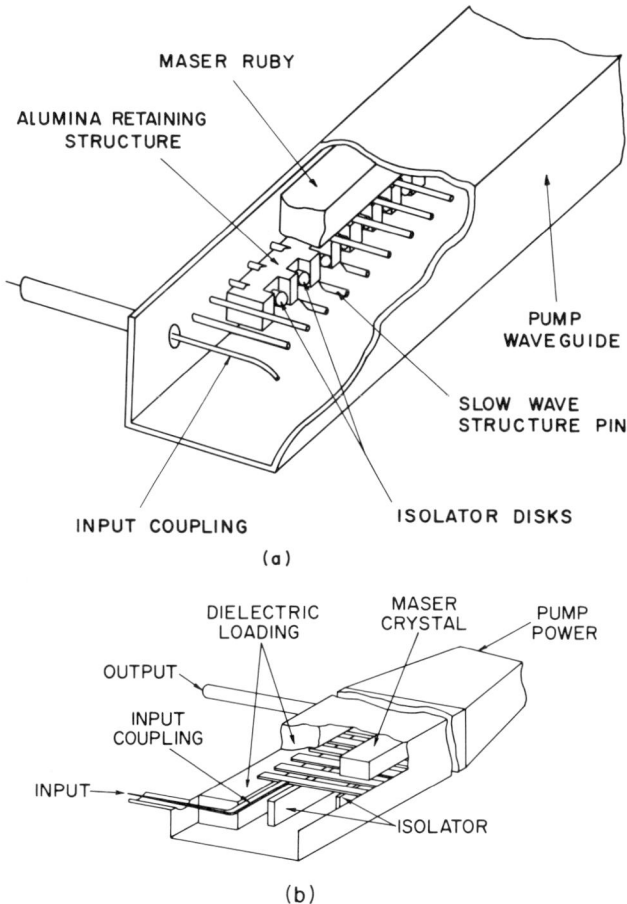

FIG. 7. (a) Sketch of the original comb-structure design; (b) Sketch of modified comb structure design for high dielectric constant materials.

with a passband at frequencies for which the pins are approximately $\lambda_0/4$ long. By arranging the correct dielectric loading effect of the ruby on the comb, one can obtain slowing as large as 400. Large slowing will, in general, mean a narrower passband of the structure, which in turn determines the frequency range over which the TWM is tunable. The theory and design principles for this comb structure may be obtained from several publications.[17-19] We should also remark that the pump wave usually propagates as an approximate TE_{10} mode in the waveguide—although in some designs a dielectric taper is used to match the pump power directly into the ruby rod.

A feature which is common to all periodic structures is of major importance for the operation of a TWM: there are regions of rf magnetic fields with good circular polarization but with reversed sense on either side of the comb. This enables the designer to position the active maser material on one side, giving a good filling factor for gain in the forward direction, as well as a ferrimagnetic material on the opposite side (at the position of optimum circular polarization) which contributes a reasonably small loss in the forward (or gain) direction while strongly attenuating any backward-traveling (reflected) wave. If such an isolator were not incorporated, the TWM would easily build up enough loop gain to fall into oscillation due to waves reflected between the input and the output. A TWM is usually designed with enough reverse attenuation to prevent it from oscillating even when input and output are short circuited.

Typical ferrimagnetic isolator materials that have been used are yttrium–iron–garnet (YIG), substituted YIG, and several different ferrites. The ferrimagnetic resonance absorption can be made to occur at the same magnetic field, which is required for the maser material, by changing the demagnetization factors, i.e., adjusting the shape of the material.[3] Since ferrimagnetic linewidths are relatively large at low temperatures (at least several hundred gauss), it is usually possible to make the two resonances track each other over an extended frequency range.

The high dielectric constant of rutile (250 parallel to the *c*-axis) makes it difficult to use the comb structure just described for rutile masers. A modified comb-structure design,[20,21] ideal for use in connection with high dielectric constant materials, is available, however, for which a detailed theory and

---

[18] S. M. Petty and R. C. Clauss, *IEEE Trans. Microwave Theory Tech.* **MTT-16**, 47 (1968).

[19] F. S. Chen, *Bell Syst. Tech. J.* **43**, 1035 (1964).

[20] E. L. Kollberg and K. S. Yngvesson, *Int. Conf. Microwave Opt. Generation Amplification, Cambridge, 1966* (*IEE Conf. Publ. 27*) 384 (1966).

[21] O. E. H. Rydbeck and E. L. Kollberg, *IEEE Trans. Microwave Theory Tech.*, **MTT-16**, 799 (1968).

design procedure[22-24] has been worked out. An example of this structure is shown in Fig. 7b. Achievable slowing and tunable bandwidths are comparable for the two types of comb structure whereas the latter gives a larger filling factor. A more recent modification of the comb structure, used in a 12-mm TWM design, produces the comb by photoetching of a thin copper or gold film, deposited directly on the ruby.[25] This structure is particularly well suited for use at millimeter waves, and possesses a remarkably high volume filling factor, about 0.9 as compared to 0.3 to 0.45 for the above two structures (assuming that active material is loaded only on one side of the comb). The theory of this structure has also been worked out.[26,27] Typical slowing is $10 \times (\varepsilon_r)^{1/2}$ where ε_r is the relative dielectric constant of the material, whereas tunable bandwidths of about 25% may be obtained independent of slowing.

3.3.2. Systems and Operational Considerations

The noise temperature of the maser proper has two main contributions[3]: one from the spins and one from losses in the microwave structure

$$T_m = \frac{T_{op}}{|I|} + \frac{L_{dB}}{G_{dB}^e} T_{op}, \qquad (3.3.10)$$

where T_{op} is the operating temperature and L_{dB} the ohmic losses (in decibels) in the structure. Equation (3.3.10) is an approximate form, valid if $L_{dB} \ll G_{dB}^e$. Typical ohmic losses are from 5 to 10 dB. Normally T_m is less than the operating temperature, which is usually either 2 or 4.2°K. This is so low that it is almost always a negligible fraction of the total system noise temperature, with the main contributions to the latter arising from antenna side- and backlobes and spill-over, atmospheric attenuation, and attenuation in waveguide components connecting the maser input to the antenna feed: thus the lowest system noise temperature reported is 13°K (at 2388 MHz) at the Jet Propulsion Laboratory (JPL), Goldstone 64-m (210-ft) antenna. In order to discuss these contributions to the system noise, we will analyze a specific system, that of the 1.6-GHz TWM at the Onsala Space Observatory[28] (Sweden). A

[22] E. L. Kollberg, Res. Rep. No. 72, Res. Lab. of Electron., Chalmers Univ. Tech. (1966).
[23] E. L. Kollberg, *Electron. Lett.* **3**, 294 (1967).
[24] E. L. Kollberg, Res. Rep. No. 98, Res. Lab. of Electron., Chalmers Univ. Tech. (1970).
[25] A. C. Cheung, M. F. Chui, A. G. Cardiasmenos, S. Y. Wang, C. H. Townes, and K. S. Yngvesson (to be published).
[26] M. H. H. Van Dijk, C. E. Hagstrom, and E. L. Kollberg, Res. Rep. No. 106, Res. Lab. of Electron., Chalmers Univ. Tech. (1972).
[27] M. H. H. Van Dijk, C. E. Hagstrom, and E. L. Kollberg, *Electron. Lett.* **8**, 70 (1972).
[28] E. L. Kollberg, *Proc. IEEE*, **61**, 1323 (1973).

FIG. 8. Packaged TWM connected to antenna feed. Note the rotatable waveguide-to-coax transition and the large diameter coax, used to minimize loss at the input side of the maser (courtesy of Onsala Space Observatory).

photograph of the maser and its connection to the feed horn is shown in Fig. 8. Note that the waveguide–coaxial transition is rotatable and allows the entire maser package to rotate in one plane as the antenna moves.

The separate contributions to system noise are cataloged in Table I. This particular system is frequency switched, which eliminates the need for a front-end switch, thus minimizing the total noise.

TABLE I. Noise Catalog for 1.6-GHz TWM System at Onsala Space Observatory

System component	Noise temperature(°K)
Maser package coaxial line	4
Waveguide–coaxial transition + waveguide polarization filter + calibration coupler	4
Second stage of receiver	1
Antenna	14
Total system noise temperature (measured separately)	24 ± 1

3.3. MASER AMPLIFIERS

The rms noise level of the system may be determined from[9]

$$(\Delta T)_{rms} = \frac{\pi}{2} \frac{T_s}{(B\tau)^{1/2}} + \frac{(\Delta G)_{rms}}{G_0}(T_a - T_{comp}). \qquad (3.3.11)$$

Here, τ is the integration time, T_a the antenna temperature on the source (beam switching) or on the line (frequency switching), T_{comp} is the comparison temperature in the switching scheme used, and $\Delta G_{rms}/G_0$ is the rms value of the random fractional fluctuations in system gain.

In a well-designed maser system, the second term in Eq. (3.3.11) should be negligible compared to the first one. This requires both that $T_a - T_{comp}$ is kept small and that the maser gain does not have large variations on the time scale of a switching period. A typical $(\Delta G)_{rms}/G_{net}$ of 0.6% with $G_{net} =$ 20 dB was found over a period of 10 min for a cavity maser in a beam-switched system,[8] whereas 0.5% (rms) over 6 hr[24] was obtained with a TWM in the laboratory at a 30-dB net-gain level. Gain fluctuations did not contribute to the rms noise level significantly in these systems. In the Jet Propulsion Laboratory (JPL) TWM system which operates at frequencies around 8 GHz an rms gain variation of 3% (0.12 dB) is obtained over a 12-hr period, with typical antenna motions.[29]

While a scheme involving rapid switching is important for continuum observations, spectral line work requires only that the shape of the bandpass curve of the system remains stable. This has led to another popular technique of observation in which spectral data are recorded alternately on-source and off-source for a typical period of a few minutes and then subtracted (see Chapter 4.3). Masers have been employed successfully with this technique, i.e., the noise fluctuations are consistent with only the first term of Eq. (3.3.11). One advantage of the method is that no switch is required, thus the system noise temperature is lower than in the beam-switched case.

In interferometer applications, one is also interested in the excellent phase stability of the maser. For the JPL 8-GHz maser system,[29] the rms changes were estimated to be 1.1° in phase; with an 0.08-nsec group delay, over a 12-hr period, including effects of antenna motion. The improved stability of TWM over a CM obviously is a desirable feature; in particular it is not possible to operate a CM at higher net gain than 20–23 dB, which will result in some second-stage noise contribution. Also the CM requires a circulator, which introduces extra noise and this works to its disadvantage. Further, most CM designs have not been easily tunable, whereas TWMs are easily tuned (one changes magnetic field and pump frequency). We have already compared bandwidths of CMs and TWMs in Fig. 3. After TWM designs

[29] R. C. Clauss, E. Wiebe, and R. Quinn, JPL Techn. Rep. 32-1526, Vol. XI, p. 71 (1972).

were brought forth it became clear that the TWM had enough advantages, and a transition from CMs to TWMs took place.

An important operational consideration is how the maser should be cooled. The most convenient way of doing this is to use a closed cycle refrigerator.[30] One such system[31] at the Jet Propulsion Laboratory was operated continuously for three years with only 24 hr down-time. Yet, operational radio-astronomy maser receivers have tended to use liquid-helium dewars that must be refilled with about 10 to 20 liters of liquid helium, typically every 24 hr, and with a typical down-time of 1 hr every time the dewar is refilled. It is also quite common to make use of the higher gain, typically three times higher, available at pumped liquid-helium temperatures (2°K or less and typical pressures of 5 to 20 mm Hg) by equipping the installation with a vacuum pump. The pump down adds at least an extra half hour of down-time.

The magnetic field required is most conveniently provided by a superconducting solenoid,[32,33] usually with an iron circuit and superconducting shields to improve the homogeneity. The size and typical design may be seen in Fig. 9. The required field is first set up with the help of an external supply, whereupon the magnet is shorted with a short section of superconducting wire and runs in the persistent mode without requiring any external supply. This very effectively eliminates any gain or phase instability that might be caused by variations in the magnetic field. By rapidly switching the superconducting magnet, one can change the frequency by as much as 80 MHz in a rutile maser.[28] This technique has applications in broadband VLBI work as it is not necessary to change the pump frequency.

Reflex klystrons are generally used to supply the pump power for a maser. The klystron may be temperature stabilized by cooling it in a bath of Fluorinert or transformer oil, although pump frequency drifts are less critical because of the common practice of frequency-modulating the pump at a rate of 10 KHz or higher. It is important to note that noise is not transferred from the pump to the signal (compare parametric amplifiers). Thus IMPATT solid-state oscillators, now becoming available, should be ideal pump sources.

3.3.3. Data for Specific Maser Amplifiers and Systems

To illustrate the preceding general discussion, we present in Table II data for some specific maser amplifiers and systems. Possibly of greatest interest are the total system noise temperature, and the rms noise fluctuations, normalized to $\tau = 10$ sec, given in the last two rows. System noise temperature

[30] W. H. Higa and E. Wiebe, *Cryog. Technol.* **3**, 47 (1967).
[31] R. C. Clauss, private communication (1970).
[32] P. P. Cioffi, *J. Appl. Phys.* **33**, 875 (1962).
[33] E. L. Hentley, *Cryogenics* **7**, 33 (1967)

3.3. MASER AMPLIFIERS

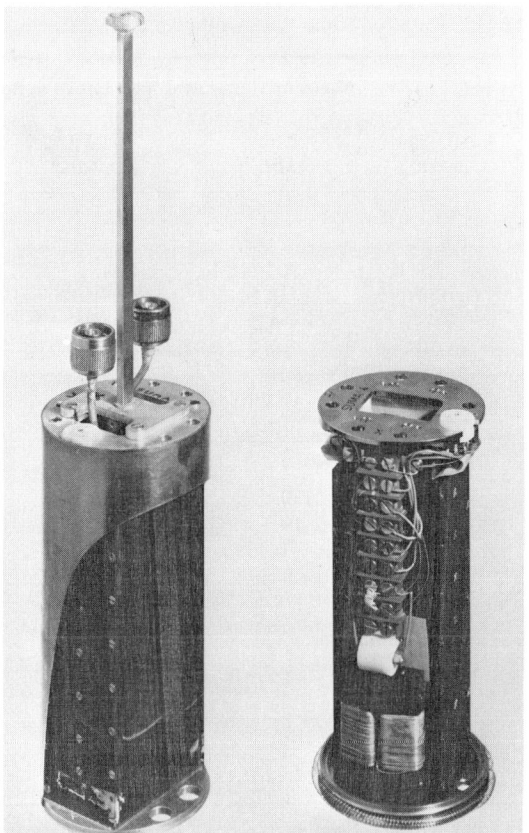

FIG. 9. Typical superconducting magnets. The superconducting switch can be seen on the magnet to the right (courtesy of Onsala Space Observatory).

and bandwidth have also been plotted in Figs. 10 and 11. With these data available, it is possible to evaluate the system performance in a particular application. One should remember that an improvement by a factor of 10 in system noise temperature represents a shortening of the observing time by a factor of 100.

A general comment may be in order: Given a basic TWM design, it is relatively easy to scale it over quite a wide frequency range. Thus the Jet Propulsion Laboratory has developed three TWMs, at frequencies around 2,380, 8000 and 15,000 MHz, respectively, all with the same basic comb structure and ruby as active material, whereas Onsala Space Observatory uses a series of eight interchangeable masers in bands from 0.96 to 8.5 GHz all with the modified dielectric-loaded comb structure and $Cr-TiO_2$ or $Fe-TiO_2$ as active material. Considering the wide frequency coverage and

TABLE II. Data for Some Representative TWM Systems

Type of data	Maser material and location of system			
	Cr–TiO$_2$ Onsala[a]	Ruby JPL[b]	Fe–TiO$_2$ Onsala[a]	Ruby JPL[b]
Material Data[c]				
$N \times 10^{-19}$ cm^{-3}	0.9	2.0	1.3	2.0
Operating point	c-axis, 1.9°K	$\theta = 90°, 4.5°$K	c-axis, 1.7°K	$\theta = 90°, 4.5°$K
$(\Delta n_s)_0$	0.0153	0.0064	0.0194	0.028
I	−9.4	−4.2	−11.5	−3.2
$|\sigma|^2$	1.50	1.44	1.85	1.53
Δf_L (MHz)	20	60	55	60
f_p (GHz)	48	12.7	60	19,23
m (MHz)	2.5	1.0	7.0	3.5
Structure Data[c]				
Slowing factor	155	130	52	62
Insertion loss (dB)	20	10	7	12
Effective η	0.31	0.40	0.18	0.14
Length (cm)	5.6	30.5	5.6	15.2
System Data[c]				
Tunable frequency range (GHz)	1.55–1.78	2.23–2.43	5.25–6.1	7.75–8.75
G_{net}(dB)	30	45	27	45
B (MHz)	6	45	20	17
T_s (°K)	24	15	49	23
$(\Delta T)_{rms}$ (°K)	0.005[d]	0.001[d]	0.005[d]	0.0025[d]

[a] Onsala Space Observatory, Onsala, Sweden.
[b] Jet Propulsion Laboratory, Goldstone, California.
[c] Symbols appearing here are defined in the text.
[d] ΔT_{rms} is calculated under the assumption that one is interested in continuum measurements, i.e., the whole bandwidth of the maser is utilized. Also, $\tau = 10$ sec and Dicke switching are assumed.

low noise temperature of such a system, it represents the most economical choice of receiver system, especially for spectral-line observations at a series of frequencies. As an illustration, a picture of a maser head with two "plug-in units" for different bands is given in Fig. 12.

3.3.4. Millimeter Wave Masers

At the present time radio-astronomy usage of millimeter wave masers is limited to three systems, one with $T_s = 200°$K at 8 mm[34] in the USSR and

[34] V. I. Zagatin, G. S. Misezhnikov, and V. B. Shteynshleyger, *Radio Eng. Electron. Phys.* **12**, 501 (1967).

3.3. MASER AMPLIFIERS

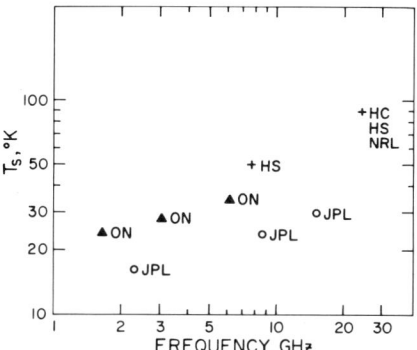

FIG. 10. Total system noise temperature in maser systems actively used for radio astronomy at Onsala (▲ON), the Jet Propulsion Laboratory (○JPL), Hat Creek(+HC), Haystack (HS) and Naval Research Laboratory (NRL).

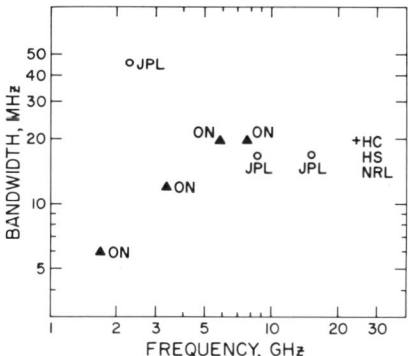

FIG. 11. Instantaneous bandwidth in maser systems actively used for radio astronomy at Onsala (▲ON), the Jet Propulsion Laboratory, (○JPL), Hat Creek (+HC), Haystack (HS), and Naval Research Laboratory (NRL).

two with $T_s \simeq 100°K$ at 12 to 15 mm[25] in the USA. Applications of millimeter wave masers to radio astronomy[35] are certain to be greatly extended in the near future, however. The two latter masers are installed at the University of California at Berkeley, Hat Creek Observatory, and the NEROC Haystack Observatory, respectively. Typical data are net gains at 1.8°K bath temperature in excess of 30 dB over a tunable range of 20 to 24 GHz with 18-MHz instantaneous bandwidth. Ruby is used as the active material at the push–pull operating point. The slow-wave structure is the photoetched comb

[35] K. S. Yngvesson, *1971 IEEE Int. Conv. Digest*, New York 160 (1971).

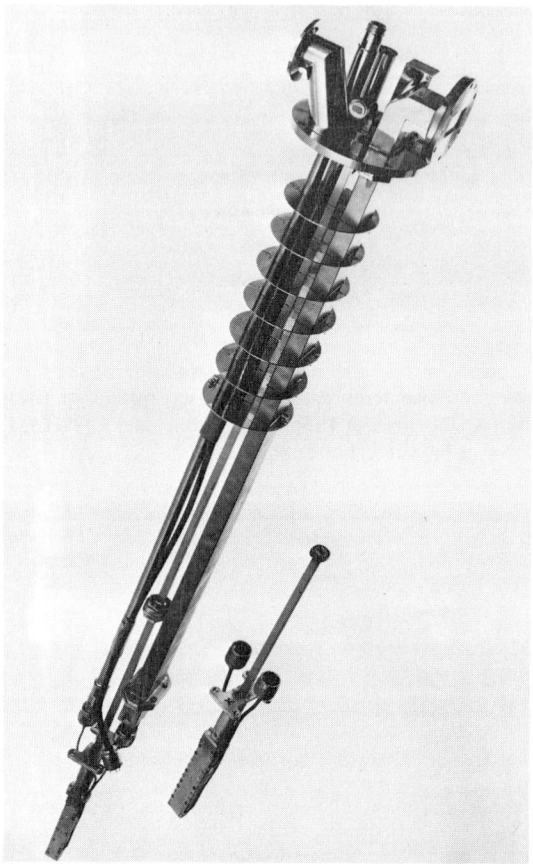

Fig. 12. Waveguide header for TWM with two exchangeable active maser units for different frequency band (courtesy of Onsala Space Observatory).

mentioned previously. An 8-mm TWM-design due to Arams and Peyton[36] uses the purely dielectric slowing obtained when rutile is loaded into a waveguide which is small enough to only support the TE_{10} mode, taking into account the high dielectric constant of the rutile. Neither this maser, nor two similar designs[37,38] have so far been used for radio astronomy, while a fourth maser of this type is likely to be taken into active use at Onsala Space Observatory in the near future.[39] Data for the rutile masers and ruby masers

[36] F. Arams and B. Peyton, *Proc. IEEE* **53**, 12 (1965).
[37] Y. de Coatpont and A. Robert, *Onde Elec.* **47**, 165 (1967).
[38] P. Swanson and J. Hagen, *Nature (London)* **218**, 158 (1968).
[39] K. S. Yngvesson, A. G. Cardiasmenos, and E. L. Kollberg, *IEEE Int. Microwave Symp.*, *Boulder, Colorado* (June, 1973).

are roughly comparable in this wavelength range. At shorter wavelengths, materials such as Fe–TiO$_2$ with larger zero-field splitting are called for. Cavity masers with this material have been operated in the laboratory at 5 to 6 mm[40] and at 4 mm,[41] but have not been taken into operational use. Another cavity maser at 3 mm, uses natural andalusite.[42] Further development is necessary; in particular an investigation of suitable structures for TWMs.

Photoetching of the structure[25, 35] should be advantageous due to its very small size, particularly when high dielectric constant crystals such as TiO$_2$ are to be used. Such development should make available maser amplifiers up to at least 100 GHz in the near future, with system temperatures in the range of 50 to 100°K (excluding the atmospheric-O$_2$ absorption region from 50 to 70 GHz). The increase in system noise temperatures with increased frequency is due to increased atmospheric absorption and greater attenuation of the waveguides and waveguide devices which connect the maser with the feed horn. The most serious limitation on the upper frequency is in the availability of pump sources at frequencies higher than the signal frequency. Schemes involving variants of push–pull, push–push, or optical pumping[43–44] show great promise for eventually getting around this limitation.

---

[40] D. L. Carter, *J. Appl. Phys.* **32**, 2541 (1961).
[41] A. Molé and M. Soutif, *Onde Elec.* **47**, 183 (1967).
[42] I. I. Eru, S. A. Peskovatskiy, and A. N. Charnets, *IEEE J. Quantum Electron.* **QE-4**, 723 (1968).
[43] E. S. Sabisky and C. H. Anderson, *IEEE J. Quantum Electron*, **QE-3**, 287 (1967).
[44] L. Mollenauer, D. Pan, and K. S. Yngvesson, *Phys. Rev. Lett.* **23**, 683 (1969).

3.4. Multichannel-Filter Spectrometers*

3.4.1. General Description

In 1951 when Ewen and Purcell, at Harvard University, made the initial detection of the 21-cm emission from interstellar neutral atomic hydrogen gas, they used a crystal-mixer superheterodyne receiver with a single-frequency selective channel bandwidth of 17 kHz.[1] The intermediate frequency of the receiver was switched at a 30-Hz rate between two frequencies which were displaced from each other by about 75 kHz. Thus any difference in spectral density between these two frequency bands would appear as a 30-Hz modulation component in the detected output of the receiver. A synchronous detector was used to convert this modulation component to a dc level which gave a direct measure of the difference in spectral density. In order to observe the hydrogen line signal, the receiver local oscillator was slowly tuned through the hydrogen frequency and a symmetrical bipolar output response was obtained.

Following the discovery of the 21-cm hydrogen line, there was great interest in constructing more sophisticated spectral-line receivers. The need for improved frequency resolution and better efficiency in obtaining detailed spectral-line profiles led to the development of multichannel receivers in which a number of adjacent or closely spaced frequency bands are processed simultaneously. Ideally, the multichannel receiver provides frequency coverage as illustrated in Fig. 1. Each individual filter channel passes only those frequencies which line inside the range of $\pm\frac{1}{2}B$ about the center frequency f_j, and the center frequencies of the filters are spaced in sequence at an interval equal to the filter bandwidth B. The total frequency coverage, therefore, is NB where N is the number of channels in the receiver. The receiver output contains N separate values, each of which is the integrated power density over the frequency band covered by the particular filter.

A typical multichannel spectral line receiver utilizes a superheterodyne configuration with the i.f. (intermediate frequency) portion of the receiver being split into the separate bandpass filter channels.[2] Figure 2 gives the block diagram of a typical channel in a multichannel-filter receiver. The i.f.

---

[1] H. I. Ewen and E. M. Purcell, *Nature (London)* **168**, 356 (1951).
[2] R. A. Batchelor, J. W. Brooks, and M. W. Sinclair, *Proc. IRE (Aust)*. **30**, 39 (1969).

Chapter 3.4 is by **Hays Penfield**.

3.4. MULTICHANNEL-FILTER SPECTROMETERS

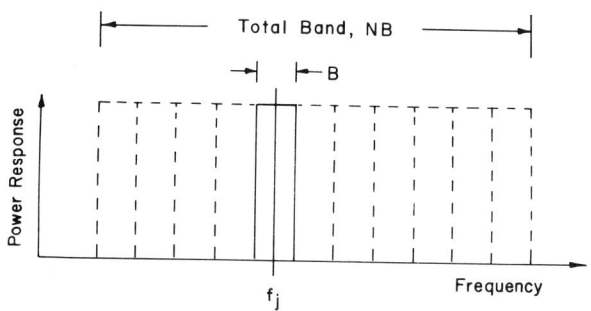

FIG. 1. Idealized multichannel-filter receiver-frequency coverage.

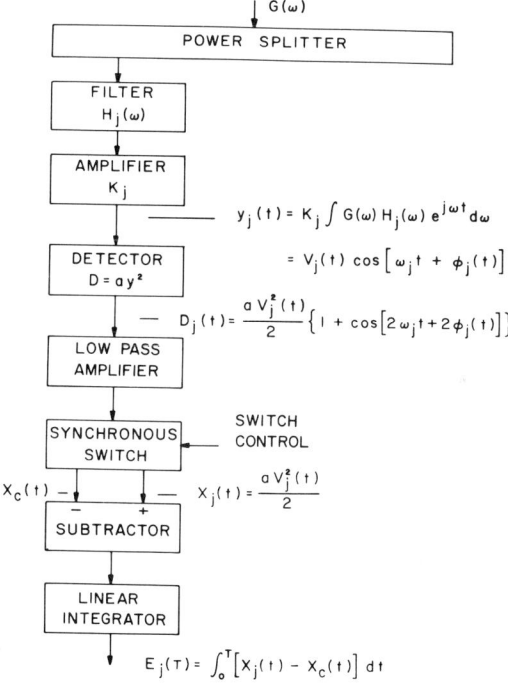

FIG. 2. Block diagram of typical channel in a multichannel-filter receiver. (See text for discussion.)

signal is coupled through a power splitter to the input of the jth channel filter. The filter output is amplified and then square-law detected. A low-pass amplifier following the square-law detector preserves the average or "envelope" term and rejects the high frequency-terms. When the receiver is operated as a switched system (i.e., input switching or frequency switching) a synchronous switch connects the low-pass amplifier output alternately to the two input terminals of a subtractor in the proper sense to produce an output which is the difference between the "signal" and "comparison" level. A linear integration process is performed on the subtractor output to produce the final channel output voltage $E_j(T)$.

Thus the switched multichannel-filter receiver provides a set of values representing the difference spectrum between the signal input obtained from the antenna and a "flat" comparison spectrum. This comparison spectrum is normally obtained either by switching the receiver input to a matched source such as a sky horn, off-axis feed or cold load or by switching the frequency of the receiver so that the filters are placed in a "flat" spectral region adjacent to the region containing the signal spectrum.

In principle, a multichannel-filter receiver can be operated in an unswitched or total power mode. For example, if we eliminated the synchronous switch and provided a uniform comparison level equal to the average of all the channels at the input to each subtractor, then we would cancel a large portion of the system noise while still preserving the spectral feature contained in the signal. Unfortunately, a total-power system of this type requires extreme flatness in the receiver bandpass prior to the multichannel-filter section and also extreme stability in the gain of each filter channel. These stability limitations have made it impossible, thus far, to achieve excellent long-term performance with the total power multichannel-filter receiver.[3] The use of a small digital computer operating directly on the low-pass amplifier output might, however, provide the necessary "bookkeeping" capacity to compensate for these inherent weaknesses of the total-power multichannel-filter receiver. Further description of the total-power multichannel-filter receiver is beyond the scope of this section, and all subsequent material will concern the switched multichannel-filter receiver.

Referring to Fig. 2, we see that the individual channel filter with its frequency characteristic transfer function $H_j(\omega)$ determines the response of each channel to any given input. The amplifier, detector, and low-pass amplifier operate on the filter output to establish a fluctuating dc level $X_j(t)$, which is proportional to the integrated power density over the bandpass of the filter. The synchronous switch, subtractor, and integrator measure the difference between the signal and comparison intervals in this fluctuating dc level and present an analog output voltage $E_j(T)$, which is the time integral of the difference between the signal level $X_j(t)$ and comparison level $X_c(t)$. Output spectral data is then obtained by direct plotting of these analog output voltages or by storing the voltages, applying calibration and baseline corrections to them, and combining or averaging a number of similar sets of spectral-

[3] F. J. Kerr, J. V. Hindman, and C. S. Gum, *Aust. J. Phys.* **12**, 270 (1959).

output data to obtain very long integration time. A digital computer is normally used to provide this data reduction function with the degree of sophistication in the final spectral-line data presentation being determined by the imagination of the programmer and the financial support available.[4,5]

In any spectral-line receiving system there are certain basic qualities which are important; that is, the system should provide the following:

(1) low system noise temperature,
(2) excellent gain and frequency stability,
(3) flexibility in frequency tuning and resolution,
(4) precise measurement of the received frequency,
(5) a true picture of the spectral power-density distribution in the received signal and,
(6) adequate total frequency coverage for efficient system utilization.

The system noise temperature, gain stability, frequency stability and flexibility in frequency tuning are determined completely or to a large degree by the low-noise preamplifier and local oscillator sections of the receiver system (see Chapters 3.1–3.3). The frequency resolution, total frequency coverage, and the degree to which the output data presents a "true" picture of the spectral power-density distribution in the received signal is determined by the multichannel signal-processing section of the receiver system. In particular, the filter, detector, and integrator portions of each channel are critical in their influence on the output data.

3.4.2. Filter Characteristics

The filter in each channel defines the frequency transfer function $H_j(\omega)$ for that particular channel. The frequency resolution and the degree to which the multichannel-filter receiver presents a true picture of the spectral power-density distribution in the received signal is determined by this channel frequency transfer function. Since the parameter which must be determined is the power-density distribution, it is the power-density transfer function $[H_j(\omega)]^2$, that must be considered. Many different theoretical and practical forms may be utilized for the power-density transfer function. The circuits used in the multichannel filter receivers to obtain the frequency transfer function $H_j(\omega)$ may all be analyzed as tuned, bandpass networks containing one or more resonant elements.

A simplified form for the power-density transfer function which may be obtained with tuned, bandpass filter circuits is written as

$$[H_j(x)]^2 = 1/(1 + x^{2n}), \qquad (3.4.1)$$

[4] B. G. Clark, *Annu. Rev. Astron. Astrophys.* **8**, 115 (1970).
[5] R. Vance, Nat. Radio Astron. Observ., Comput. Div. Intern. Rep. No. 7 (1966).

where n is the number of tuned circuits, $x = 2[(\omega - \omega_0)/(\omega_2 - \omega_1)]$, ω_0 the center frequency, ω_1 the lower half-power frequency, and ω_2 the upper half-power frequency.

The channel output level is proportional to the total power at the filter output which is given by the integral

$$\int_0^\infty [H_j(\omega)]^2 \, d\omega \qquad (3.4.2a)$$

or, expressing this in terms of the variable x,

$$\int_0^\infty [H_j(x)]^2 \, dx. \qquad (3.4.2b)$$

The ideal filter, as depicted in Fig. 1 passes all frequencies between ω_1 and ω_2 with equal response and completely rejects frequencies less than ω_1 and greater than ω_2. This ideal filter has the following form of power-density transfer function:

$$[H_j(\omega)]^2 = \begin{cases} 1, & \omega_1 \leq \omega \leq \omega_2, \\ 0, & \omega_1 > \omega, \quad \omega_2 < \omega \end{cases} \qquad (3.4.3a)$$

and in terms of the variable x

$$[H_j(x)]^2 = \begin{cases} 1, & -1 \leq x \leq +1, \\ 0, & -1 > x, \quad +1 < x. \end{cases} \qquad (3.4.3b)$$

The total power at the output of the ideal filter is therefore unity. Various practical and theoretical filter-transfer functions may be evaluated on the basis of the ratio of their total output power to the output power from an ideal bandpass filter with the same half-power bandwidth. This ratio is, in fact, equal to the integral $[H_j(x)]^2 \, dx$, and is sometimes known as the "equivalent noise bandwidth of the filter." Table I lists several types of filter-transfer functions giving the power-density function and the equivalent noise bandwidth.

The Gaussian power-density distribution is used extensively in analyzing spectral-line data and appears to represent a close approximation to the actual shape of radio astronomical spectral lines. Therefore it is of interest to compare the practical filters used in multichannel-filter receivers with this Gaussian response. The double-tuned filter matches the Gaussian shape most closely and is reasonably easy to obtain. As the number of tuned circuits increases, the filter shape approaches that of the ideal filter and multipole-crystal filters are used extensively to obtain a response closely matching that of the ideal filter.

3.4. MULTICHANNEL-FILTER SPECTROMETERS

TABLE I. Comparison of Filter Types

Type	$[H_J(x)]^2$	Equivalent noise bandwidth factor
Gaussian	$e^{-(0.833x)^2}$	1.06
Cosine squared	$\cos^2 \frac{\pi}{4} x, \quad x \leq 2$	1.0
Single-tuned	$\dfrac{1}{1+x^2}$	$\dfrac{\pi}{2} = 1.57$
Double-tuned	$\dfrac{1}{1+x^4}$	$\left(\dfrac{\pi}{2}\right)^{1/4} = 1.12$
Triple-tuned	$\dfrac{1}{1+x^6}$	$\left(\dfrac{\pi}{2}\right)^{1/9} = 1.05$

One of the inherent problems in the multichannel-filter receivers is the nonuniformity in the bandwidth and filter-response shape from channel to channel. This nonuniformity distorts the spectral-line output data obtained from the receiver and therefore must be minimized as much as possible. In general, the susceptibility of the receiver to this nonuniformity in channel response gets worse as the number of tuned circuits in the filter increases. One notable exception to this would be the case of crystal filters all constructed identically at the same center frequency and utilizing separate frequency conversion prior to each filter to obtain the required channel center-frequency distribution.[6]

Multichannel-filter receivers utilizing single-tuned channel filters offer a simple means for achieving uniform channel to channel response. Because of the relatively long "tail" on the single-tuned filter response beyond the half-power response points, however, these filters "smear" or broaden the spectral-line signals. When the single-tuned filter is used in multichannel receivers and the channel center frequencies are spaced by an amount equal to the filter half-power band width, 28.5% of the power output in any one channel is contributed by signal energy outside the half-power band. This results in a behavior that is similar to that obtained with multipole filters which are partially overlapped in frequency. This overlap produces a "smoothed" pattern in the output data since adjacent channels are not independent, but are correlated to an extent related to the degree of overlap.

[6] D. L. Thacker and L. Beale, Nat. Radio Astron. Observ. Electron. Div. Intern. Rep. No. 88 (1969).

No particular filter type may be characterized as being the best for multichannel-filter receivers. What is important is the uniformity and the stability of the individual filter response characteristics, and that the data output be analyzed with these filter characteristics clearly understood.

3.4.3. Detectors

Radio-astronomy signal intensity is normally expressed in terms of noise temperature with the units of degrees Kelvin (°K). Noise temperature T, for any given bandwidth B is linearly related to the average power level P by the expression

$$T = P/2kB, \qquad (3.4.4)$$

where k is the Boltzmann constant (1.38×10^{-23} W/Hz °K). Hence, the detector utilized in the typical multichannel filter receiver is designed to provide an average output voltage (dc level) that is directly proportional to the input power. This type of detector is generally called a *square-law detector* since it provides an output signal in which the average value is proportional to the square of the input voltage.

The square-law-detector characteristic is obtained by operating a point-contact diode in the low-current range (1–10 μA) and selecting a detector load resistance (normally 1000 ohms or less) which gives the greatest range for the square-law behavior. For spectral-line systems a range of 10 dB centered on the normal operating level is adequate and a range of 20 dB for the square-law characteristic is often obtained.

3.4.4. Integrators

The signal present at the output of the low-pass amplifier in each channel contains an amplitude-modulation component at the receiver switch rate. In order to obtain the final channel output it is necessary to measure this difference between the average level in the signal and comparison portions of the switch cycle. Two common methods utilized are the analog synchronous detector followed by a linear or rc integrator and the digital computer which samples, stores, and averages the signal and comparison data for each channel and then processes this information to produce the spectral-line output data.

The analog synchronous detector and integrator combination provides a means for obtaining a direct analog-voltage output for each channel. The synchronous detector operates directly on the output from the low-pass filter and utilizes mechanical contacts (i.e. relay or magnetic reed devices) or solid-state elements (i.e., diodes or transistors) to extract the difference between the signal and comparison level. This difference is then amplified and integrated for the duration of the observation. A typical analog synchronous de-

3.4. MULTICHANNEL-FILTER SPECTROMETERS

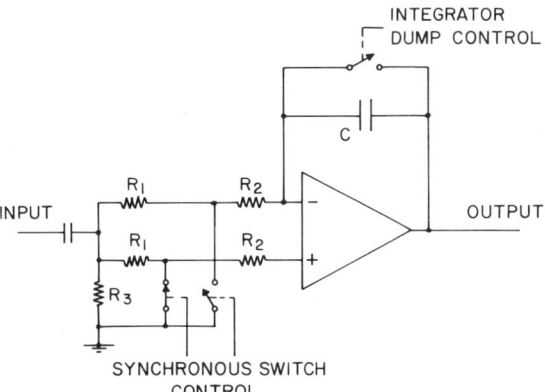

FIG. 3. An example of an analog synchronous detector and integrator utilizing a high-gain operational amplifier.

tector and integrator circuit is shown in Fig. 3. The signal from the low-pass amplifier is coupled through a capacitor to remove the dc component and split into two separate paths which connect to the differential inputs of a high-gain operational amplifier. In each of the input paths a set of contacts permits the signal to be removed by shorting it to ground. When these contacts are opened and closed, alternately, in synchronism with the receiver-switching cycle, the differential input to the operational amplifier is equal to the amplitude of the switch rate-modulation component divided by two. A capacitor connected from the operational amplifier output to the inverting input constrains the output voltage to be the linear time integral of the differential input voltage. Assuming that the switching cycle is symmetrical with no dead time, the expression for the output voltage at any time interval T may be written as

$$E_0(T) = \frac{1}{2RC} \int_0^T e_{\text{in}}(t) \, dt, \qquad (3.4.5)$$

where R is the average resistance from the inverting input to ground, C the value of the capacitor connected around the operational amplifier, and $e_{\text{in}}(t)$ the amplitude of the switch rate-modulation component in the input. The value of the integration constant $1/2RC$ is typically of the order of unity and has the units of volts per volt-seconds. Thus a 0.1-V amplitude-modulation component applied to the input for 10 sec would produce an output of the order of 1 V. If the integrator continued to operate, the output voltage would reach a maximum value or saturation value determined by the operational amplifier. This saturation level will, in many instances, be the determining factor for selecting the value of integration time.

Prior to any integration cycle, the integrator output must be returned to zero. This is accomplished by closing both of the input contacts and dumping the charge on the integrating capacitor by shorting it with another set of contacts. At the end of the integration cycle the integrator output may be "frozen" or placed in a "hold" condition by closing both of the input contacts.

Aside from the saturation limitation already mentioned, the analog synchronous detector and integrator has another limit imposed by the drift characteristic of the operational amplifier. Any change in the integrator output which is not the direct result of an input signal may be attributed to drift in the operational amplifier. If this drift is to have negligible effect on the output data, it must be small compared to the statistical noise fluctuation in the output. Therefore, drift becomes a more important limitation on integration time as the channel-filter bandwidth becomes larger.

A typical saturation level for analog integrators is ± 10 V and drift rates of 1×10^{-5} V/sec or less may be achieved with nonchopper stabilized operational amplifiers.

The use of small digital computers to perform the synchronous detection and integration process in the multichannel-filter receiver circumvents the normal saturation and drift problems associated with the analog system.[7] In this case the output from the low-pass amplifier following the square-law detector is connected to an rc integrator with a time constant that is small compared to the half-period of the receiver-switching cycle. The output level at each of these rc integrators is then sampled and converted to digital form at the end of each signal and comparison interval in the switch cycle. The computer sorts and stores this digital information thereby accumulating separate signal and comparison data for each channel. During certain of the switch cycles, a calibration noise level is added to the signal portion of the switch cycle and the computer sorts and stores this data separately for each channel to be used in the data reduction process within the computer. The final spectral-line data output obtained from the computer operating in conjunction with the multichannel-filter receiver will be processed to the extent that each channel is corrected for its individual calibration scale factor and a linear baseline slope is computed and removed from the data. Further processing of the data is usually available with such options as subtraction of "OFF" source from "ON" source observations, averaging of many individual observations and least square fitting of standard Gaussian line shapes to the reduced data.

[7] S. Weinreb, Progress in radio science 1963–1966, *Proc. General Assembly URSI, 15th Munich 1966* 1284 (1967).

3.4.5. Special Operating Features

Special techniques which involve the automatic stabilization of receiver gain and gain balance are used in multichannel-filter receivers to improve the quality of the output data. This stabilization of gain and balance is applicable primarily in receivers which utilize analog synchronous detectors and integrators.

Both of these stabilization techniques operate on a square law detected signal which is obtained from a filter whose bandpass is essentially equivalent to the total band covered by the multichannel filter set. The gain stabilization control level is obtained by gating that portion of the total band detected output which corresponds to the signal interval of the switch cycle.[8] The dc level of this gated signal is then compared with a fixed reference voltage and any difference is amplified and applied as a correction input to an electronically controlled attenuator in the intermediate frequency portion of the receiver. Thus the receiver noise level during the signal portion of the receiver switch cycle in the total band covered by the multichannel filter set is maintained constant throughout the observation.

The gain-balance stabilization-control level is obtained by synchronously detecting the total band-detected output. Any difference between the noise level during the signal and comparison intervals of the receiver switch cycle is then amplified and applied as a correction voltage to an electronically controlled gain-balance attenuator in the intermediate frequency portion of the receiver. This gain-balance attenuator is controlled by a switched input signal which is in synchronism with the receiver switching cycle. During the signal portion of the switch cycle, the attenuator-input control voltage is held at a fixed level. During the comparison portion of the switch cycle an adjustable input voltage obtained either from a manual control or the automatic balance-control circuit is applied to the attenuator-input control. Thus under the condition of automatic gain-balance operation, the receiver noise level during the comparison portion of the receiver switch cycle will be constrained to be the same as the noise level during the signal portion.

The effect of the automatic gain stabilization is to normalize the output data in terms of the total system noise temperature and to maintain a constant total noise level into the multichannel-filter set.

The effect of the automatic gain-balance stabilization is to maintain an average difference between the signal and comparison noise level across the multichannel-filter set, which is equal to zero. This results in an output spectrum that is balanced or centered about zero. In the multichannel-filter receiver which utilizes the analog synchronous detector and integrator this is

---

[8] T. V. Seling, *IEEE Trans. Antennas Propagat.* **AP-12**, 636 (1964).

particularly important since it minimizes the influence of integrator saturation and thereby allows longer integration time.

In receivers which use a digital computer as the synchronous detector and integrator, automatic gain and gain balance are not required because the calibration is applied at regular intervals throughout the observing period and the computer does not have any saturation limit.

3.4.6. Output Devices

Most multichannel-filter receivers are equipped with at least one of each of the following types of output devices:

(1) Real-time oscilloscope display giving continuous spectral-line output during the observation;

(2) printer and plotter to record output data at the conclusion of each observation; and

(3) punched card, punched paper tape, or magnetic tape output for permanent data storage at the conclusion of the observation.

A multiplex switch that connects the output from each channel in sequence to a common output bus provides the link between the multichannel output and each of the output devices.

In a typical multichannel-filter receiver using the analog synchronous detector and integrator, each channel output would be connected through an element of the multiplex switch to the common output bus. During the observation time the multiplex switch sequences rapidly through the channels and the common output bus connects to the vertical input of an oscilloscope. A synchronous pulse coincident with the start of each multiplex scan triggers the horizontal sweep, and intensity-modulation pulses are applied during each step of the multiplex switch. When the oscilloscope sweep-time and intensity-level controls are properly adjusted, this provides a real-time picture of the spectral-line output with one dot on the oscilloscope for each channel.

At the end of the integration time, the analog integrators are placed in a "hold" condition and the multiplex switch sequences through the channels (usually at a slower rate). The common output bus is connected to an analog-to-digital converter or digital voltmeter, and a data output pulse is generated during each step of the multiplex switch to trigger the data conversion. This output data is then printed on a paper tape printer, plotted on a point plotter or X–Y recorder and stored in digital form on punched cards, punched paper ape or magnetic tape.

In the multichannel-filter receiver utilizing a digital computer for the synchronous detection and integration the spectral-line data is continually being updated in the computer and the data display and output may occur in a wide variety of forms. Normally, there will be an oscilloscope display which can provide either real-time spectral-line output or the processed spectral-line data. The computer will also normally be equipped with a teletype for input of data-processing instructions and for printed and plotted output data. Digital data output from the computer will usually be in the form of magnetic tape; however, punched cards and punched paper tape are sometimes used.

3.4.7. Calibration

The temperature scale factor for each of the channels in the multichannel-filter receiver is determined by injecting a known level of broadband Gaussian noise at the receiver input. This calibration signal is normally provided by a temperature-limited diode, gas-discharge tube, avalanche diode, or temperature-controlled load.

In the case of a switched input system, the calibration noise is injected in the line connecting the antenna feed to the signal-input terminal of the input switch. In a frequency-switched system, the calibration noise is injected through a modulation element into the line connecting the antenna feed to the receiver input. In either case, the calibration noise will appear as a modulated signal with equal relative intensity in all channels.

Differences in the temperature scale factor for the different channels, as determined by the response to the calibration signal, will be due to actual gain differences in the channels and to the "noise" variation associated with the total system noise temperature. This "noise" variation in the channel calibration factors is a false indication and must be made small compared to the actual gain difference factor in order to make the calibration valid. The rms value of the "noise" variation is given by the expression

$$\Delta T_{\rm rms} = \frac{2\,(T_{\rm sys} + T_{\rm CAL})}{(Bt)^{1/2}}, \qquad (3.4.6)$$

where $T_{\rm sys}$ is the total system noise temperature, $T_{\rm CAL}$ the calibration noise temperature, B the channel noise bandwidth, and t the integration time. The value of calibration noise temperature required to achieve a signal-to-noise ratio R in the calibration process is

$$T_{\rm CAL} = \left(\frac{2R}{(Bt)^{1/2} - 2R}\right) T_{\rm sys}. \qquad (3.4.7)$$

A reasonable value for R is 100. In order to achieve this value of signal-to-noise ratio for a particular filter bandwidth B it is necessary to select the proper calibration noise temperature and integration time.

In multichannel-filter receivers that utilize the analog-type integrators, the calibration signal is usually applied for a relatively short time prior to obtaining observational data on any particular source. In the case where a computer is utilized for the synchronous detection and integration process, the calibration signal is applied for short intervals (say, ten switch-cycle intervals) periodically throughout the time of the observation. In either case the total time during which the calibration signal is applied determines the integration time.

Table II lists the minimum calibration noise temperature required to obtain

TABLE II. Calibration Level as a Function of Bandwidth to Achieve Signal-to-Noise Ratio of 100 in 30-sec Integration Time

Bandwidth	Minimum calibration
2 MHz	$0.027 T_{sys}$
1 MHz	$0.038 T_{sys}$
0.3 MHz	$0.072 T_{sys}$
0.1 MHz	$0.13 T_{sys}$
30 kHz	$0.27 T_{sys}$
10 kHz	$0.58 T_{sys}$
3 kHz	$2.0 T_{sys}$

a calibration signal-to-noise ratio of 100 when the total calibration integration time is 30 sec. Normally the calibration noise temperature will be less than $0.25 T_{sys}$, therefore with the small bandwidths it will be necessary to increase the total calibration integration time. In the case of multichannel-filter receivers that utilize analog-type synchronous detectors and integrators there is a minimum bandwidth limit for any particular calibration signal-to-noise ratio. This minimum bandwidth limit is a function of the integration constant $1/2RC$, the integrator saturation voltage E_{sat}, the calibration noise temperature relative to the system noise temperature, and the low-pass amplifier output voltage, E_{dc}, corresponding to the system noise temperature. The expression for this minimum bandwidth limit may be written as

$$B_{min} = \frac{4R^2(1+\beta)^2 E_{dc}}{2RC\beta^2 E_{sat}}, \qquad (3.4.8)$$

where β is the ratio of T_{CAL} to T_{sys} and the other terms are as previously defined. For a typical system where $\beta = 0.25$, $E_{dc} = 1$ V, $2RC = 1$, and $E_{sat} = 10$ V, the minimum bandwidth for attaining a calibration signal-to-noise ratio of 100 is

$$B_{min} = 100 \quad \text{kHz}.$$

Hence, it is obvious that for the narrower bandwidths a compromise must be made and lower values of calibration signal-to-noise ratio must be accepted.

In the multichannel-filter receivers that utilize computer-type synchronous detection and integration, there is no saturation problem and the total calibration integration time may be adjusted to give the desired signal-to-noise ratio.

3.5. Autocorrelation Spectrometers*

3.5.1. Introduction

Early spectral-line work in radio astronomy was, of necessity, done with the aid of frequency-scanning receivers or with multichannel receivers of the type described in the previous chapter. With the rapid development of digital electronics in the early 1960s it became possible to exploit the precision and stability of digital techniques for spectrum analysis.

The digital correlation approach makes use of the Fourier transform relationship that exists between the autocorrelation function of a signal and its power spectrum, and becomes relatively simple to implement when one-bit quantization of the input signal is employed.

The one-bit approach was first used by Goldstein[1] to analyze the spectrum of radar echoes from Venus, and shortly afterwards Weinreb[2] introduced the method to radio astronomers in an attempt to detect the deuterium line at 327 MHz. More recently the method has been used in very long baseline interferometry.[3]

Although the correlation approach requires ready access to a computer to carry out a Fourier transformation, a small computer is now no more expensive than the multichannel analyzer that is normally used at the "back end" of filter-type spectrometers. Within the bandwidth limitations imposed by the speed of available digital logic the digital approach is now to be preferred to the multifilter approach, especially for high-resolution spectrometry, and the ability to vary the spectral resolution simply by changing the clock frequency makes the digital correlator equivalent to a multiplicity of filter banks.

Some hydrogen-line and OH profiles show spectral detail on a scale down to almost a thousandth part of the total profile width, making it desirable to have as many as 1000 channels if all the spectral detail is to be measured without resorting to piecemeal observations. This aspect is particularly important when the optical depth of a narrow absorption feature in a broad profile is to be gauged accurately, since the alternative of building up a profile

---

[1] R. M. Goldstein, *IRE Trans. Space Electron. Tele.* **SET-8**, 170 (1962).
[2] S. Weinreb, MIT Res. Lab. of Electron. Rep. 412 (1963).
[3] B. F. Burke, *Phys. Today* **22**, 54 (1969).

---

*Chapter 3.5 is by B. F. C. Cooper.

in sections can give serious accumulated errors. Again a correlator with 1000 channels can be subdivided into sections which, for example, will allow the four Stokes parameters of a spectral line to be measured simultaneously with good resolution.

The autocorrelation spectrometer has been demonstrated to have superior stability for long integrations. By locking the i.f. conversion oscillators and the digital clock system to an atomic standard, the frequency calibration of the spectrum can also have atomic accuracy, whereas in a filter-type spectrometer expensive crystal filters are needed to guarantee freedom from frequency drifts.

The cost of digital logic has now decreased to the point where a multichannel digital correlator can be built for a fraction of the cost of a comparable filter-type spectrometer, and it now appears that the filter approach will in future be restricted to wide-bandwidth applications.

In what follows a basic knowledge of correlation functions and Fourier transforms as applied to signal processing is assumed. The reader is referred to one of the standard texts on the subject, for example, the one by Blackman and Tukey.[4]

3.5.2. Sampling and Quantizing Considerations

In a digital correlator the signals $x_1(t)$ and $x_2(t)$ which are to be correlated are first converted to video signals and are sampled at a rate at least twice as high as the video cut-off frequency in accordance with the Nyquist criterion. The samples are quantized, delayed, and multiplied to form an estimate of the correlation function

$$R(n\,\Delta t) = (1/K) \sum_{m=0}^{K-1} x_1(t_0 + m\,\Delta t)\, x_2(t_0 + (m+n)\,\Delta t), \quad n = 0, 1, \ldots, N-1 \tag{3.5.1}$$

where K is the number of products averaged and Δt a delay increment usually made equal to the sampling interval. When $x_1(t)$ is equal to $x_2(t)$ an autocorrelation function is generated.

For noise signals having Gaussian amplitude statistics, which is the case for naturally occurring signals, it is known from the work of Van Vleck[5] that the quantizing can be carried to the extreme of recording merely the sign of the samples. The penalty paid for deleting the amplitude information in such a one-bit correlator is a loss of sensitivity by a factor of approximately $2/\pi$ relative to a continuous or many-bit correlator.

[4] R. B. Blackman and J. W. Tukey, "The Measurement of Power Spectra." Dover, New York, 1958.
[5] J. H. Van Vleck and D. Middleton, *Proc. IEEE* **54**, 2 (1966).

In a one-bit correlator in which the samples are represented by normalized values ± 1, the products required for (3.5.1) take the value $+1$ when the samples coincide in sign and -1 if they differ. Hence the estimate of the normalized one-bit correlation function is given by

$$\rho_1(n\,\Delta t) = [K_c(n) - K_a(n)]/K, \tag{3.5.2}$$

where K_c and K_a are, respectively, the number of polarity coincidences and anticoincidences observed in K products at a delay $n\,\Delta t$. Alternatively

$$\rho_1(n\,\Delta t) = 1 - 2[K_a(n)/K] = 2[K_c(n)/K] - 1. \tag{3.5.3}$$

The expressions of (3.5.3) are more suited to one-bit multipliers which employ either exclusive-or or exclusive-nor logic functions. The exclusive-or function between two quantities A and B is expressed as $A \oplus B = A\bar{B} + \bar{A}B$ and yields a logical "1" only when A and B are unlike quantities. The exclusive-nor function $AB + \bar{A}\bar{B}$ yields a "1" only when A and B are alike. And-or-invert logic gates will perform either function.

Having generated the one-bit correlation function the continuous correlation is recovered using the Van Vleck relationship

$$\rho_c = \sin \tfrac{1}{2}\pi\rho_1. \tag{3.5.4}$$

When quantization with two or more bits is employed, the sensitivity rapidly approaches that of a continuous correlator.[6] Two bits give 88% and three bits give 95% of the continuous correlator sensitivity. The complexity of the digital logic, however, also increases quite rapidly for such correlators.

In a two-bit correlator[7] an extra bit is used to specify whether the modulus of the sampled amplitude is in a high state or a low state, with optimum performance being obtained when the high–low transition is set close to the rms voltage. A high-level product is generated when both multiplier and multiplicand are in the high state, an intermediate product is generated when one is high and one is low, and a low product is generated when both are low. Since the signs of the products can be either positive or negative, there are six different products to handle. These are accumulated with weighting factors of n^2, n, and 1 assigned, respectively, to the high, intermediate, and low products, where the optimum value of n is either 3 or 4. A significant saving in logical complexity with a sensitivity loss of only 1% can be effected by deleting the low-level products. A scheme which combines this mode of operation with an optional one-bit mode is described later.

[6] T. Cole. *Aust. J. Phys.* **21**, 273 (1968).
[7] B. F. C. Cooper, *Aust. J. Phys.* **23**, 521 (1970).

The correction function for recovering the continuous correlation function from the two-bit correlation function cannot be given in an exact analytical form. A relatively simple formula giving a good approximation for the degrees of correlation, usually encountered in radio astronomy is available, however.[7]

A three-bit correlator appears to involve much more logical complexity than the expected sensitivity improvement would justify. A hybrid scheme in which one of the input signals is kept in analog form and the other is quantized with three bits has been used in a commercial correlator.[8] It has also been shown[9] that the sensitivity of a one-bit correlator can be significantly improved by sampling at a rate higher than the Nyquist rate. For example, sampling at four times the video cut-off frequency gives a theoretical improvement in the sensitivity of a one-bit correlator to 74% of the ideal figure and an experimental test gave a factor of 78%.

Quantizing schemes having two, three, four, five, or eight levels have also been considered by Bowers,[10] who reaches the following conclusions:

(a) The two voltages being multiplied need not be quantized with the same number of levels. Certain hardware advantages can be gained by quantizing them with different numbers of levels.

(b) The sensitivity degradation due to the quantizing can be considered separately for the two inputs and expressed as a product $D_1 D_2$, where D_1 applies to the signal $x_1(t)$ and D_2 to the signal $x_2(t)$.

(c) For minimum degradation the discrete values assigned to the quantizer outputs should be symmetrical about zero, but not spaced uniformly. Multiplication is, however, much easier with equidistant quantizer levels, which can then be assigned integer values. The degradation penalty for doing so is extremely small.

(d) For equidistant quantizer levels the optimum decision levels are also equidistant and symmetrical about zero.

(e) Deviations of the decision levels by up to 20% from the optimum values are not serious.

Table I reproduces values of the degradation factor D for correlators having two to eight quantizer levels. The quantity D^2 is the inverse of the sensitivity factor used earlier in describing one- and two-bit correlators. The two, four, and eight-level schemes of Table I correspond, respectively, to one-bit, two-bit, and three-bit correlators, and σ denotes the rms input voltage.

[8] B. LuBow, *Electronics* **39**, 75 (1966).
[9] W. R. Burns and S. Yao, *Radio Sci.* **4**, 431 (1969).
[10] F. K. Bowers, 1971 *IEEE Int. Conv. Digest, New York*, 1971, 156 (1971).

TABLE I. Degradation Factors for Quantized Correlators

Number of levels	Choice of integral values	Decision level spacing	D	D^2
2	$-1, +1$	—	1.253	1.571
3	$-1, 0, +1$	1.224σ	1.112	1.236
4	$-3, -1, +1, +3$	0.995σ	1.065	1.135
5	$-2, -1, 0, +1, +2$	0.884σ	1.043	1.087
8	$-7, -5, \ldots, +5, +7$	0.585σ	1.019	1.039

3.5.3. Computation of the Power Spectrum

The N points on the autocorrelation function generated by the correlator can be transformed into N points on a power spectrum using the discrete Fourier transform (DFT) relationship,[11]

$$P_j = \frac{1}{N}\left[\rho_0 + 2\sum_{n=1}^{N-1} \rho_n \cos\left(\pi \frac{nj}{N}\right)\right], \quad j = 0, 1, \ldots, N-1, \quad (3.5.5)$$

where P_j represents the jth point on the power spectrum and ρ_n is the nth point on the normalized autocorrelation function corrected according to (3.5.4). A cosine transformation is used, since an autocorrelation function is a symmetrical function of delay. The transformation normalizes the power spectrum in such a way that

$$\sum_{j=0}^{N-1} P_j = 1.$$

To denormalize the spectrum in terms of antenna temperature the procedure of Section 3.5.5. is used.

When N is large considerable time is saved by making use of the (FFT) algorithm for computing the (DFT). Here it is advantageous for N to be a power of 2 to exploit fully the speed of the FFT. The time taken for the FFT is proportional to $N \log_2 N$ and with an average computer using an assembly language program is of the order of $\frac{1}{2}$ to 1 sec for $N = 1024$. Such a speed is adequate for updating a spectrum during typical radio-astronomy observations.

Because of the truncation of the autocorrelation function the transformed spectrum is convolved with a scanning function whose width is inversely related to the maximum autocorrelation delay $\tau_{\max}$. If the autocorrelation

[11] E. Oran Brigham, "The Fast Fourier Transform." Prentice-Hall, Englewood Cliffs, New Jersey, 1974.

function is transformed without modification, the scanning function has a sin f/f shape and exhibits large sidelobes. The sidelobes can be reduced at the expense of broadening the scanning function by applying tapered weighting functions to the autocorrelation function before transformation. The problem of optimum-weighting functions† encountered here is closely related to the problem of optimum illumination of linear antenna arrays, as treated, for example, by Taylor.[12]

Table II lists a few of the many possible weighting functions together with their Fourier-transformed scanning functions and some useful parameters of the scanning functions; also see Section 4.3.5.

3.5.4. Standard Deviation of Spectral Estimate

The standard deviation of the spectral estimate obtained from a correlation spectrometer operating in a total-power mode is given by[2]

$$\langle \Delta T^2 \rangle^{1/2}/T_N = \alpha\beta(\Delta f\,\tau)^{-1/2}(1 - \Delta f/B)^{1/2}, \qquad (3.5.6)$$

where T_N is the system noise temperature, $\alpha = (\Delta f/B_N)^{1/2}$ as defined in Table II, β the clipping or quantization loss factor, Δf the frequency resolution, τ the integration time, and B the bandwidth under analysis. For the usual multichannel correlator ($N = 100$), the quantity $\Delta f/B$ will be very small so that (3.5.6) can be approximated by

$$\langle \Delta T^2 \rangle^{1/2}/T_N = \beta(B_N\,\tau)^{-1/2}. \qquad (3.5.7)$$

As might be expected, (3.5.7) differs from the corresponding expression for a filter-type spectrometer only by the presence of the clipping factor β. The value of β is nominally $\pi/2$ for a one-bit correlator or 1.14 for a two-bit correlator. Weinreb,[2] however, found experimentally a value 1.39 for a one-bit correlator.

3.5.5. Spectrum Denormalization

The spectrum generated by the discrete Fourier transform represents the true spectrum of the input noise modified by the receiver bandpass shape and

[11a] R. T. Lacoss, *Geophysics* **36**, 661 (1971).
[12] T. T. Taylor, *IRE Trans. Antennas Propagat.* **AP-3**, 16 (1955).

---

† Since preparing this chapter, the author's attention has been drawn to recent work by Lacoss[11a] and others on data-adaptive spectral-analysis methods which can give much better spectral resolution than the naive "windowing" functions listed in Table II. It is likely that such data-adaptive methods will come into wide use in processing correlation functions, and the reader is urged to follow up further developments in the technical literature.

TABLE II. Weighting Functions

Shape	Weighting function[a] $w(\tau)$	Scanning function $W(f)$	Scanning function width[b] Δf	Noise bandwidth[c] B_N	$(\Delta f/B_N)^{1/2}$	Max. sidelobe level (%)
Uniform	1	$2\tau_{max} \operatorname{sinc} 2f\tau_{max}$[d]	$0.605/\tau_{max}$	$0.5/\tau_{max}$	1.099	21.7
Triangular	$1 - |\tau|/\tau_{max}$	$\tau_{max} \operatorname{sinc}^2 f\tau_{max}$[d]	$0.885/\tau_{max}$	$0.667/\tau_{max}$	1.16	4.7
Cos²	$0.5(1 + \cos \pi\tau/\tau_{max})$	[e]	$1/\tau_{max}$	$0.75/\tau_{max}$	1.155	2.6
Cos² on pedestal	$\dfrac{1+a}{2} + \dfrac{1-a}{2} \cos\dfrac{\pi\tau}{\tau_{max}}$	[f]	$0.85/\tau_{max}$[g]	$0.635/\tau_{max}$[g]	1.155^g	1.7^g

[a] For $|\tau| \leq \tau_{max}$; $w(\tau) = 0$ elsewhere; $\tau_{max} = N\Delta t$.
[b] Measured at half-maximum of $W(f)$.
[c] $B_N = [1/W^2(0)] \int_{-\infty}^{\infty} W^2(f) \, df = [1/W^2(0)] \int_{-\infty}^{\infty} w^2(\tau) \, d\tau$.
[d] Here sinc $x = \sin x / x$.
[e] $W(f) = \tau_{max}[\operatorname{sinc}(2f\tau)_{max}) + 4f\tau_{max}\sin(2\pi f\tau_{max})/2\pi(1 - 4f^2\tau^2_{max})]$.
[f] $W(f) = \tau_{max}[(1 + a\operatorname{sinc}(2f\tau_{max}) + 4(1-a)f\tau_{max}\sin(2\pi f\tau_{max})/2\pi(1-4f^2\tau^2_{max})]$.
[g] $a = 0.15$, approximate optimum value.

multiplied by a normalizing factor. When the input is a comparison spectrum produced by pointing the antenna at a region of sky which is known to have a flat spectrum or by switching the input to a matched comparison load we can write for the normalized value of point j on the transformed spectrum

$$P_{jc} = \alpha_1(T_r + T_c)\, G_j,$$

where T_r and T_c are, respectively, the receiver noise temperature and the comparison source temperature, G_j the receiver power gain at the point j, and α_1 a normalizing factor. When the antenna is pointed at a source showing spectral detail we can similarly write

$$P_{js} = \alpha_2(T_r + T_{js})\, G_j,$$

where T_{js} is the antenna temperature at the frequency corresponding to point j on the spectrum.

If the integrated noise for the two input conditions is equalized across the bandpass, it is easy to show that $\alpha_1 = \alpha_2$, leading to the result

$$T_{js} = \frac{(P_{js} - P_{jc})(T_c + T_r)}{P_{jc}} + T_c,$$

which can be written as

$$T_{js} = \frac{\Delta P_j T_n}{P_{jc}} + T_c. \qquad (3.5.8)$$

Here $T_n = T_c + T_r$ is the system noise temperature, which can be determined by injecting a known calibration noise step into the feed system and measuring the resultant power increase in an auxiliary total-power channel. The difference spectrum ΔP_j can be obtained after transformation of the signal and reference autocorrelation functions, or it can be obtained by transforming the difference autocorrelation function as will be described later.

3.5.6. Prefilters and Video Converters

Since the digital correlator can analyze unambiguously only a band extending from dc to one-half of the sampling frequency, considerable care must be taken to ensure that the video converters and prefilters exploit the useful bandwidth to the best advantage without allowing significant aliasing errors to occur. Figure 1 illustrates a section of a conversion chain designed for a multiband digital correlator. Here the spectrum generated by a previous conversion stage is first heterodyned by mixer M1 to a band centered somewhat above the desired video band. A bandpass filter BPF1 following M1 selects the center portion of this band. A second mixer M2 fed by a local oscillator of frequency f_2, which is located on the upper edge of the passband of BPF1,

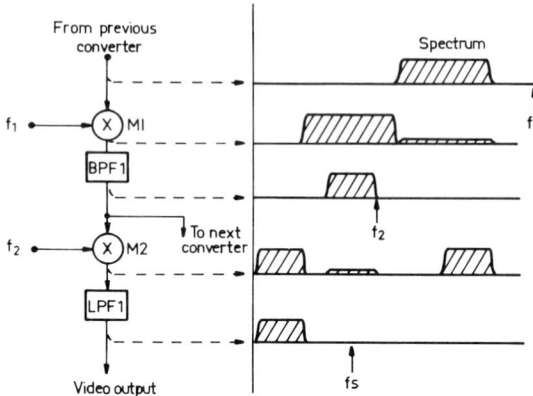

FIG. 1. Video conversion chain.

generates the desired video band. The filter LPF1 following M2 rejects the upper sideband output of M2 and any direct signal which may leak through M2.

Placement of f_2 at the upper 20-dB insertion-loss point of BPF1 will reduce to an acceptable level the residual signal that is converted from just above f_2 to a low video frequency. Again by making the width between the 20-dB points of BPF1 exactly equal to one-half of the sampling frequency for the video output, the aliasing distortion will also be reduced to an acceptable level.

It is apparent that the skirts of BPF1 should be as steep as possible if maximum use is to be made of the video band which it generates, and the use of a high-order Chebyshev–Cauer[13] filter is indicated. In practice a filter of the seventh or eighth order will ensure that at least 95% of the video spectrum will lie between the 3-dB loss points. It is also easier to provide steep skirts to the video spectrum if BPF1 is centered not too far above the desired video band, thus making its fractional bandwidth as large as possible. It may also be possible to relieve some of the demands on the bandpass filters by adopting image-canceling mixers, but this could lead to considerable increase in cost and complexity where there are many conversion stages.

3.5.7. Digital Correlator Logic

Figure 2 shows the basic logical organization of a one-bit correlator. Here the clipped and sampled video signal is passed down a shift register chain of length equal to the desired number of points on the autocorrelation function.

[13] A. I. Zverev, "Handbook of Filter Syntheses." Wiley, New York, 1967.

3.5. AUTOCORRELATION SPECTROMETERS

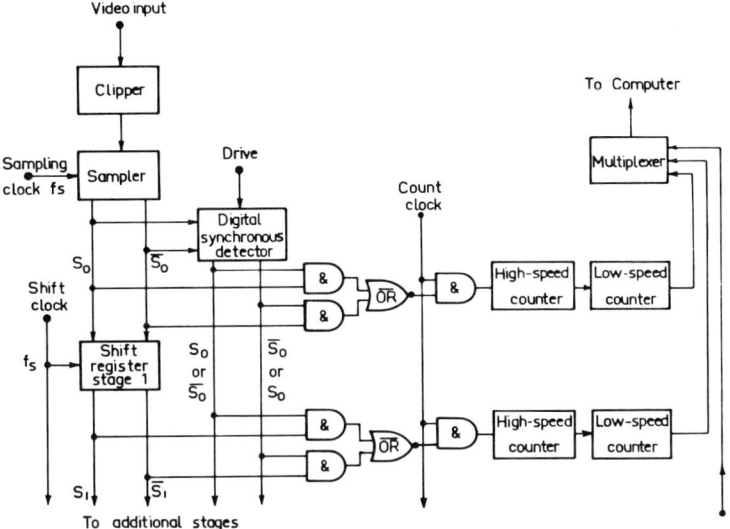

FIG. 2. Block diagram of one-bit correlator.

One-bit products are formed for each unit of delay by and-or-invert gates which multiply the undelayed signal S_0 and the delayed signals $S_1, S_2, S_3, \ldots$. Products are accumulated in counters of length appropriate to the rate at which the computer can handle the counter readouts, typically a few times a second.

If K samples are handled during each integration period a binary counter of length not less than $M = \log_2 K$ is required to handle the one-bit products. However, not all M bits are useful. To see this write (3.5.3) in the form

$$\rho_1 = 1 - 2p,$$

where $p = K_a/K$ is the probability of a count being generated by the exclusive-or multiplier. The counts resulting from accumulation of K products have a binomial distribution about the mean value K_a with rms deviation

$$\sigma_K = [p(1-p)K]^{1/2};$$

whence

$$\sigma_K = [(1 - \rho_1^2)K]^{1/2}/2.$$

In practice $\rho_1 = 1$ for zero delay, giving a fully determined count, or $\rho_1 \ll 1$ for other delay stages. For these stages $\sigma_K \approx K^{1/2}/2$. For economy in read-out hardware the least significant counter bits can be truncated at a point where the truncation error is insignificant compared to σ_K. If T bits are truncated, the rms truncated error is given[14] by $\sigma_T = 2^T/(12)^{1/2}$. Since σ_T adds quadratically to σ_K, a value $\sigma_T \leq 0.14\sigma_K$ will increase the effective σ by no more than 1%. Hence we require

$$2^T/(12)^{1/2} \leq 0.07 \times 2^{M/2},$$

or

$$T \leq (M/2) + \log_2 0.245 \approx (M/2) - 2.$$

In practice it is usual to discard no more than $(M/2) - 3$ bits.

Provision may be made to reverse periodically the sign of the undelayed data bits in synchronism with the front-end comparison switch, giving a form of digital synchronous detection. Since the one-bit system is amplitude insensitive, the comparison switching is necessary only if the receiver bandpass shape changes significantly in the time required to complete a pair of on-source–off-source observations. Observational experience suggests that "total-power" operation is usually quite satisfactory and gives a valuable improvement in observing efficiency.

When the digital synchronous detector is in operation, the cycle time is adjusted so that an integral number of signal–reference cycles is completed before the counter contents are transmitted to the computer. It may then be shown by manipulation of (3.5.3) that the counter outputs correspond to points on a difference autocorrelation function given by

$$\Delta\rho_1(n\,\Delta t) = [\rho_1(\text{sig}) - \rho_1(\text{ref})]_{n\Delta t}$$
$$= 2(C_n/K) - 2,$$

where C_n is the accumulated count for stage n and K the number of samples. The first counter stage operates with zero time delay and therefore counts at the full sampling rate or at zero rate depending on the state of the synchronous detector switch. It therefore registers one-half of the sample count over a full integration period. Fourier transformation of the difference autocorrelation function gives directly the quantity ΔP_j required in (3.5.8), while the reference spectrum must be measured as a separate operation.

Having outlined the general logical structure we should now consider some specific circuit details.

3.5.7.1. *Zero-Crossing and Amplitude Discriminators.* An ideal zero-crossing discriminator should be a bistable device which makes an instantaneous transition between its two discrete output levels whenever the input passes through zero. As a practical compromise it may be noted that a noise voltage spends 99% of its time in the amplitude range between 0.02 to 3 times the rms value. Hence if the output saturates over a range of input amplitude of

[14] W. R. Bennett, "Electrical Noise," p. 48, McGraw-Hill, New York, 1960.

at least several hundred to one, the number of erroneous samples should be acceptably small. The bandwidth of the discriminator should also be considerably greater than the signal bandwidth, and the time delay of the output transitions relative to the input zero crossings should be as nearly as possible amplitude independent. A requirement closely related to the previous one is that the input dc error should be small and independent of input amplitude.

Figure 3 illustrates a discriminator circuit using the Motorola MC1035

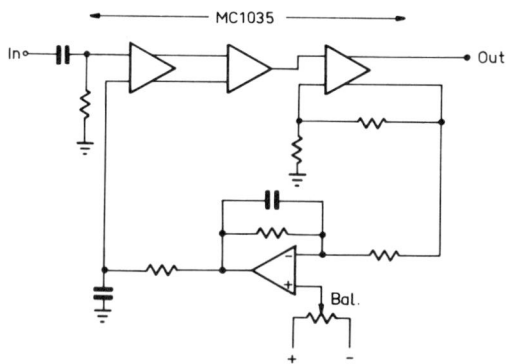

FIG. 3. Zero-crossing discriminator. (See text for discussion.)

emitter-coupled logic package which works well for signal bandwidths up to at least 10 MHz. It employs two emitter-coupled amplifier stages followed by a Schmidt trigger and gives a saturated output for inputs in the range of a few millivolts to greater than 1 V. By application of a dc feedback loop through an operational amplifier the dc error is kept to a small fraction of a millivolt. Slewing time is 4 nsec and time delay is 15 ± 2 nsec over this range of input voltages.

The circuit can be readily adapted for amplitude discrimination by appropriate adjustment of the input reference and the dc feedback voltage to achieve the desired duty factor of the output waveform.

3.5.7.2. Samplers. Sampling gates of the type used in sampling oscilloscopes have been used in some digital correlators, but for a two-level input the function of sampling the value of the input signal at regular intervals is very conveniently carried out with the aid of an integrated circuit type D flip–flop. Such a flip–flop transfers to the Q output terminal the value of the D input signal coexisting with the leading (or trailing) edge of the clock waveform, with a small uncertainty defined by the sharpness of the leading edge and the internal time delays. For a high-speed transistor–transistor logic

(TTL) flip–flop the width of the sampling aperature is about 6 nsec making it suitable for sampling signals of up to 10 MHz bandwidth. Sampling apertures down to about 2 nsec can be realized with emitter-coupled logic (ECL) flip–flops.

3.5.7.3. *Shifting, Multiplication, and Accumulator Logic.* Integrated circuit logic capable of performing the functions just listed is available in many forms. Of these, transistor–transistor logic combines versatility, speed, noise immunity, and low cost to the best advantage. High-speed TTL with a maximum rating of 50 MHz is suitable for clock rates up to a little over 20 MHz, when allowance is made for the propagation problems encountered in a large correlator. For higher clock speeds emitter-coupled logic is suitable, or alternatively, a technique of interleaved sampling may be employed, as illustrated in Fig. 4. This scheme has been devised for use in a 50-MHz correlator at the

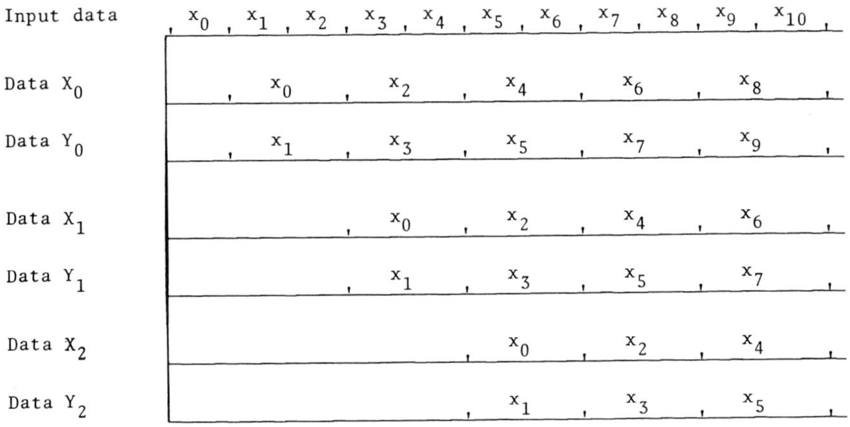

FIG. 4. Interleaved sampling system.

Haystack Observatory† in Westford, Massachusetts. In Fig. 4 the original data samples $x_0, x_1, x_2, \ldots$, generated at a rate f_s, are separated into parallel streams X_0 and Y_0 of odd and even numbered samples clocked at a rate $f_s/2$. The bit streams are propagated down two shift-register chains, also clocked at $f_s/2$, producing delayed streams $X_1, X_2, X_3, \ldots$, and $Y_1, Y_2, Y_3, \ldots$. It may then be seen by inspection that points on the one-bit autocorrelation function may be generated according to the rules

$$\begin{aligned} \rho_1(2n) &= \langle X_0 \cdot X_n + Y_0 \cdot Y_n \rangle, \\ \rho_1(2n+1) &= \langle X_0 \cdot Y_{n+1} + Y_0 \cdot X_n \rangle, \end{aligned} \qquad n = 0, 1, 2, \ldots, \qquad (3.5.9)$$

---

† J. I. Levine. Private communication.

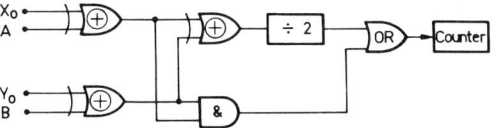

FIG. 5. One-bit multiplication logic for interleaved sampling system. $A = X_n$ for channel $2n$ or Y_{n+1} for channel $2n + 1$. $B = Y_n$ for channel $2n$ or X_n for channel $2n + 1$.

where the · represents one-bit multiplication and the + represents arithmetical addition. Equations (3.5.9) may be implemented by standard digital logic blocks, as shown in Fig. 5, with (3.5.3) being applied to compute the correlation function. With this scheme it may be seen that once the data has been separated into the two parallel bit streams all subsequent logic is only required to work at a rate of $f_s/2$ or less.

At the low speed end of the accumulators it is advantageous to change from conventional flip–flop counters to some form of serial accumulation employing circulation of a digit pattern in a shift register or delay line. The readout from each accumulator to the computer multiplexer then only requires a single line instead of the multiplicity of wires that are required for readout from a conventional counter. For serial accumulation at clock speeds of about 1-MHz MOS shift registers provide a most economical and compact storage unit requiring very low dc power.

Fast data-transfer channels are usually available for servicing computer peripherals, but a magnetic core buffer storage unit can be employed to advantage between the correlator and the computer, allowing the accumulators to be read out, cleared, and immediately begin a new accumulation cycle. The computer is then free to interrupt its main program whenever convenient to service the buffer core store. Increasing use of IC memories may also be expected in new correlator designs.

3.5.7.4. **Constructional Considerations.** The layout of a correlator operating at the highest clock speeds demands special care to minimize unwanted reflections and propagation delays. Most of the precautions required in assembling integrated circuits on a large scale are covered in manufacturer's application notes, but particular emphasis should be placed on the provision of liberal ground planes in the printed-circuit boards and of an extensive low-impedance grounding mesh between the circuit boards. Power supplies must be well decoupled and transmission lines properly terminated.

Multilayer circuit boards offer the best scope for providing an effective ground plane as well as low-impedance conductors for the signal paths. For maximum packing density the "flat pack" package has advantages over the more popular dual in-line package, although costing considerably more.

3.5.8. A Scheme for Optional One-Bit or Two-Bit Correlation

Figure 6 shows a logic scheme designed for optional one-bit correlation or for the modified form of two-bit correlation described in Section 3.5.2 in which the low-level products are deleted. In Fig. 6, $L = 0$ for the low-level

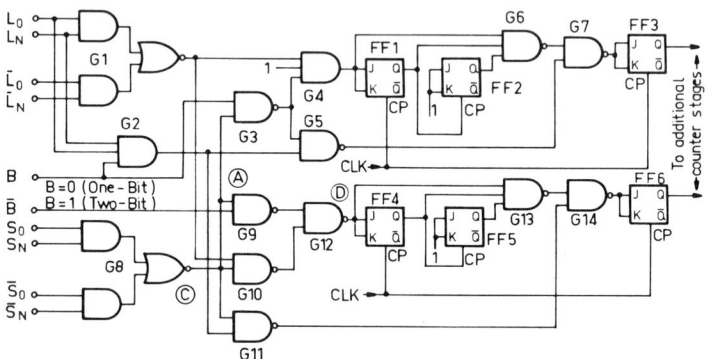

FIG. 6. Logic for optional one-bit or two-bit correlation.

state and 1 for the high-level state, $S = 0$ for a negative sign and 1 for a positive sign. Separate shift registers are employed for the sign and level bits and the subscripts 0 and N denote undelayed and delayed bits, respectively.

In the one-bit mode gates G2 and G3 are disabled, and the upper half of the circuit is disconnected from the lower half. Sign bits are substituted for the level bits L_0 and L_N as well as for S_0 and S_N. The and-or-invert gate G1 performs the exclusive-or function on the (sign) bits L_0 and L_N. Whenever its output is a "1" a count is clocked into FF1 and thence to the remainder of the upper counter chain. The timing of the clock pulses applied to the first three counter stages is adjusted to give optimum discrimination against the spurious breakthrough pulses which are generated during the gate transitions.

In the lower part of the circuit the gates G9, 10, 12 perform the logic function

$$D = C(A + \bar{B}) = C \quad \text{when} \quad \bar{B} = 1.$$

Thus the anticoincidences between S_0 and S_N are counted by the lower counter chain which can function independently of the upper chain. In the one-bit mode the outputs of G5 and G11 are always in the high state, thus keeping gates G7 and G14 open to the passage of carries from the earlier counter stages.

In the two-bit mode ($B = 1$), six conditions can be distinguished:

(a) $S_N = S_0$, $L_N = L_0 = 1$. Both inputs have the same sign and are in the high-level condition. Of the four gates (G4, 7, 12, and 14) controlling the

JK inputs of the counters, only G7 has its output in the high-level state. A count is therefore passed into FF3 at the next clock transition.

(b) $S_N = S_0$, $L_N \neq L_0$. This condition defies a positive intermediate-level product and causes the output of G4 to go high, whereupon a count is propagated into FFI at the next clock pulse transition. Since there is a division by 4 before FF3 is reached, this condition has one-quarter of the weight of a high-level product.

(c) $S_N = S_0$, $L_N = L_0 = 0$. This condition corresponds to a positive low-level product, and is ignored because all J–K inputs are at a low level, thus inhibiting any counting action.

(d) $S_N \neq S_0$, $L_N = L_0 = 1$. This condition defines a negative high-level product and causes a count to be transmitted into FF4.

(e) $S_N \neq S_0$, $L_N \neq L_0$. An intermediate-level negative product causes a count to be transmitted into FF4.

(f) $S_N \neq S_0$, $L_N = L_0 = 0$. No count is propagated.

The modified two-bit correlation function is proportional to the difference of the counts accumulated in the upper and lower counter chains and is normalized by dividing by the count accumulated in channel 0. For channel 0, $L_0 = L_N = 1$ and $S_0 = S_N$ and counts are accumulated only in the upper counter chain via G7. The number of counts accumulated in channel 0 is related to the total number of samples correlated by the cumulative probability distribution for random noise, and provides an independent check on the correct adjustment of the high–low transition level.

In the preceding scheme and in one-bit correlators in general, it is advantageous to clock the products into the first counter stages at a constant rate equal to the highest sampling rate used with the correlator. This causes the most significant bits to occupy the same position in the low-speed counter chain after a standard integration time, irrespective of the clock rate that is used for sampling and bit-shifting.

3.5.9. Extension to Cross-Correlation Spectrometry

So far attention has been restricted to autocorrelation spectrometry, but there are extensive applications for digital correlators in generating cross-power spectra with polarimeters and interferometers. An advanced form of digital cross correlation will also be used with the Westerbork aperture synthesis telescope for spectral-line observations.†

In general a cross-correlation function of the type defined in (3.5.1) will be an unsymmetrical function of delay so that the transformed cross spectrum will be complex. In this case it is convenient to generate the function $R(-n\,\Delta t)$

---

† J. L. Casse, Private communication.

by interchanging the delay parameter in (3.5.1). Then the cross-correlation function can be separated into its even and odd components according to the relationships

$$R^{\text{even}}(n\,\Delta t) = [R(n\,\Delta t) + R(-n\,\Delta t)]/2,$$
$$R^{\text{odd}}(n\,\Delta t) = [R(n\,\Delta t) - R(-n\,\Delta t)]/2.$$
(3.5.10)

The real and imaginary parts of the discrete Fourier transform become, after normalization,

$$\text{Re}(P_j) = (1/N)[\rho_{0\text{even}} + 2\sum_{n=1}^{N-1} \rho_{n\text{even}} \cos(\pi n j/N)],$$

$$\text{Im}(P_j) = (2/N)\sum_{n=1}^{N-1} \rho_{n\text{odd}} \sin(\pi n j/N).$$
(3.5.11)

In practice, using the FFT, (3.5.11) are computed together as a complex transform.

Figure 7 shows a possible arrangement of a digital cross correlator. Two shift-register chains are employed to allow accumulation of polarity coincidences for progressively increasing delays, both positive and negative. The correlation functions for positive and negative delays are calculated according to (3.5.3) and (3.5.4) and are combined in the computer to obtain the even

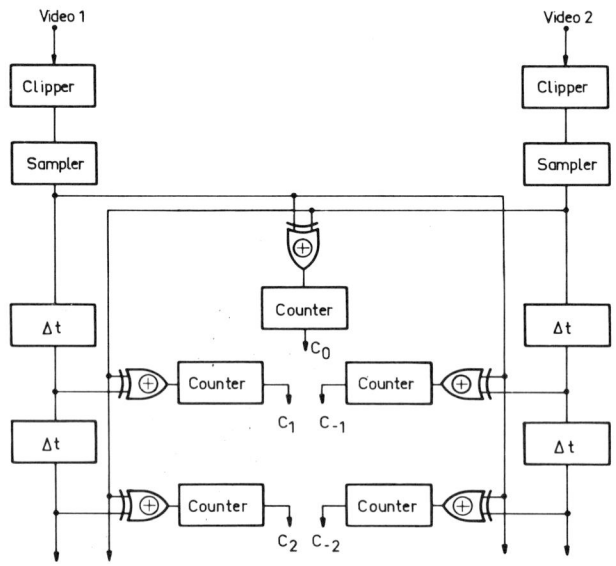

FIG. 7. One-bit cross-correlator block diagram.

3.5. AUTOCORRELATION SPECTROMETERS

and odd parts. For polarimetry with a single antenna the correlator is preceded by receivers of exactly matched phase delays connected to orthogonal linearly polarized probes in the feed. The cosine transform of the even part of the cross-correlation function then gives the spectrum of the Stokes U parameter, while sine transformation of the odd part gives the spectrum of the V parameter.

Where only a single autocorrelator is available it is possible to generate the even part of the cross-correlation function of the two input signals by combining them at i.f. in a sum and difference hybrid. Autocorrelating the sum and difference signals individually and differencing the two autocorrelation functions yields the desired result, since

$$\langle [x_1(t) + x_2(t)] [x_1(t + \tau) + x_2(t + \tau)] \rangle$$
$$- \langle [x_1(t) - x_2(t)] [x_1(t + \tau) - x_2(t + \tau)] \rangle$$
$$= 2\langle x_1(t)x_2(t + \tau)\rangle + 2\langle x_2(t)x_1(t + \tau)\rangle. \quad (3.5.12)$$

Sensitivity is, however, reduced by a factor of $\sqrt{2}$ relative to the full cross correlator of Fig. 7 because of the use of time-sharing.

In the interferometry application the even and odd cross correlation functions correspond to the cosine and sine fringe components. Here it is necessary to slow down the fringe motion so that the fringe phase does not drift significantly during an integration period. The identification of the particular Stokes parameters which are being measured depends in a rather complicated way on the polarization of the individual antenna feeds and the phase of the source relative to the central fringe maximum and will not be discussed further in the present section.

3.5.10. Some Examples of Correlation Spectrometers

A 413-channel correlator operating at a maximum clock rate of 20 MHz has been in operation at the National Radio Astronomy Observatory, Green Bank, West Virginia, for several years. This correlator has 384 main channels which can be split into two independent banks of 192 channels. Clock frequency is selectable in the range of 39 kHz to 20 MHz by factors of 2. An auxiliary set of 29 channels is available to set a wide-bandwidth baseline for the 384-channel correlator. A duplicate model of the NRAO correlator is also being put into service at the time of this writing by the Max Planck Institute for Radio Astronomy, Bonn, West Germany.

A correlator working at a maximum clock rate of 50 MHz and using the interleaved sampling scheme of Section 3.5.7 has been built at the Lincoln Laboratory of M.I.T., replacing an earlier 100-channel, 10-MHz correlator. A 50-MHz correlator is also being built at the U.S. National Radio Astronomy Observatory. In England, the Jodrell Bank Observatory is equipped with a 256-channel, 10-MHz digital correlator.[15] A 1024-channel correlator with 20-MHz maximum clock frequency is being built at the C.S.I.R.O. Radiophysics Laboratory for use with the Parkes 64-m telescope. In the University of British Columbia a dual 128-channel correlator capable of operating in either a one-bit mode or a 3 × 5 level mode is being built for use with a super synthesis telescope.[10] Its bandwidth is 16 MHz in the one-bit mode or 4 MHz in the 3 × 5 level mode.

A new correlator system designed by James I. Levine of Haystack Observatory uses a scheme that permits frequency doubling or quadrupling. The system consists of 16 separate correlator modules each with 64 channels. When these modules are interconnected in the usual way, one obtains a correlation function with 1024 lag intervals. For this case the sampling rate is 25 MHz and the spectrum width is 10 MHz. In the frequency-doubling mode a sampling rate of 50 MHz is processed by interconnecting the modules so alternate samples are processed in parallel to obtain a 20 MHz spectrum. In the frequency-quadrupling mode the sampling rate is 100 MHz, and the data are multiplexed into four streams. Then the sixteen modules form autocorrelations and crosscorrelations between the four data streams to obtain a bandwidth 40 MHz wide with the equivalent of 256 channels. Systems of this design have been constructed in 1975 at the Haystack Observatory, Westford, Massachusetts, and at the Owens Valley Radio Observatory, Big Pine, California.

[15] R. D. Davies, J. E. B. Ponsonby, L. Pointon, and G. DeJager, *Nature* (*London*), **222**, 933 (1969).

INDEX FOR VOLUME 12, PART B

A

Absorption
  atmospheric gas, 142-176
  ionospheric, 125-126
Absorption coefficient, *see also specific gases*
  atmospheric, 149
  general expression for, 145
Albedo, single scattering of rain, 182-183
Ammonia
  absorption coefficient of, 171-172
  atmospheric concentration, 170
Amplifiers, *see specific types*
Antenna arrays, *see* Array antennas
Antenna calibration, 82-96
  pointing, 96
  radio sources for, 83, 86-93
  radio sources, classes of, 84
Antenna patterns
  definition of, 3
  for paraboloidal reflectors, 34
Antennas, *see also specific types*; Reflectors
  characteristics of, 7
  types of, 7-28
Antenna temperature, 203
  definition of, 2
Aperture blocking, in paraboloidal reflectors, 38-44
Aperture efficiency, *see* Efficiency, aperture
Aperture synthesis, 24-28
Arecibo Antenna, 19
Argon, atmospheric concentration, 170
Array antenna patterns, 101-102
Array antennas, 98-118
  building problems, 104-111
  cables and lines for, 114-116
  cost of, 104, 117
  design of, 111
  elements for, 108-110
  feeder networks for, 104-108
  frequency sensitivity of, 105

  gain measurements of, 112-114
  groundscreen for, 110
  materials, 118
  phasing techniques for, 104-108
  power line noise in, 118
  site selection for, 104
Atmosphere
  complete microwave spectrum of, 172-176
  gaseous constituents, 170
  neutral structure of, 136-141
  propagation in, 140-141
Atmospheric effects, 119-200
Atmospheric gases, *see specific gases*
Attenuation, *see* Absorption
Attenuation computations, models for
  rain, 180
Autocorrelation, *see* Spectrometers
Autocorrelation spectrometers, 280-298
  amplitude discriminators in, 290-291
  constructional considerations for, 293
  examples of, 297-298
  prefilters for, 287-288
  quantizing considerations, 281-284, 294-295
  samplers for, 291-293
  spectrum renormalization, 285-287
  standard deviation of spectral estimate, 285
  three-bit, 283
  two-bit, 282
  video converters for, 287-288
  weighting function for, 286
  zero-crossing in, 290-291

B

Balloons, use in antenna calibration, 113
Balun, 117
Bending angle, computation of, 191
Blackbody radiation, 2
  equivalent temperature, 144
Bonn Telescope, 13

Boundary value solutions, references for, 62
Brightness temperature, definition of, 2

C

Calibration, *see* Antenna calibration
Carbon dioxide, atmospheric concentration, 170
Carbon monoxide
 absorption coefficient of, 169-171
 atmospheric concentration, 170
 microwave lines of, 171
Cassegrain antenna, 13-14, 48
Chlorine, atmospheric concentration, 170
Circular array antenna, 24
Clark Lake Radio Observatory, 21
Collisional broadening of lines, 146-147
Compound-grating antenna, 23
Conical spiral feed, 76, 79
Continuum signals, 202
Correlation receivers, 218-219
 many-channel, 219
 two-channel, 218
Correlator, *see* Autocorrelation spectrometers
Correlator logic, 288-293
Cross antenna, 21
Cross-correlation spectrometry, 107
Crossed-grating antenna, 22
Culgoora Radio Heliograph, 24

D

Defocusing, *see* Feed defocusing
Degenerate parametric amplifiers, 227
Detectors, multichannel spectrometer, 272
Dicke receivers, 214-217
Diffraction, geometrical theory of, references for, 63
Dimers, water vapor, 156, 173
Dipole arrays, 98-101, 116, 117
Dipole feeds, 71, 78
 two parallel, 79
Directivity function, 2
Doppler broadening of microwave lines, 146
Double-sideband receivers, 212
Drop size distribution, 178-179
 Laws and Parsons, 181
 Marshal and Palmer, 181

E

Effective area, definition of, 3
Efficiency
 aperture, 5
 in paraboloidal reflectors, 30-33
 radiation, 3

F

Faraday rotation, ionospheric, 126-127
Feed defocusing
 axial, in paraboloidal reflectors, 45
 lateral, in paraboloidal reflectors, 44
Feeds, *see also specific types*
 characteristics, 78-79
 use on paraboloidal reflectors, 64-81
Filled aperture antennas, 11-21
Flat spiral feed, 77, 79
Flux density, units of, 4
Flux scale, for antenna calibration, 84
Formaldehyde
 absorption coefficient of, 172
 atmospheric concentration, 170

G

Gain, antenna, 3
Gain modulation, use in receivers, 216
Geometrical optics, references for, 61
Grating lobes, use in array antennas, 102-103
Green's function, dyadic, references for, 61
Gregorian antenna, 48-50
Group delay, ionospheric, 127-128

H

Haystack Radio Observatory, 16
Helicopters, use in antenna calibration, 113
Helium, atmospheric concentration, 170
Helix feed, 74, 79
Homologous design principle, 12
Horn antennas, simple, 11
Horn feeds, 65-71
 corrugated, 68, 78
 diagonal, 67, 78
 Love, 67
 multimode, 70, 78
 rectangular, 78
 ridge, 70

ridged square, 78
scalar, 70
Horn-reflector antenna, 13
Hydrogen, atmospheric concentration, 170
Hydrogen sulfide
  absorption coefficient of, 172
  atmospheric concentration, 170

I

Index of refraction, atmospheric, 186
Integrators, use in multichannel spectrometers, 272-274
Interferometer receivers, 221
Interferometers
  multielement, 27
  two-element, 25
Iodine, atmospheric concentration, 170
Ionosphere, 119-135
  irregularities, large scale in, 133-134
  irregularities, small scale in, 131-133
  propagation in, 120-125
Ionospheric measurements, 134-135

J

Jansky, 4
Jicamarca Array, 21

K

Kraus-type antenna, 17
Krypton, atmospheric concentration, 170

L

Line-shape function, 146
Line shapes
  Ben-Reuven, 148
  Van Vleck-Weisskopf, 148
Linewidth parameter, 148
Log-periodic dipole array feed, 79
Log-periodic feed, 73
Loop feed, 73, 79

M

Manley-Rowe equation, 226
Maser amplifiers
  basic properties of, 246-265
  cryogenic cooling of, 260
  millimeter wave, 262-265
Maser systems, 257-260
Methane, atmospheric concentration, 170
Method of moments, references for, 61
Mie scattering theory, 177
Minimum detectable signal, 208-210
Model atmosphere
  exponential (CRPL), 189
  U.S. Standard, 139, 188
Modulated receivers, 214-217
Mountings, antenna, 14-16
Multichannel-filter spectrometers, 266-279
Multichannel spectrometer, 220
Multiple scattering, 183-184
Multireflector antennas, 48-60

N

Neon, atmospheric concentration, 170
Nitrogen dioxide
  absorption coefficient of, 172
  atmospheric concentration, 170
Nitrous oxide
  absorption coefficient of, 169-171
  atmospheric concentration, 170
  microwave lines of, 171
Noise-adding receivers, 216-217
Noise figure, 208
Noise temperature
  cascaded stage, 208
  definition of, 5
  equivalent, 203
Numerical techniques, use in antenna design, 62

O

Opacity, see Absorption
Orthomode transducers, 67
Output devices, use in multichannel spectrometers, 276-277
Oxygen, thermal emission by, 172
Oxygen molecule
  absorption coefficient of, 58
  microwave lines of, 166-168
Ozone
  absorption coefficient of, 164-169
  atmospheric concentration, 170
  atmospheric lines, measured, 176

P

Parabolic cylinder antenna, 19
Paraboloid antennas, 12-14
Parametric amplifiers, 225-241
  cryogenic cooling of, 238-239
  design considerations for, 233-241
  equivalent circuits for, 228
  for millimeter wavelengths, 239
  noise in, 231-233
  wide-band tunable, 237
Partition function
  linear molecule, 145-146
  nonlinear molecule, 145-146
Path length, atmospheric, 196-200
Path length correction, 197
Pencil-beam antennas, 10-24
Petzval surface, 46
Phase delay, ionospheric, 127-128
Physical optics, references for, 61
Planets, size and mean disk temperature of, 94
Pointing calibration, *see* Antenna calibration, pointing
Pointing correction, atmospheric refraction, 193
Pointing error, atmospheric refraction, 190
Polarization receivers, 222
Power spectrum, computation of, 284-285
Prime-focus feed, in paraboloidal reflectors, 30-47
Pulsar receivers, 221
Pulsar signals, 202
Pressure distribution
  atmospheric, 138-139
  ionospheric, 138-139

Q

Quarter-wave plate, use with horn feeds, 67

R

Radiation Pattern, paraboloidal reflector, 34-38
Radiative transfer, equation of, 142-145
Radiometer designs, considerations in, 223
Radiometers, 201-224, *see also specific types*
Radiometric measurements, 1-6
Radiorefractivity, atmospheric, 186-190
Radiosonde data, 141
Radio telescopes, comparison with optical, 8-9
Radomes, 16
Rain
  attenuation coefficient of, 179
  measured attenuation by, 184-185
  radar measurements of, 185
  scattering by, 177-178
Rayleigh limit, 8
Receivers, *see also specific types*; Spectrometers
  basic types of, 210-212
  special purpose, 219-223
Reciprocity, references for, 63
Reflectors, *see also* Antennas
  dual, 48-51
  dual, shaped, 55-58
  mechanical measurements of, 94-96
  paraboloidal, 29-63
  surface measurements of, 94-95
Reflex feed system, 58
Refraction
  ionospheric, 128-130
  neutral atmospheric, 186-200

S

Sampling, autocorrelation spectrometer, 281-284
Schelkunoff gain, 85
Selove receivers, 213
Servo control receivers, 217
Shaped dual reflectors, *see* Reflectors
Signals, *see also specific types*
  measurement of in radio astronomy, 202-210
  types of in radio astronomy, 201-202
Single cavity maser amplifiers, 250
Single-channel tunable spectrometer, 220
Single-sideband receivers, 212
Solar radio-astronomy receivers, 221
Source mapping, 26
Spectral-line receivers, 219-221
Spectral-line signals, 202
Spectrometers, *see also specific types*
  autocorrelation, 220-221
  basic qualities of, 269
  calibration of multichannel, 277-278
Spherical antennas, 18-19

Stationary-phase evaluation, use in
 radiation integrals, 62
Subreflectors
 ellipsoidal, 51
 hyperboloidal, 50-51
 nonoptical, 52
Sulfur dioxide
 absorption coefficient of, 172
 atmospheric concentration, 170
Superheterodyne receivers, 212-213
System noise temperatures, 204-207
System temperature, 203
 definition of, 5

T

Temperature distribution, atmospheric,
 136-138
Tertiary reflector antennas, 77, 79
Thermal broadening of microwave lines,
 146
Three-reflector feed system, 58
Total power receivers, 212-216
 stabilization of, 78
Traveling-wave maser amplifiers, 251
 representative systems, 262
Tuned radio-frequency receivers, 213
Turnstile feed, 79
Two parallel dipole feed, 79

U

Unfilled apertures, 21-24

V

Varactors, 230
Very long baseline interferometry, *see*
 VLBI
VLBI, definition of, 26
VLBI receivers, 221

W

Water, extinction by condensed, 177-185
Water vapor
 absorption coefficient of, 150-257
 distribution in atmosphere, 139-140
 lower frequency spectral lines of, 153
 thermal emission by, 172

X

Xenon, atmospheric concentration, 170

Z

Zeeman effect, in molecular oxygen,
 162-163

INDEX FOR VOLUME 12, PART C

A

Aberration, 280
Allan variance, 201
Angular separation
 VLBI measurement of, 268
Antenna temperature, 1
Aperture synthesis, 149
 earth rotation in, 170
 example of, 173
 one-dimensional, 166–167
 two-dimensional, 170
 VLBI observations for, 252–256
ASLEP, 273
Astrometry, 161–166
 VLBI applications in, 261, 269
Atomic clocks
 types of, 208
Atomic frequency standards
 comparison of, 225–227
Atomic hydrogen maser
 description of, 214–219
Atomic-hydrogen-maser oscillator, 208
Atomic time, 263
 definition of, 182
Atmospheric fluctuations
 compensation for, 14
Autocorrelation spectrometers, 22
 measurement techniques with, 46–57
 resolution in, 55–57
 sidelobes in, 55
 weighting function in, 55

B

Background sources, 17
Baseline
 interferometer determination of, 139, 162–163
 VLBI measurement of, 229, 261–276
Baseline curvature
 spectra with, 48
Baseline fitting
 spectra with, 51
Beam switching measurements, 14

C

Cassiopeia A, 173
Cesium 133
 resonant frequency of, 210
Cesium-beam clock, 178
Cesium-beam resonator, 208, 219–221
CFFT2, 292, 294, 318
CHCH
 transition frequencies of, 26, 27
CHCN
 transition frequencies of, 26
CHNH
 transition frequencies of, 27
CHO
 transition frequencies of, 26
CHOH
 transition frequencies of, 26
CHS
 transition frequencies of, 26
CLEAN, 151
CN
 transition frequencies of, 27
CO
 transition frequencies of, 26
Coherence function, 175
Coherence time
 in VLBI systems, 241
Computer control
 radiometer use of, 20–21
Computer programs
 use in radio astronomy, 277–295
Confusion noise
 from weak sources, 16
Continuum sources
 measurements of, 7–18
Contour mapping
 mechanical plotter programs for, 296
 nonmechanical plotter programs for, 299
CONTR, 296

continuous contours, 322-324
contours in segments, 325-326
Convolution integral
 solution of, 2-7
Correlation function
 VLBI use of, 182, 196
Cross-correlation function
 clipped signals, 233
 interferometer use of, 145
 VLBI use of, 182
Cross-power spectrum, 234
 VLBI use of, 196
Cross-spectrum function, 144, 148, 231
Crystal clock, 179
Crystal-controlled oscillators, 212-214
CS
 transition frequencies of, 27

D

Data presentation
 techniques for, 296-307
DCN
 transition frequencies of, 27
Delay, see Interferometer delay
Delay ambiguity function, 147
DFT, 285, 289
Diffraction
 by extended scattering medium, 124-125
 by thin layer, 119-121
Dispersion
 in pulsar signals, 80-81
DOP, 283, 308-314
Doppler effect
 history of, 277
 reference frame for, 280-282
 relativistic, 278
 velocity components in, 282
Doppler line widths, 29
Drift-scan measurements, 7-10
Dual-beam technique
 measurements with, 14

E

Earth rotation
 nonuniform rate of, 182
 VLBI measurement of, 261-276
Earth tides, 264
 VLBI measurement of, 275
Earthquake prediction
 VLBI use in, 276

F

Faraday rotation
 in pulsar signals, 82
Fast Fourier Transform, see FFT
FFT
 accuracy of, 293
 fixed-indexing, 290
 implementation of, 295
 in-place, 290
 memory requirements of, 293
 multidimensional data, 292
 natural, 289
 principles of, 284-287
 real data, 291
 real symmetric data, 293
 testing programs for, 295
 types of, 284
 zeros in data, 291
Filter-type spectrometers, see also
 Multichannel filter spectrometers
Filter-type spectrometers, 22
Five-kilometer Cambridge telescope,
  165-166
Flux density measurements
 complex sources, 7
 Gaussian source-corrections in, 5
 small-diameter sources, 3
 uniform-disk corrections in, 5
FORS1, 294, 320-321
Fortran, 284
FOUR1, 290, 294, 316
FOURG, 289, 294, 315
Fringe amplitude
 effects of noise on, 238-247
Fringe-frequency spectrum
 VLBI measurement of, 233
Fringe phase
 VLBI measurement of, 265-266
Fringe rate
 VLBI measurement of, 266
Fringe-visibility spectrum, 234
Fringes, see Interferometers, fringes in
Frequency stability
 characterization of, 198
 frequency-domain measures of, 200-202
 time-domain measures of, 200-202
Frequency standards
 bibliography of, 227
Frequency standards, 198-227
Frequency switching, 31-33, 49
FXRL1, 292, 294, 317

G

Gain modulation
  spectral-line measurements with, 50
Galactic hydrogen spectra
  data presentation for, 65
Gaussian fitting
  drift-scan use of, 7
General relativity
  tests of, 270
Geodesy
  VLBI applications in, 261
Geodetic latitude
  calculation of, 257
Geophysical measurements
  VLBI techniques for, 274-276
Goldstack interferometer, 267, 269
GRAYM, 304, 331
GRAYMP, 332-333
Gray-scale mapping
  programs for, 303-307
Green Bank
  three-element interferometer at, 164-165

H

HCN
  transition frequencies of, 27
HCNO
  transition frequencies of, 26
HCOOH
  transition frequencies of, 26
HFFT2, 292, 294, 319
HNC
  transition frequencies of, 26
HO
  transition frequencies of, 26
HS
  transition frequencies of, 27
Hydrogen 1
  transition frequencies of, 26, 210
Hydrogen absorption
  interferometer measurements of, 73
  single antenna measurements of, 72-74
  spectrometer baseline determination with, 74-77
Hydrogen clouds
  emission surveys of, 58-64
  high velocity, 61
  measurements of, 58-77
  smaller region, 62-64
  standards regions in, 36
Hydrogen cloud measurements
  scale calibration in, 69-72
  spectrometer baseline determinations for, 66-69

I

Interferometer delay, 147
Interferometers
  connected-elements, 158-173
  data inversion techniques for, 151-152
  detection schemes for, 157
  fringes in, 139
  noise analysis for, 154-157
  observing techniques for, 163
  phase correction for, 160
  phase measurements with, 158
  position measured with, 161
  signal analysis for, 141-147
  theory of two-element, 139-157
  white fringes in, 142, 158
International Radio Consultative Committee, 182

L

Line widths
  calculation of, 29-31
Load switching, 33-34, 48-49
Lobe rotation
  VLBI use of, 184
Local oscillators for VLBI
  coherence function of, 175
Local standard of rest, 281
Loran C stations, 180-181
Loran C system, 179
Lunar diffraction, 94
Lunar occultation
  bandwidth considerations in, 109-111
  finite beam effects in, 104-105
  measurements, 92-117
  methods of observations, 92-93
  noise effects in, 106-109
  occurrence of, 92
  point source in, 95-96
  refraction effects in, 115-117
  restoration technique for, 101-104, 107-108, 111
  source model fitting in, 100-101
  source-size effects in, 98

surveys by, 114-115
time scale of, 96-97

M

Mapping sources
  interferometer techniques for, 149, 166-173
  VLBI techniques for, 248-256, 270-271
Maser action
  spectral-line sources, 30
Maximum entropy method
  interferometer use of, 150-151
Moon observations
  VLBI techniques for, 273
Multichannel filter spectrometers
  applications in millimeter range, 44
  filter spacing in, 42
  measurements with, 38-45
  use in survey work, 44
Multiradix index, 287

N

NH
  transition frequencies of, 26
NHHCO
  transition frequencies of, 27
Noise fluctuation
  effects of, 22
Nonthermal distribution, 28-29
Nutation
  VLBI measurement of, 274

O

Observing techniques, *see also source types*
Occultation times
  duration of, 98-99
OCS
  transition frequencies of, 26
OH
  transition frequencies of, 26
OH masers
  VLBI measurement of, 251
On-off measurements, 15, 48-50, 52
One-mile telescope, 162

P

Path length
  atmospheric effects in, 264
  ionospheric effects in, 263
Phase prediction
  in frequency standards, 207-208
Plate tectonics
  VLBI measurement of, 275
Polar motion, 274
Polarization measurements
  continuum source, 10-12
  use of interferometers in, 152-153
Polarization switching, 34
Position measurements
  lunar occultation in, 99-100
  VLBI techniques for, 251, 261-276
Precession
  VLBI measurement of, 274
Proper motion
  VLBI measurement of, 269
Principal solution, 2
Pulsar receivers, 86
Pulsar signals
  dispersion measurements, 87
  dispersion removal, 84-87
  intensity variations of, 80
  interferometer observations, 90-91
  limitations to measurement of, 83-84
  observing techniques for, 78-91
  period measurement of, 87-88
  periods of, 78
  polarization measurements of, 79, 88
  propagation effects with, 80
  search techniques for, 91
  spectral measurements of, 89-90
  spectral-line absorption of, 90
  spectra of, 79

R

Recombination lines
  examples of, 20
  frequencies of, 23-24
Redshift
  optical spectra, 279
  radio spectra, 279-280
Refractions
  lunar ionosphere as cause of, 116-117
  solar corona as cause of, 116-117
RFFT2, 292, 294, 318
Ripple
  spectrometer baselines with, 33, 48
Royal Radar Establishment
  interferometer at, 165

Rubidium, 87
  resonant frequency of, 210
Rubidium clock, 179
Rubidium-gas-cell resonator, 208
  description of, 221-223
RULED, 300, 327-330
Ruled-surface mapping
  programs for, 299-307

S

Satellite orbits
  VBLI measurement of, 271-273
Searching techniques
  for spectral lines, 49
Servo systems
  frequency lock in, 224-225
Schwarzchild coordinates, 262
Scintillation
  bandwidth effects in, 135-137
  by interstellar medium, 118
  in pulsar signals, 81, 135-137
  interplanetary, 118, 126
  measurements, 118-137
Single antenna
  spectral-line measurements with, 20
SiO
  transition frequencies of, 27
Sky-horn switching, 33, 49
Small-diameter sources
  observation of, 1-18
Smoothing
  spectral data with, 47
Source size measurements, 4-5
  complex sources, 6-7
Source structure
  scintillation effects of, 126
Source survey
  4C, 170
Source shape,
  measurements of, 4-7
Source position
  measurements of, 7
Spatial frequency, 3
Spectral-density models, 202-207
Spectral distribution
  flicker of frequency, 204
  flicker of phase, 205
  random walk of frequency, 204
  white frequency noise, 205
  white phase noise, 206

Spectral lines
  frequencies of, 23-24
  strengths of, 24-29
Spectral-line measurements
  equipment requirements for, 20-23
  fundamentals of, 19-44
  history of, 19-20
  instrumental correlation of, 30
  noise in, 52
  requirements of, 19
  use of, 31
Spectral-line narrowing, 30
Spectral-line observations
  radial-velocity corrections to, 277-283
Stokes parameters
  interferometer measurement of, 152
  measurement of, 35
Switching schemes, *see also specific types*
Switching schemes, 48-51
Switching techniques, 31-35
  hydrogen cloud observations wtih, 66

T

Time delay
  ionospheric, 263
Time prediction
  in time standards, 207-208
Time scales, 182
Time standards, 198-227
  bibliography of, 227
Turbulent line widths, 29

U

Universal time
  definition of, 182
  variations in, 274
UTC
  definition of, 182
UT0
  definition of, 182
UT1
  definition of, 182
UT2
  definition of, 182

V

Van Vleck relation, 183
VBLI data reduction, 228-260

continuum fringe visibility in, 235
delay tracking in, 232
fringe phase in, 233
fringe rate in, 232
incoherent averaging in, 241-247
misidentification of signal in, 247-248
shift correction in, 232
VLBI measurements
  coordinate systems for, 262
  dual antenna techniques for, 269
  examples of, 266-276
  observables in, 261
  observational techniques in, 266
  relativistic effects in, 262
VBLI observations, 228-260
  checklist for, 258-260
  fringe amplitude estimation, 228
  fringe identification from, 228
  measurable qualities from, 228
  measurement of fringe amplitude, 229-238
  measurement of fringe phase, 229-238
  operational problems, 228-229
VBLI systems, 174-197
  block diagram of, 174
  Canadian, 174, 187, 192-196
  comparative sensitivity of, 197
  correlators for, 184

data processor for, 184
data recording for, 183
digital recording in, 186
frequency conversion in, 177
frequency standards for, 176
Mark I, 183-184, 186-187
Mark I, recorder in, 174
Mark II, 183-184, 187, 189-192
Mark II, recorder in, 174
playback system for, 184
storage capacity of, 187
synchronization of, 179
tape recorders for, 187
timing required for, 178

W

Water vapor masers
  VLBI measurement of, 251, 267

X

X-ogen
  transition frequencies of, 26

Z

Zeeman effect
  hydrogen cloud measurements of, 76-77